Administrative Records for Survey Methodology

Administrative Records for Survey Methodology

Edited by

Asaph Young Chun
Statistics Research Institute
Statistics Korea, Republic of Korea

Michael D. Larsen
Department of Mathematics and Statistics
Saint Michael's College, United States

Gabriele Durrant
Department of Social Statistics and Demography
Southampton University, UK

Jerome P. Reiter
Department of Statistical Science
Duke University, United States

Registered Office
John Wiley & Sons, Inc., 111 River Street, Hoboken, NJ 07030, USA

Editorial Office
111 River Street, Hoboken, NJ 07030, USA

For details of our global editorial offices, customer services, and more information about Wiley products visit us at www.wiley.com.

Wiley also publishes its books in a variety of electronic formats and by print-on-demand. Some content that appears in standard print versions of this book may not be available in other formats.

Library of Congress Cataloging-in-Publication Data

Names: Chun, Asaph Young, editor. | Larsen, Michael D., 1977- editor.
Title: Administrative records for survey methodology / edited by Asaph
    Young Chun, Statistics Research Institute | Statistics Korea, Republic of Korea, Michael D. Larsen,
    St. Michael's College, Colchester, United States, Gabriele Durrant, UK, Jerome P.
    Reiter, United States.
Description: First edition. | Hoboken, NJ : Wiley, 2021. | Series: Wiley
    series in survey methodology
Identifiers: LCCN 2020030571 (print) | LCCN 2020030572 (ebook) | ISBN
    9781119272045 (cloth) | ISBN 9781119272052 (adobe pdf) | ISBN
    9781119272069 (epub)
Subjects: LCSH: Surveys–Methodology. | Surveys–Quality control.
Classification: LCC HA31.2 .A36 2021 (print) | LCC HA31.2 (ebook) | DDC
    001.4/33–dc23
LC record available at https://lccn.loc.gov/2020030571
LC ebook record available at https://lccn.loc.gov/2020030572

Cover Design: Wiley
Cover Image: © PopTika/Shutterstock

Set in 9.5/12.5pt STIXTwoText by SPi Global, Chennai, India

Contents

Preface

Sample surveys are used by governments to describe the populations of their countries and provide estimates for use in policy decision making. Surveys can focus on individuals, households, businesses, students and schools, patients and hospitals, plots of land, or other entities. For surveys to be useful for official purposes they must cover the target population, represent the entirety of the population, collect information on key variables with accurate measurement methods, and have large enough sample sizes so that estimates are sufficiently precise at national and subnational levels. Achieving these four goals in a nationwide sample survey with a limited budget while being conducted in a short time interval is very challenging. The purpose of this book is to explore developments in the use of administrative records for improving sample surveys.

Sample surveys aim to gather information on a population. The target population is the specific part of the population that one aims to survey. Some parts of the broader population typically are excluded from the target population based on contact mode, data collection mode, the survey frame or list, or convenience. Individuals without a regular address, residing in some forms of group quarters, or without phone or Internet access, for example, might be effectively ineligible to serve as respondents. Survey frames record contact information and some other variables on members of a population, but of course they do not necessarily include all members of the population and have up-to-date information on everyone. Some individuals with accurate contact information in the frame will prove harder than others to contact or even refuse to participate. Surveys then are potentially limited to reporting about respondents and the population to which they are similar. Surveys cannot be overly long or else they risk deterring potential respondents and costing a lot of money per respondent. As a result, surveys can accommodate only so many questions. Self-report and less detailed questions, with their inherent limitations, for sensitive and complex items, often must be used for expediency. Budgets for national surveys compete with other government interests. Even large surveys typically have smaller-than-desired sample sizes in local areas and in subsets of the population. Despite these significant challenges, official statistical agencies around the world gather critically useful data on a myriad of topics.

The conditions for conducting sample surveys have changed immensely in the past 100 years. There is little chance that change will slow down. In-person surveys have been replaced and augmented by surveys by mail, by phone, and by Internet. Contact and data collection via multiple modes now are standard. The social environment, too, has evolved. Response rates are lower. Despite technological advances, people are increasingly busy. Official government surveys compete for attention with ever-more marketing and polling. Concerns over privacy and confidentiality have been elevated, rightly so, in the public consciousness. Simultaneously, government, researchers, and the public want more from data and surveys. Official surveys contribute to identifying challenges and to improvements in society. It is not practical, or maybe even possible, to get more out of old ways of conducting surveys.

Administrative records in a general sense are records kept for administrative purposes of the government. Administrative records can pertain to almost all aspects of life, including taxes, wages, education, health, residence, voting, crime, and property and business ownership. Does an individual have a license for a dog, for fishing at public lakes, to drive a car or motorcycle, or to own a gun? Does an individual receive public assistance through a government program? Administrative records, essential for government operations, contain a wealth of information on large segments of the population, but there are limitations. The records contain information on only some variables on subsets of the overall population. Information is collected so that a government can execute its program, but not typically for other purposes. Additional variables that might be interesting for study purposes likely are not recorded. Methods of recording variables might not be those that would be used in a scientific study. Those included in an administrative data file are not a random sample from the population. Some administrative records are collected over the course of several months or years, instead of only during a succinct time interval.

The use of administrative records has been part of the survey process for many decades. Survey textbooks since at least the 1960s (Cochran 1977; Kish 1967; Hansen, Hurwitz, and Madow 1953; Särndal, Swensson, and Wretman 1992) present methods for using auxiliary variables. It typically is assumed that values of auxiliary variables are available for all members of the population without error, or at least that aggregate totals are known. They might have come from a census, from a large survey at a previous time, or as part of the sample frame. Auxiliary variables are used for stratified surveys, probability proportional to size sampling, difference estimation, and ratio estimation. Often, they are treated in classic literature as known, fixed values.

Despite the limitations of administrative records, researchers, including the authors in this book, have been exploring how "adrecs" can be used to improve sample surveys in today's world and build on the record of past successes. They have examined new possibilities for using administrative record information to address four goals (coverage, response, variables, and accuracy) of official surveys. Increasing timeliness and decreasing costs through use of administrative records also are of continuing interest.

The book is organized into four sections. The first section contains two chapters. Chapter 1, by Li-Chun Zhang, presents fundamental challenges and approaches to integrating survey and administrative data for statistical purposes. The chapter focuses on administrative data, also called register or registry data, as a source for proxy variables. The proxy variables obtained from administrative sources can, for example, enhance a survey by providing additional information, be used for quality assessment of responses, and provide substitutes for missing values. Chapter 2, by John Marion Abowd, Ian Schmutte, and Lars Vilhuber addresses confidentiality protection and disclosure limitation in linked data. Linking data on population elements is an essential step for many uses of administrative records in conjunction with survey data. If individuals from a survey can be located uniquely in administrative records, then variables in those administrative records can be meaningfully associated with their originating units, thereby generating useful proxy variables. Data files from surveys, both from those linked to administrative information and those not, are made available to researchers and policy analysts. In standard practice, values of personally identifying information, such as names, fine-level geographic information including addresses, birthdates, and identification numbers, are suppressed. A data file containing a rich set of variables for analysis, however, increases the chance that someone could identify a unique individual from the survey in the population based on the values for several variables. The concern is that such an identification violates legal promises of confidentiality, causes harm to individuals who view their survey responses and administrative information as sensitive, and endangers future survey operations. Chapter 2 describes three applications, traditional statistical disclosure limitation methods, and new developments. The article includes discussion of how researchers access data (access modalities) and the usefulness (analytic validity) of data made available after modification for enhanced disclosure limitation.

Section 2 groups together five chapters on data quality and record linkage. Chapter 3, by Piet Daas, Eric Schulte Nordholt, Martjin Tennekes, and Saskia Ossen, examines the quality of administrative data used in the Dutch virtual census. A challenge in assessing quality of a data source is having better information on some variables for at least a subset of the population. Coen Hendriks, in Chapter 4, reports on improving the quality of data going into Norwegian register-based statistics. In Chapter 5, William Winkler considers a wide range of topics from initial cleaning of data files, record linkage, and integrated modeling, editing, and imputation. The impact of cleaning data files through standardizing variables, parsing variables such as addresses into separable components, and checking for logical errors cannot be overstated. Various approaches are in use for linking records from two files on the same population. Dr. Winkler reviews several enhancements, including variations in string comparator metrics and memory indexing, that have been put into practice at the U.S. Census Bureau. Jerry Reiter writes about assessing uncertainty when using administrative records in Chapter 6. Along with survey estimates, one typically needs to provide estimates of standard error. How do the quality of administrative records and the performance of the linkage to the survey impact the accuracy of estimates? Multiple imputation (Rubin 1986, 1987) could be one area for further exploration. In Chapter 7, Joseph Sakshaug addresses the specific question of measuring and controlling non-consent bias when surveys and administrative data are linked together. It is increasingly common for surveys that plan to link respondents to administrative data to ask for permission to do so. Some individuals refuse to give permission for linkage or cannot be linked due to other reasons, such as refusing to provide information on key linkage variables. Those whose records are not linkable can be different in many ways from those whose records are. Bias due to non-consent to linkage and failed linkage is therefore a novel contributing factor to total survey error.

Section 3 contains four articles on uses of administrative records in surveys and official statistics. Chapter 8 by Ingegerd Jansson, Martin Axelson, Anders Holmberg, Peter Werner, and Sara Westling describes experiences in the first Swedish register-based census of the population. In a register-based census, the population is counted and characteristics are gathered directly from administrative records, which, in this case, are referred to as population registers. Chapter 9 by Vincent Tom Mule and Andrew Keller of the U.S. Census Bureau presents research on administrative records applications for the U.S. 2020 Decennial Census of the population. In the U.S., there is no universal population register and the census involves enumerating and gathering basic information on every person in the country. Administrative records have been used to improve the data gathering process in the past. This chapter describes expanded options for improved design, quality and accuracy assessment, and dealing with missing information. Chapter 10 by Andrea Erciulescu, Carolino Franco, and Partha Lahiri concerns methods for improving small area estimation using administrative records. Surveys are designed to provide accurate estimates at a national or large subnational level, but not typically for small geographic areas or groups. Small area estimation uses models that provide a rationale for borrowing strength of sample across small areas for local estimation. The methodology relies on an advantageous bias–variance trade-off and estimation admissibility ideas (e.g. Efron and Morris 1975). Administrative records can provide key variables for use in such models.

Section 4 looks beyond statistical methodology for use of administrative records with surveys and provides three articles about using administrative data in evidence-based policymaking. The applications are in health, economics, and education. Chapter 11, by Cordell Golden and Lisa Mirel, focuses on enhancement of health surveys at the U.S. National Center for Health Statistics, through data linkage. Chapter 12, by Bruce Meyer and Nikolas Mittag, concerns economic policy analysis, with an emphasis on using administrative records to improve income measurements. Chapter 13, by Peter Siegel, Darryl Creel, and James Chromy, discusses combining data from multiple sources in the context of education studies.

The book is intended for a diverse audience. It should provide insight into developments in many areas and in many countries for those conducting surveys and their partners who manage and seek to improve administrative records. Several articles present theory as well as application and advice based on practical experience. Many chapters in the book include exercises for reflection on the material presented. The book could be of interest to students of statistics, survey sampling and methodology, and quantitative applications in government. Certainly, the book will have useful chapters for a variety of courses.

Data science has emerged as a term for an integration of statistics, mathematics, and computing and their integration in the effort to solve complex problems. Administrative records along with large-scale sample surveys provide a setting for the best applications in data science. This book hopefully will motivate those in the data science community to learn about survey sampling, official statistics, and a rich body of work aiming to utilize administrative records for sample surveys and survey methodology.

23 May 2020

Asaph Young Chun
Statistics Research Institute
Statistics Korea, Republic of Korea

Michael D. Larsen
Department of Mathematics and Statistics
Saint Michael's College, United States

Gabriele Durrant
Department of Social Statistics and Demography
Southampton University, UK

Jerome P. Reiter
Department of Statistical Science
Duke University, United States

References

Cochran, W.G. (1977). *Sampling Techniques*, 3e. Wiley.

Efron, B. and Morris, E. (1975). Data analysis using Stein's estimator and its generalizations. *Journal of the American Statistical Association* 70 (350): 311–319.

Hansen, M.H., Hurwitz, W.N., and Madow, W.G. (1953). *Sample Survey Methods and Theory, Volume 1: Methods and Applications; Volume 2: Theory.* Wiley.

Kish, L. (1967). *Survey Sampling*, 2e. Wiley.

Rubin, D.B. (1986). Statistical matching using file concatenation with adjusted weights and multiple imputations. *Journal of Business and Economic Statistics* 4: 87–94.

Rubin, D.B. (1987). *Multiple Imputation for Nonresponse in Surveys.* New York: Wiley.

Särndal, C.-E., Swensson, B., and Wretman, J. (1992). *Model Assisted Survey Sampling.* Springer Series in Statistics: Springer.

Acknowledgments

The origin of this book can be traced to the 2017 meeting of the European Survey Research Association and the session "Administrative Records for Survey Methodology" (https://www.europeansurveyresearch.org/conference/programme2017? sess=81). Dr. Asaph Young Chun (then of the U.S. Bureau of the Census). was the lead organizer and chair. Additional coordinators of that session included Drs. Michael Larsen (then at George Washington University, Washington, DC), Ingegerd Jansson (Statistics Sweden), Manfred Antoni (Institute for Employment Research, IAB, Germany), and Daniel Fuss and Corinna Kleinert (Leibniz Institute for Educational Trajectories, Germany). Papers presented at the conference included "Evaluation of the Quality of Administrative Data Used in the Dutch Virtual Census" (Schulte et al. 2017), "Evaluating the Accuracy of Administrative Data to Augment Survey Responses" (Berzofsky, Zimmer, and Smith 2017), and "Assessing Administrative Data Quality: The Truth is Out There" (Chun and Porter 2017).

Dr. Chun with Dr. Larsen proposed the book entitled *Administrative Records for Survey Methodology* to Wiley publishing. The intent of the book was to follow on the conference and reach further into topics and applications in additional countries and disciplines. Dr. Jerry Reiter (Duke University) and Dr. Gabriele Durrant (University of Southamptom) joined the team as assistant editors. Since the inception of this book, Dr. Chun has joined Statistics Korea and Dr. Larsen has moved to Saint Michael's College in Vermont.

The topics described by authors in this book have been described by these authors and others at international conferences since the 2017 ESRA meeting. Dr. Chun organized panel sessions at the Joint Statistical Meetings in 2019 entitled "Linked Data to Advance Evidence Building in Public Policy" (https://ww2.amstat .org/meetings/jsm/2019/onlineprogram/ActivityDetails.cfm?SessionID=218399) and in 2018 entitled "Administrative Records for Survey Methodology and Evidence Building" (https://ww2.amstat.org/meetings/jsm/2018/onlineprogram/

ActivityDetails.cfm?SessionID=215012). Some contributors to the current book participated in these panels.

We wish to thank individuals who have contributed toward bringing this volume to fruition. Many people have worked together to make this book possible. First, editors made comments and suggestions to improve the several chapters in this book. Second, a few individuals served as anonymous reviewers on individual chapters. Third, eight anonymous reviews on the overall scheme of the book were provided by the publisher Wiley. Fourth, the individual authors have been attentive to comments and suggestions from the editors and reviewers and generous with their time in improving their contributions. Fifth, authors and reviewers have contributed to these efforts with the support of their government agencies, educational institutions, sponsored funding organizations, and companies. Together all involved have made the present work a reality. We apologize if we have failed to mention any contributors.

Finally, we wish to thank individuals at Wiley who agreed to publish this manuscript and who have helped us along the way. Their reminders of deadlines and encouragements have kept us going through some transitions. Specifically, we wish to thank Associate Editor Kathleen Santoloci, Project Editors Blesy Regulas and Linda Christina, support person Mindy Okura-Marszycki, Managing Editor Kimberly Monroe-Hill, and Content Refinement Specialist Viniprammia Premkumar of Wiley Knowledge & Learning.

We hope you find the chapters in this book interesting and useful. We look forward to new developments with the use of administrative records and other data sources with sample surveys.

References

Berzofsky, M., Zimmer, S., and Smith, T. (2017). Evaluating the accuracy of administrative data to augment survey responses. Presentation at the 7th Conference of the European Survey Research Association (ESRA).

Chun, A.Y., and Porter, S. (2017). Assessing administrative data quality: the truth is out there. Presentation at the 7th Conference of the European Survey Research Association (ESRA).

Schulte, E., Daas, P., Tennekes, M., and Ossen, S. (2017). Evaluation of the quality of administrative data used in the Dutch virtual census. Presentation at the 7th Conference of the European Survey Research Association (ESRA).

List of Contributors

John M. Abowd
U.S. Census Bureau
4600 Silver Hill Road
Washington, DC 20233
USA
Cornell University
Ithaca, NY 14853
USA

Martin Axelson
Statistics Sweden
Box 24300
Stockholm SE-104 51
Sweden

James Chromy
RTI International
Research Triangle Park, NC 27709
USA

Darryl Creel
RTI International
Rockville, MD 20852
USA

Piet Daas
Statistics Netherlands
CBS-weg 11
Heerlen
the Netherlands, 6412 EX

Andreea L. Erciulescu
Westat
Rockville, MD 20850
USA

Carolina Franco
U.S. Census Bureau
4600 Silver Hill Road
Washington, DC 20233
USA

Cordell Golden
U.S. National Center for Health
Statistics (NCHS)
3311 Toledo Road
Hyattsville, MD 20782
USA

Coen Hendriks
Statistics Norway
Akersveien 26
0177 Oslo
Norway

Anders Holmberg
Statistics Sweden
Box 24300
Stockholm SE-104 51
Sweden

Ingegerd Jansson
Statistics Sweden
Box 24300
Stockholm SE-104 51
Sweden

Andrew Keller
U.S. Census Bureau
4600 Silver Hill Road
Washington, DC 20233
USA

Partha Lahiri
University of Maryland
College Park
Maryland 20742
USA

Bruce D. Meyer
U.S. Census Bureau
4600 Silver Hill Road
Washington, DC 20233
USA
University of Chicago
1307 E. 60th Street
Chicago, IL 60637
USA

Lisa B. Mirel
U.S. National Center for Health
Statistics
3311 Toledo Road
Hyattsville, MD 20782
USA

Nikolas Mittag
CERGE-EI
Politických vězňů 7
111 21 Prague 1
Czech Republic

Vincent T. Mule Jr.
U.S. Census Bureau
4600 Silver Hill Road
Washington, DC 20233
USA

Eric S. Nordholt
Statistics Netherlands
CBS-weg 11
Heerlen
the Netherlands, 6412 EX

Saskia Ossen
Statistics Netherlands
CBS-weg 11
Heerlen
the Netherlands, 6412 EX

Jerome P. Reiter
Duke University
Durham, NC 27708
USA

Joseph W. Sakshaug
Institute for Employment Research
Regensburger Str. 104
90478 Nuremberg
Germany
Ludwig Maximilian University of
Munich
Ludwigstr. 33
80539 Munich
Germany

Ian M. Schmutte
University of Georgia
Athens, GA 30602
USA

Peter Siegel
RTI International
Research Triangle Park, NC 27709
USA

Martijn Tennekes
Statistics Netherlands
CBS-weg 11
Heerlen
the Netherlands, 6412 EX

Lars Vilhuber
Cornell University
Ithaca, NY 14853
USA

Sara Westling
Statistics Sweden
Box 24300
Stockholm SE-104 51
Sweden

William E. Winkler
U.S. Census Bureau
4600 Silver Hill Road
Washington, DC 20233
USA

Li-Chun Zhang
University of Southampton
Southampton SO17 1BJ
UK

Part I

Fundamentals of Administrative Records Research and Applications

1

On the Use of Proxy Variables in Combining Register and Survey Data

Li-Chun Zhang

S3RI/Department of Social Statistics and Demography, University of Southampton, Southampton SO17 1BJ, UK

1.1 Introduction

In this chapter, we present an overview of the uses of proxy variables when combining data from multiple sources. In the remaining of this introductory section, we will explain what we mean by register and survey data, how the multisource data perspective differs from the survey-data centric view, and the concept of proxy variables in the context of multisource data. In Section 1.2, we consider the many and various instances of proxy variable, based on a systematic examination of the processing steps of data integration and associated error sources. In Section 1.3, we classify and outline estimation methods in the presence of multiple proxy variables. It is seen that the traditional role of auxiliary data from administrative sources can be greatly extended under the multisource data perspective. A short summary and discussion of future research is given in Section 1.4.

1.1.1 A Multisource Data Perspective

Under the presumption that the target units and measures are collected in survey data, register data traditionally have two principal uses: to provide the frames for sampling and estimation, to provide the auxiliary data for reducing both sampling and non-sampling survey errors (Särndal, Swensson, and Wretman 1992). The term auxiliary data conveys that register data play a helpful supporting role but is ultimately not indispensable. A broader view is necessary in order to cover the full range of approaches for combining register and survey data, where the two types of data are on an equal footing to each other.

Let us first clarify what we mean by register and survey data. We shall simply refer to statistical data arising from administrative sources as register data. On

Administrative Records for Survey Methodology, First Edition.
Edited by Asaph Young Chun, Michael D. Larsen, Gabriele Durrant, and Jerome P. Reiter.

the one hand, this extends the narrow interpretation of the term register as an authoritative list of objects; on the other hand, it implies that generally some processing may be required in order to transform "raw" administrative data into a state that permits them to be utilized for statistical purposes. Next, we shall simply refer to statistical data collected from samples and censuses as survey data. Our usage of the term survey here is conventional and more limiting, e.g. compared to that of Statistics Canada (2015), where it is used generically to cover any activity that collects or acquires statistical data, including administrative records and estimated data. We do not wish to contend the general interpretation, but we adopt the convention to facilitate the discussion that follows. A central distinction between what we call register and survey data is that the survey data are purposely designed and collected for statistical uses, whilst the register data are originally generated and recorded for purposes other than making statistics. This is also the reason why we refer to both survey sampling and census data as survey data, rather than taking on an even narrower interpretation which equates survey data with survey sampling data.

Brackstone (1987) characterizes the uses of administrative records, i.e. register data, into (i) direct tabulation, (ii) indirect estimation, (iii) survey frames, and (iv) survey evaluation. To appreciate what we shall refer to as the multisource data perspective and by way of introduction, let us consider the following question: *Are the four uses (i)–(iv) of register data equally applicable to survey data?*

Direct tabulation refers to the situation where statistics are produced based on the relevant register data without any explicit use of survey data. The scope of such register-based statistics has increased greatly in the past decades. A prominent example is the latest round of register-based census-like statistics in a number of European countries (UNECE 2014). See Wallgren and Wallgren (2014), for many other examples. As Zhang and Giusti (2016) point out and illustrate, sometimes relevant survey data are available and used implicitly to define the processing rules or to assess the accuracy of the register data, but are not part of the statistics directly. Clearly, in this sense, one can equally speak of direct tabulation based on survey data, such as the use of the Horvitz–Thompson estimator in survey sampling, or direct census enumeration of the population size.

Brackstone (1987) includes, under indirect estimation, the cases where register data "comprise one of the inputs into an estimation process." In the split-population or split-data approach (UNECE 2011), register and survey data supplement each other literally. A practical example of the split-population approach is the Unified Enterprise Survey at Statistics Canada, where register data are used for over half of the smaller enterprises with simple structures, and survey data are collected from the remaining units with more complex structures. Under the split-data approach, register data would provide some but not all of the required variables for the whole population, which otherwise would have to be collected in survey questionnaires. For example, at Statistics Norway, it is possible to derive income and education level data from statistical registers, so that these variables are not collected in the European Union Statistics on Income and Living Conditions (EU-SILC) and other social surveys. Imputation for survey nonresponse using register data can be viewed as a hybrid approach, where the units and variables to be substituted are determined post hoc after survey data collection. Indirect estimation beyond the split-population/split-data approach will be discussed in details later on, after we have explained the concept of proxy variables in Section 1.1.2.

Regarding the use of register data to create, supplement, or update frames for sample surveys and censuses, it takes only a moment of reflection to realize that exactly the same can be said of survey data. For instance, a census can be used to create, supplement, or update frames for postcensal sample surveys. The yearly Structural Business Survey and specific quality assurance surveys are used to proof or update the Business Register. As a noteworthy special case, one may include here population size estimation based on Census and Census Coverage Surveys (Nirel and Glickman 2009).

Survey evaluation covers the use of register data for checking, validating, or assessing survey data, whether they are collected in a sample or census. This may be done at both individual and aggregate levels. Reversely, using survey estimates for external validation of register-based statistics has been a natural approach from early on (Myrskyla 1991). Quality survey in a census year is another common approach in Scandinavia (Axelson et al. 2020), which is usually not directed at the population coverage errors of the Central Population Register in those countries, but at the various classification and measurement errors in the register data. Or, as mentioned above, survey data are commonly used implicitly to define the processing rules or to assess the accuracy of the register data.

In summary, one can speak of a multisource data perspective for combining register and survey data on at least two different levels. In the wider sense, it is possible to characterize equally the uses of both register and survey data into four broad categories: (i) single-source estimation, (ii) multisource estimation, (iii) frames, and (iv) evaluation. Both can be treated as statistical data and used as such. In a narrower sense, one can greatly extend the scope of "indirect estimation" under the multisource data perspective, where register and survey data each may comprise part of the inputs on an equal footing provided the proxy variables are present. Indirect estimation will be discussed in more details in Section 1.3. But first we shall explain below what we mean by proxy variables.

1.1.2 Concept of Proxy Variable

According to Upton and Cook (2008), a proxy variable is "a measured variable that is used in place of a variable that cannot be measured." We make two observations.

Firstly, one may distinguish between the cases where the ideal measure is unobservable in principle and where it is unavailable by chance. For example, per-capita gross domestic product (GDP) is sometimes used as a proxy measure of living standard, where it seems reasonable to acknowledge that the latter is unobservable in principle. For a contrasting example, country of birth can generate a proxy to mother tongue, by referring to the official language in that country. One should think that in this case the ideal measure is unavailable only due to circumstances. Secondly, in order for a proxy to be used in place of the ideal measure, the two should have the same support. Taking the previous example, it is not the birth country that is a proxy to the true mother tongue, but the official language in that country, and the common support of the proxy and ideal measures being all the existing languages in this case.

Zhang (2015a) defines a proxy variable as one that is *similar in definition and has the same support* as the target variable. It follows that one can regard two variables as proxy to each other, *without* having to specify one of them to be the target (or ideal) measure. Variables such as age, sex, education, income can be useful auxiliary but not proxy variables for the binary International Labour Organization (ILO) unemployment status. In particular, sex is not a proxy despite it being binary and thus have the same support as the unemployment status, because they do not have similar definitions. The binary register-based job-seeker status is a proxy, and the ILO unemployment status does not have to be the ideal measure for every conceivable purpose. But the job-seeker status is not a proxy variable for the activity status defined as (employed, unemployed, and inactive) because the two have different support.

Proxy variables can arise from survey data. For example, indirect interview yields proxy measures (Thomsen and Villund 2011), where household members respond on behalf of the absentees. Data collected in different modes can be proxy to each other. A variable collected in a census can be proxy to the same variable or a similarly defined one in the postcensal years. Synthetic datasets released for research can contain proxy variables for the target measures, based on which the synthetic ones are modeled and generated. Register data are perhaps the richest source of proxy variables. It is often possible to have both complete coverage and concurrency, or nearly so. As some common examples of register proxy variable one can mention economic activity status, education level, income, family and housing condition, etc. in social statistics; value-added tax (VAT) based turnover, export and import, house price, animal holding, fishing and hunting figures, arable soils, vegetation, etc. in economic and environmental statistics.

Finally, it is useful to reflect on the relationship between a proxy variable and one that can be affected by measurement errors, since one can always envisage a proxy variable as an attempt to measure the target variable, whether the effort is real or imaginary. Measurement errors are commonly decomposed into two components:

random errors and systematic errors. By definition random errors occur by chance and has zero expectation. Insofar as one considers random measurement errors to be unavoidable and omnipresent, any measured variable can only be a proxy of the ideal measure. In contrast, many proxy variables will remain the case even when it is acceptable to disregard the potential random errors for practical purposes. Systematic errors due to discrepancy in definition, instrument, time point, etc. are then the cause of imperfect measure, including when the ideal measure is unobservable in principle. Notice that this interpretation of systematic errors differs from the usage of the term in statistical data editing (de Waal, Pannekoek, and Scholtus 2011), where a systematic error is regarded as an error for which a plausible cause can be detected and knowledge of the underlying error mechanism enables then a satisfactory treatment in an unambiguous deterministic manner. Some examples of such systematic errors are typographical, measurement unit or sign errors. In summary, regardless of whether proxy variables may arise due to measurement errors, we are concerned here with the proxy variables that cannot be corrected by data editing methods.

1.2 Instances of Proxy Variable

Zhang (2012) presents a two-phase life cycle model of integrated statistical micro data, which provides a total-error framework for combining data from multiple sources. The first phase concerns the respective input data before integration takes place. Here, we consider the instances of proxy variables in relation to the various processing steps and associated error sources at the second, integration phase (Figure 1.1).

1.2.1 Representation

We start with the Representation side in Figure 1.1, which concerns the target population and units. Let us consider coverage error first. For instance, one may have a Population Register that is not sufficiently accurate to allow for direct tabulation of census-like population counts at detailed aggregation levels, so that Population Coverage Surveys are carried out in order to obtain the desired population estimates. The Population Register and Coverage Survey enumerations are proxies of the true population enumeration. This is the situation in Switzerland 2000 (Renaud 2007) and Israel 2008 (Nirel and Glickman 2009). Other instances may involve one or several register enumerations, Census enumeration and Census Coverage Survey enumeration. Capture–recapture methodology is a commonly used estimation approach that combines two or more proxy enumerations subjected to under-counts (Fienberg 1972; Wolter 1986; Hogan 1993). Adjustment

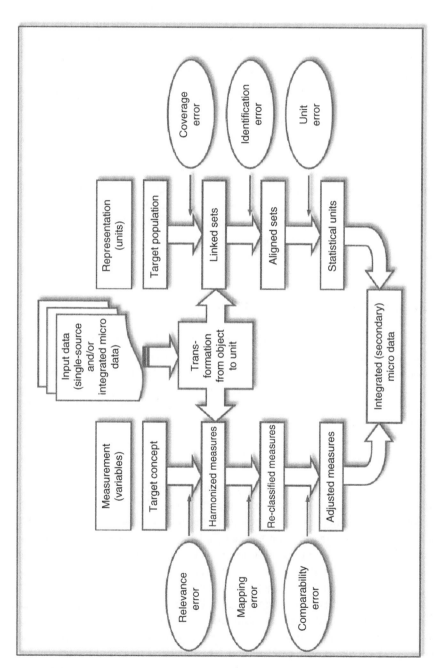

Figure 1.1 Second phase of integrated statistical micro data. Source: Zhang (2012).

of erroneous over-counts has attracted increasing attention recently, in situations where one does not have a Population Register and over-coverage errors are found to be large in the available register enumerations (ONS 2013). See, e.g. Zhang (2015b), for an extension of the capture–recapture modeling approach, Zhang and Dunne (2017) for trimmed dual-system estimation, and Di Cecco et al. (2018) for a latent class modeling approach.

All the methods mentioned above require the matching of records in separate sources. In reality, linkage errors may be unavoidable, unless a unique identifier exists in the different files and facilitates exact matching. A linkage error is the case if either a pair of linked records actually do not belong to the same entity or if the records that belong to the same entity fail to be linked. Both multi pass deterministic and probabilistic record linkage procedures are common in practice and often used in tandem. See, e.g. Fellegi and Sunter (1969) and Herzog, Scheuren, and Winkler (2007). The records in different files are compared to each other in terms of key variables such as name, birth date, address, etc. One can regard a concatenated string of key variables as a proxy of the true identifier, insofar as the key variables involved in principle could lead to unique combinations. Distortion of the key variables would then result in erroneous proxy identifiers and potentially cause linkage errors. For population size estimation, see, e.g. Di Consiglio and Tuoto (2015) for a study of linkage errors and dual system estimation, and Zhang and Dunne (2017) for a discussion regarding the trimmed dual-system estimation. More generally, since record linkage is a prerequisite for combining multisource data at the individual level, the matter of linkage errors due to imperfect proxy identifiers can be relevant in many other situations.

In a frame that is constructed from combining multiple population datasets, one can often find several related classification variables. Identification errors (Zhang 2012) arise if the classification variables or the relationships between them are mistaken based on the input datasets. For instance, the variable address is central for population and household statistics. Multiple addresses can be collected by combining the Population Register with resident address, the Post Register with postal address, the Higher-Education Student Register with term-time address, the various Utility datasets with occupant address, etc. Each person may be assigned a unique de jure address based on all these sources, in a way that is judged to be most appropriate, which would then yield a proxy variable for the de facto address that is of interest in many social-economic statistics.

The economic activity classification, e.g. NACE in Europe, is a well-known example in business statistics. The NACE code in the Business Register is generally a proxy of the target "pure" economic classification that has its root in the System of National Accounts. Several issues contribute to this fact, such as inconsistent operational rules of the Business Register, misreporting, lack of updation, etc. It is common in sample surveys to observe that for some units the

NACE code based on the updated survey returns will differ from the existing one in the Business Register. Such domain classification error is a kind of identification error. See, e.g. Brion and Gros (2015) for an example of how the matter is dealt with in the French Structural Business Surveys, and Van Delden, Scholtus, and Burger (2016) for an analysis of the NACE-classification errors in the Dutch context.

For survey data, the statistical unit can be identified in fieldwork. Based on register data, however, it is sometimes necessary to construct proxy statistical unit of interest, in which case unit errors may be unavoidable even if all the input data are error-free. For instance, consider register-based household. Provided all dwelling (or address) in the Population Register are correct, one may define a dwelling household to consist of all the persons who de jure share the same dwelling. We do not consider such a dwelling household to be a constructed statistical unit, precisely because it can be obtained from error-free input data directly. The perfection is another way of saying that there are no identification errors. An example of a constructed unit in this context is living household, which does not have to include everyone registered at the same dwelling nor be limited to these. Errors in a constructed living household is the case if two persons in different living households are placed in the same constructed living household, or if two persons in the same living households are placed in different constructed living households.

Constructed or not, unit error can be the case whether it results from lack of data or errors in data. Zhang (2011) devises a mathematical representation of unit error. It is assumed that each statistical unit of interest can consist of one or several so-called base units, but never cuts across a base unit. For example, person can be the base unit for household. The mapping from the set of base units to the set of statistical units can then be specified in terms of an allocation matrix, where each element takes value 1 or 0 depending on whether or not the corresponding base unit (arranged by column) belong to the statistical unit (arranged by row). In the case where a base unit can be assigned to one and only one statistical unit, such as a person can only belong to one household, the column sum of the allocation matrix is always equal to 1. Zhang (2011) develops a unit error theory for household statistics. Despite the unit error clearly being one of the most fundamental difficulties in business statistics, a statistical theory has so far been lacking. This may be partly due to the prominence of the identification error mentioned above. Another important reason may simply be the lack of a commonly acknowledged choice of base unit in business statistics.

1.2.2 Measurement

Consider now the measurement side in Figure 1.1. Relevance error refers to the discrepancy between the target measure that may be a theoretical construct and the measure that is achievable based on the available data. In a widespread scenario for combining register and survey data, the survey variable is treated as the target measure and the register proxy an auxiliary variable, which can be used either to adjust the survey sampling weights or to build a prediction model of the survey variable.

Sometimes, however, all the available measures entail relevance error, regardless of the source of the data, and there does not exist a way in which they can be combined to derive the target measure directly. For instance, Meijer, Rohwedder, and Wansbeek (2012) adopt such a viewpoint and study earnings data in register and survey using a mixture model approach, whereas Pavlopoulos and Vermunt (2015) apply latent class models to analyze income-based labor market mobility. It is also possible to formulate an adjusted measure as the solution of an appropriately defined constrained optimization problem, without explicitly introducing a model that spells out the relationship between the true measure and the observed proxy measures. For instance, Mushkudiani, Daalmans, and Pannekoek (2014) apply such an approach to Census aggregated tables and turnover variable from different sources.

Mapping error due to reclassification of input register data is highly common, since a register proxy variable often arises by means of reclassification. For instance, inferring the mother tongue from birth country is reclassification of the input variable birth country to the outcome variable mother tongue. For another example, to classify someone receiving unemployment benefit as unemployed is to reclassify the input variable benefit or not to the outcome variable unemployed or not. Examples as such are numerous.

It is worth noting that mapping error may be caused by delays or mistakes in the administrative sources, even where reclassification has no conceptual difficulties. Register data may be progressive in the sense that the observations for a particular reference time point may differ depending on when the observations are compiled. According to Zhang and Fosen (2012) and Zhang and Pritchard (2013), let t be the reference time point of interest and $t + d$ the measurement time point, for $d \geq 0$. Let $U(t)$ and $y(t)$ be the target population and value at t, respectively. For a unit i, let $I_i(t; t + d) = 1$ if it is to be included in the target population and 0 otherwise, based on the register data available at $t + d$, and let $y_i(t; t + d)$ be the observed value for t at $t + d$. The data are said to be progressive if, for $d \neq d' > 0$, one can have $I_i(t; t + d) \neq I_i(t; t + d')$ and $y_i(t; t + d) \neq y_i(t; t + d')$. Progressiveness is a distinct feature of register data compared to survey data.

The observed proxy measures may need to be adjusted in order to satisfy micro- as well as macro-level constraints, so as to resolve incompatibility across the data sources. For instance, register data from corporate tax returns may be used to impute for the missing items in Structural Business Survey. If this results in numerical inconsistency with the items observed from the survey, then imputation or adjustment of some of the items will be necessary in order to produce a clean and coherent dataset. See e.g. Pannekoek, Shlomo, and DeWaal (2013) and Pannekoek and Zhang (2015) for relevant instances.

Imposing macro-level survey estimates as benchmarks, when micro-adjusting a register proxy variable, can be regarded as a means to achieve statistical relevance at the level where the unbiased benchmarks are introduced (Zhang and Giusti 2016), though one is unable to remove the relevance bias at the micro-level. The Norwegian register-based employment status provides an example of such uses of proxy variables. Initially, the register proxy variable is rule-processed based on several input administrative registers, covering employee benefit, self-employment, tax, military or civilian service, leave of absence, etc. This results in the tripartition of the target population: (I) the compatible part, where the register data are compatible across the sources and allow for unequivocal reclassification accordingly, (II) the resolved part, where reclassification can be determined after making room for administrative regulations and progressiveness of the data, (III) the unsolved part, where register data are either lacking or incompatible, beyond what can be rule-processed. The Labor Force Survey (LFS) estimate of the yearly total of employed is then introduced to define an income threshold in the different subsets of part (III), whereby everyone above the threshold is reclassified as employed, such that the register total of employed coincides with the LFS estimate. As shown by Fosen and Zhang (2011), the resulting adjusted register proxy variable entails smaller mean squared error at the municipality level, compared to the survey estimates where the register proxy is used as an auxiliary variable.

1.3 Estimation Using Multiple Proxy Variables

Within the context of combining register and survey data, we consider here multisource estimation methods that make use of two or more proxy variables. Deficiency of coverage, relevance, and timeliness is often the reason that register-based estimation is not viable. When the lack of coverage can be limited to specific domains or variables, the problem can be remedied by the collection of supplementary survey data using the split-population or split-data approach. There would be only one value for each variable of interest now that the data supplement each other. Different multisource estimation approaches are needed for multiple proxy variables.

We shall classify the various scenarios using two conditions summarized in Table 1.1. (i) Whether one treats one of the proxy variables as the target measure and the others as associated with relevance bias – to be referred to as the asymmetric setting; the setting is symmetric otherwise, where either none of the proxy variables is considered to be the ideal measure, or all are correct measures

Table 1.1 Indirect estimation using register and survey proxy variables.

Linked data	One target measure and relevance bias in the others?	
	Yes (asymmetric)	**No (symmetric)**
Yes (linked)	Survey weighting	Capture–recapture methods
	Prediction modeling	Structural equation modeling
No (unlinked)	Benchmark adjustment	Constrained optimization

which nevertheless do not have perfect population coverage, (ii) Whether it is necessary to have linked data at the individual or cell level – to be referred to as the linked setting; the setting is unlinked otherwise. Each of the approaches listed in Table 1.1 covers a variety of methods with an extensive body of literature. The following elaboration aims merely to provide a brief accessible overview, and the references given serve only as points of departure for further exploration.

1.3.1 Asymmetric Setting

The two most common approaches under the asymmetric-linked setting are survey weighting and prediction modeling, where the register proxy variable is used as an auxiliary variable or a covariate. See e.g. Säarndal, Swensson, and Wretman (1992), for design-based approach to survey weighting that makes use of auxiliary variables; Valliant, Dorfman, and Royall (2000) and Chambers and Clark (2012) for model-based approach to finite population prediction; Rao and Molina (2015) for relevant methods of small area estimation. We make two observations. Firstly, when the overlapping survey variable is deemed necessary despite the presence of a register proxy, the latter is typically the most powerful among all the auxiliary variables when it comes to weighting adjustment and regression modeling. See e.g. Djerf (1997) and Thomsen and Zhang (2001) for the use of register economic activity status in the LFS, and the effects on reducing sampling and nonresponse errors. Secondly, applications to remedy Representation errors are much less common. However see, e.g. survey weighting under dependent sampling for the estimation of coverage errors (Nirel and Glickman 2009), mixed-effects models for assessing register coverage errors (Mancini and Toti 2014), and different misclassification models for register NACE (Van Delden et al. 2016), and register household (Zhang 2011).

The nature of a proxy variable implies a special use that is beyond what is feasible with a non-proxy auxiliary variable, no matter how good an auxiliary it is: *provided suitable conditions, it is possible to substitute (or replace) the target measure by the proxy value.* However, substitution would only be acceptable for a subset of the units but not all since, had it been acceptable for all the units, one would have had "direct tabulation" instead.

It follows that adjustment, or imputation in the case of a rejected value, will be necessary. Macro-level survey estimates can be imposed as benchmarks to achieve statistical relevance at the corresponding level. Linked datasets are typically not necessary here – recall the Norwegian register-based employment status described earlier. This yields many methods under what may be referred to as the benchmarked adjustment approach for combining register and survey proxy variables under the asymmetric-unlinked setting.

Repeated weighting and constrained (mass) imputation are two common approaches of benchmarked adjustment; see e.g. de Waal (2016) for a discussion. Repeated weighting is a technique initially presented for sample reweighing in the presence of overlapping survey estimates (Renssen and Nieuwenbroek 1997). It has been used for the reconciliation of Dutch virtual census output tables (Houbiers 2004). But it can equally be applied to adjust register datasets so that afterward, e.g. the weighted register proxy total agrees with the valid target totals imposed. This does not require linking the register datasets and the external datasets from which the benchmark totals are obtained. An inconvenience arises in cases where there are multiple proxy variables to be benchmarked and the variables are available for different subsets of units. This may be the case due to partial missing data in a single register file or when merging multiple register files. Some imputation will then be necessary if one would like to have a single set of weights for the whole dataset.

The one-number census imputation provides an example of the alternative imputation-based benchmarked adjustment methods (Brown et al. 1999). In the case of multiple proxy variables observed on different subsets of units, imputation is applied not only to the units with partially missing data, but also to the units with no observed variables at all, or possibly the units with completely observed data. The result is a complete dataset that guarantees numerical consistency for any tabulation across the variables and population domains. Constrained imputation for population datasets are e.g. discussed by Shlomo, de Waal, and Pannekoek (2009) and Zhang (2009a). Methods that incorporate micro-data edit constraints are e.g. studied in Coutinho, de Waal, and Shlomo (2013), Pannekoek, Shlomo, and DeWaal (2013), and Pannenkoek and Zhang (2015). Chambers and Ren (2004) consider a method of benchmarked outlier robust imputation. Obviously, it may be difficult to generate a single population dataset that is fit for all possible statistical uses. de Waal (2016) discusses the use of "repeated imputation." Notice that there are many relevant works on the generation of benchmarked synthetic populations in Spatial Demography, Econometrics, and Sociology.

The distinction between weighting and imputation can be somewhat blurred when it comes to the adjustment of cross-classified proxy contingency tables, because an adjusted cell count is just the number of individuals with the corresponding cross-classification one would have in an imputed dataset. Take, e.g. a two-way table, where the rows represent population domains at some detailed level, say, by local area and sex-age group, and the columns a composition of interest, say, income class. Let X denote the table based on combining population and tax register data. Let $Y^{(r)}$ denote the known vector of population domain sizes, and let $\hat{Y}^{(c)}$ denote the survey-based estimates of population totals by income class, which are the row and column benchmarks of the target table Y, respectively. Starting with X and by means of iterative proportional fitting (IPF) until convergence, one may obtain a table $\tilde{Y}$ that sums to both $Y^{(r)}$ and $\hat{Y}^{(c)}$ marginally. The technique has many applications including small area estimation (Purcell and Kish 1980) and statistical matching (D'Orazio, Di Zio, and Scanu 2006; Zhang 2015a) – more in Section 1.3.2.

A key difference between the asymmetric-linked setting and the asymmetric-unlinked setting discussed above is that, one generally does not expect a benchmarked adjustment method based on unlinked data to yield unbiased results below the level where the benchmarks are imposed. For instance, repeated weighting of Renssen and Nieuwenbroek (1997) can yield design-consistent domain estimates subjected to population benchmark totals, because the overlapping survey variables are both considered as the target measure here and no relevance bias is admitted. However, when the same technique is applied to reweight a register dataset, e.g. with the initial weights all set to 1, one cannot generally claim design or model-based consistency below the level of the imposed benchmarks, regardless of whether the benchmarks themselves are true or unbiased from either the design- or model-based perspective. Similarly, provided suitable assumptions, the one-number census imputation can yield model-consistent estimates below the level of the imposed constraints, because the donor records are taken from the enumerated census records that are considered to provide the target measures. However, the model-consistency would fall apart when the donor pool is a register dataset that suffers from relevance bias, even if all the other "suitable" assumptions are retained. Assessment of the statistical uncertainty associated with benchmarked adjustment is therefore an important research topic. An illustration in the contingency table case will now be given in Section 1.3.2.

1.3.2 Uncertainty Evaluation: A Case of Two-Way Data

Let $a = 1, ..., A$ and $j = 1, ..., J$ form a two-way classification of interest. For example, a may stand for ethnicity (White, Black, and Others), and j election votes for party (Democratic, Republic, Others). Or, let a be the index of a large number of local areas, and j the different household types such as single-person, couple without children, couple with children, etc. Let $X = \{X_{aj}\}$ be a known register-based proxy table that is unacceptable as "direct tabulation" of the target table $Y = \{Y_{aj}\}$.

For the asymmetric-linked setting, suppose there is available an observed sample two-way classification of (a, j). For survey weighting, let s denote the sample and let $d_i = 1/\pi_i$ be the sampling weight of unit $i \in s$, where π_i is the inclusion probability. Let $y_i(a, j) = 1$ if sample unit $i \in s$ has classification (a, j) according to the target measure and $y_i(a, j) = 0$ otherwise; let $x_i(a, j) = 1$ if it has classification (a, j) according to the proxy measure and $x_i(a, j) = 0$ otherwise. Post-stratification with respect to X yields then the poststratification weight, say, $w_i = d_i X_{aj} / \hat{X}_{aj}$, where $\hat{X}_{aj} = \sum_{i \in s} d_i x_i(a, j)$.

This is problematic when there are empty and very small sample cells of (a, j). Raking ratio weight can then be given by $w_i = d_i \tilde{X}_{aj} / \hat{X}_{aj}$, where $\tilde{X}$ is derived by the IPF of $\hat{X}$ to row and column totals X_{a+} and X_{+j}, respectively. Deville, Särndal, and Sautory (1993) provide approximate variance of the raking ratio estimator, say, $\hat{Y}_{+j} = \sum_{i \in s} w_i y_i(j)$ where $y_i(j) = \sum_{a=1}^{A} y_i(a, j)$.

A drawback of the weighting approach above is that no estimate of Y_{aj} will be available in the case of empty sample cell (a, j), and the estimate will have a large sampling variance when the sample cell (a, j) is small in size. This is typically the situation in small area estimation, where, e.g. a is the index of a large number of local areas. Zhang and Chambers (2004) and Luna-Hernández (2016) develop prediction modeling approach.

The within-area composition $(Y_{a1}, Y_{a2}, ..., Y_{aJ})$ is related to the corresponding proxy composition $(X_{a1}, X_{a2}, ..., X_{aJ})$ by means of a structural equation

$$\alpha_a^Y = \beta \alpha_a^X$$

where $\alpha_a^Y = \left(\alpha_{a1}^Y, ..., \alpha_{aJ}^Y\right)^T$ is the area-vector of interactions on the log scale, i.e. $\log Y_{aj} = \alpha^Y + \alpha_a^Y + \alpha_j^Y + \alpha_{aj}^Y$ where $\sum_{a=1}^{A} \alpha_a^Y = \sum_{j=1}^{J} \alpha_j^Y = \sum_{a=1}^{A} \alpha_{aj}^Y = \sum_{j=1}^{J} \alpha_{aj}^Y = 0$, and similarly for α_a^X, and β a matrix of unknown coefficients that sum to zero by row and by column.

The structural equation can be used to specify a generalized linear model of the observed sample cell counts, or their weighted totals, which allows one to estimate β and Y. It is further possible to develop the mixed-effects modeling approach that is popular in small area estimation, by introducing the mixed structural equation

$$\alpha_a^Y = \beta \alpha_a^X + \mathbf{u}_a$$

with the same quantities and the additional random effects $\mathbf{u}_a = (u_{a1}, ..., u_{aJ})^T$, where $\sum_{a=1}^{A} u_{aj} = \sum_{j=1}^{J} u_{aj} = 0$. The associated uncertainty will now be evaluated under the postulated model. The prediction modeling approach can thus improve on the survey weighting approach in the presence of empty and very small sample cells.

For an example under the asymmetric-unlinked setting, consider the Norwegian register-based household statistics. At the time the household register was first introduced for the year 2005, there were still about 6% persons with missing dwelling identification in the Central Population Register. As the missing rate differed by local areas as well as household types, direct tabulation did not yield acceptable results compared to the Census 2001 outputs. The IPF was applied to the sub-population of households that have the dwelling identification to yield a weight for every such household. The method falls under the benchmarked adjustment approach. However, direct evaluation of the associated uncertainty is not straightforward. Zhang (2009b) extends the prediction modeling approach above to accommodate the informative missing data. By comparison with the model-based predictions, one is able to assess indirectly the benchmarked adjustment results.

Using the IPF for small area estimation is known as structure preserving estimation (SPREE, Purcell and Kish 1980). The model underpinning the SPREE is a special case of the prediction models mentioned above, i.e. by setting $\beta = 1$. It does not require linkage between the proxy data X and the data that yield the benchmarks Y_{a+} and Y_{+i}. While this is convenient for deriving the estimates, a difficulty arises when it comes to uncertainty evaluation directly under the SPREE model. See also Dostál et al. (2016) for a benchmarked adjustment method based on the chi-squared measure in this respect.

Finally, let Y by ethnicity and party votes be the table of interest. Suppose one can obtain $\hat{Y}_{a+}$ and Y_{+j} in an election, but there are no joint observations of the cells (a, j). This can be framed as a problem of statistical matching. Provided a proxy table X, say, ethnicity by party membership, the IPF can be applied to obtain an estimated table $\hat{Y} = \{\hat{Y}_{aj}\}$. Zhang (2015a) develop an uncertainty measure that combines the identification uncertainty and the sampling uncertainty in this context, which enables one to quantify the relative efficiency of the proxy data X, compared to statistical matching without X. The application of the IPF here is an example of the benchmarked adjustment approach.

1.3.3 Symmetric Setting

In the symmetric setting none of the proxy variables is ideal due to errors of relevance, measurement, or coverage. The two most common approaches under the symmetric-linked setting are capture–recapture methodology for population size estimation and Structural Equation Modeling (SEM) that covers the latent class models mentioned earlier.

Capture–recapture methods that originate from wide-life, social, and medical applications are traditionally used for under-count adjustment. Imagine catching fish in a pond on two separate occasions, where one marks and identifies the fish that happen to be caught on both occasions (i.e. the recaptures). Then, under a number of simplifying assumptions, including independent and constant-probability captures, it becomes possible to estimate the total number of fish in the pond (i.e. the target population), for which the captures on each occasion generally entail undercounts. The method can be generalized to multiple captures to allow for relaxation of the independent assumption. The capture probability can be modeled using covariates to allow for heterogeneity across different subpopulations. See, e.g. Böhning, Van der Heijden, and Bunge (2017), for some recent developments.

Combining survey and register-based enumerations for population size estimation has attracted growing interest in the recent years, under the assumption that none of the sources can yield the true target population enumeration directly. We refer to the *Journal of Official Statistics* (2015, vol. 31, issue 3) for several useful references in this regard. There is plenty of scope for developing a range of models in order to address the different problems, including erroneous enumerations that are not dealt with in the traditional capture–recapture methodology. The potential impact can be huge if it enables one to produce census-like population statistics without the traditional census.

SEM is often considered to have evolved from the genetic path modeling of Sewall Wright. See, e.g. Kline (2016) for a general introduction. The approach is popular in many social science disciplines that share a common interest in "latent constructs" such as intelligence, attitude, well-being, living standard, and so on. The postulated latent constructs cannot be measured directly and are only manifested through observable indicators. The SEM consists of two main components: the structural model showing potentially causal dependencies among the latent variables, and the measurement model relating the latent variables and their indicators. The approach can be referred to in different ways depending on the continuous-categorical nature of the variables involved, the presence of causality or stochastic process on the latent level, etc.

The SEM approach is applicable under the symmetric-linked setting, where the proxy variables are treated as the indicators of the unobserved target measure. In the context of combining register and survey data, this can serve a number of purposes, including assessing potential relevance bias of proxy measures, detecting and possible treatment of measurement errors in editing and estimation, and statistical analysis of latent relationships using proxy indicators. For examples of data types that have been studied recently, see e.g. Pavlopoulos and Vermunt (2015) for temporary employment, Guarnera and Varriale (2015) for labor cost, and Burger et al. (2015) for turnover.

Di Cecco et al. (2018) apply latent class models for population size estimation based on multiple register enumerations that entail both over and under-counts. It is intriguing to notice the connection with some recent developments in record linkage. Imagine K lists of records, where each record may or may not refer to a target population unit (i.e. latent entity). Provided the union of the lists entail only over-counts of the target population, a potential alternative approach is record linkage, also referred to as entity resolution or co-reference – see e.g. Stoerts, Hall, and Fienberg (2015). The records in the same list that refer to the same entity represent duplicated enumerations; the records in the different lists that refer to the same entity can be conceived as the target for record linkage. The errors in compiling the population total are then the potential de-duplication and record linkage errors, which are traditionally the topics of computerized record linkage.

Multiple macro-level proxy totals may need to be reconciled under the symmetric-unlinked setting. A typical example is multiple time series with different frequencies, e.g. with register-based yearly figures and survey-based sub-annual figures. Another example is the Supply-and-Use Tables for the production of GDP, where the initial estimates generally do not balance out because they are derived from different sources, or when the GDP is compiled using different approaches. Census output tables derived from fragmented data sources instead a one-number file is yet another example. See e.g. Bikker, Daalmans, and Mushkudiani (2013) and Mushkudiani, Daalmans, and Pannekoek (2014, 2015).

Reconciliation is often achieved as the solution to a constrained optimization problem. The approach requires the specification of two components. A loss function may be defined to measure the changes from the initial proxy estimates to the final reconciled estimates. The constraints that the final estimates must satisfy need to be explicitly stated, which may contain both equality and inequality constraints. Minimizing the loss function subjected to the constraints would then yield the final estimates. The approach is feasible without linked data across the sources. Notice that there are many advanced techniques of constrained optimization in Applied Mathematics, Engineering, and Computer Sciences.

Mushkudiani, Pannekoek, and Zhang (2016) develop scalar uncertainty measure of macro accounts to replace, say, the entire variance–covariance matrix of all the estimates involved. Devising simple summary statistical uncertainty measures for an accounting system such as the System of National Account can be helpful in at least two respects: (i) it can inform the choice among alternative adjustment methods that seem equally viable to start with, (ii) it can identify and assess the changes, or potential improvements, that are most effective in terms of the final estimated account directly. Implementation of the approach to the System of National Account is currently under development.

1.4 Summary

In this chapter, we have provided an overview of the uses of proxy variables when combining register and survey data. The nature of the proxy variables discussed is such that they are not subjects of data editing methods, even when they can be considered to have risen from some kind of measurement errors. The presence of proxy variables is a characteristic feature of the settings involving data from multiple sources, because in a single-source setting proxy variables can be eliminated by design. The various instances discussed in Section 1.2 demonstrate the ubiquitous presence of proxy variables in multisource statistics. Sometimes proxy variables raise challenges because the conflict between them needs to be resolved, sometimes they are a blessing – indeed statistics may be impossible without them as in the case of capture–recapture methods for population size estimation. Either way, they always represent potentially useful sources of statistical information. We believe that the appropriate conceptualization, treatment, and usage of proxy variables provide a wide-ranging perspective, which enables one to draw on insights and experiences from diverse problems. An important theme for future research is the assessment of statistical uncertainty associated with indirect estimation based on unlinked data (Table 1.1). Several methods are mentioned in Sections 1.3.2 and 1.3.3, which however do not cover all the practical situations.

References

Axelson, M., Holmberg, A., Jansson, I. et al. (2020). A register-based census: the Swedish experience (Chapter 8). In: *Administrative Records for Survey Methodology* (eds. A.Y. Chun and M. Larsen). Wiley.

Bikker, R., Daalmans, J., and Mushkudiani, N. (2013). Benchmarking large accounting frameworks: a generalized multivariate model. *Economic Systems Research* 25: 390–408.

Böhning, D., Van der Heijden, P.G.M., and Bunge, J. (2017). *Capture–Recapture Methods for Social and Medical Sciences*. Chapman & Hall/CRC.

Brackstone, G.J. (1987). Issues in the use of administrative records for statistical purposes. *Survey Methodology* 13: 29–43.

Brion, P. and Gros, E. (2015). Statistical estimators using jointly administrative and survey data to produce French Structural Business Statistics. *Journal of Official Statistics* 31: 589–609.

Brown, J.J., Diamond, I.D., Chambers, R.L. et al. (1999). A methodological strategy for a one-number census in the UK. *Journal of the Royal Statistical Society, Series A* 162: 247–267.

Burger, J., van Delden, A. and Scholtus, S. (2015). Sensitivity of Mixed-Source Statistics to Classification Errors. *Journal of Official Statistics*, 31: 489–506.

Chambers, R.L. and Clark, R. (2012). *An Introduction to Model-Based Survey Sampling with Applications*. Oxford University Press.

Chambers, R.L. and Ren, R. (2004). Outlier robust imputation of survey data. In: *JSM Proceedings, Survey Research Methods Section. Alexandria, VA*, 3336–3344. American Statistical Association.

Coutinho, W., de Waal, T., and Shlomo, N. (2013). Calibrated hot deck imputation subject to edit restrictions. *Journal of Official Statistics* 29: 1–23.

Deville, J.-C., Särndal, C.-E., and Sautory, O. (1993). Generalized raking procedures in survey sampling. *Journal of the American Statistical Association* 88: 1013–1020.

Di Cecco, D., Di Zio, M., Filipponi, D., and Rocchetti, I. (2018). Population size estimation using multiple incomplete lists with overcoverage. *Journal of Official Statistics* 34: 557–572.

Di Consiglio, L. and Tuoto, T. (2015). Coverage evaluation on probabilistically linked data. *Journal of Official Statistics* 31: 415–429.

Djerf, K. (1997). Effects of post-stratification on the estimates of the Finnish Labour Force Survey. *Journal of Official Statistics* 13: 29–39.

D'Orazio, M., Di Zio, M., and Scanu, M. (2006). *Statistical Matching: Theory and Practice*. Chichester: Wiley.

Dostál, L., Münnich, R., Gabler, S., and Ganninger, M. (2016). Frame correction modelling with applications to the German Register-Assisted Census 2011. *Scandinavian Journal of Statistics* 43: 904–920.

Fellegi, I.P. and Sunter, A.B. (1969). A theory for record linkage. *Journal of the American Statistical Association* 64: 1183–1210.

Fienberg, S.E. (1972). The multiple recapture census for closed populations and incomplete 2k contingency tables. *Biometrika* 59: 409–439.

Fosen, J. and Zhang, L.-C. (2011). Quality evaluation of employment status in register-based census. In: *Proceedings of the 58th World Statistical Congress, Dublin*, 2587–2596. International Statistics Institute.

Guarnera, U. and Variale, R. (2015). Estimation from contaminated multi-source data based on latent class models. NTTS.

Herzog, T.N., Scheuren, F.J., and Winkler, W.E. (2007). *Data Quality and Record Linkage Techniques*. Springer.

Hogan, H. (1993). The post-enumeration survey: operations and results. *Journal of the American Statistical Association* 88: 1047–1060.

Houbiers, M. (2004). Towards a social statistical database and unified estimates at Statistics Netherlands. *Journal of Official Statistics* 20: 55–75.

Kline, R.B. (2016). *Principles and Practice of Structural Equation Modeling*, 4e. The Guilford Press.

Luna-Hernández, A. (2016). Multivariate structure preserving estimation for population compositions. University of Southampton, School of Social Sciences, Doctoral Thesis, 155 pp. https://eprints.soton.ac.uk/404689/

Mancini, L. and Toti, S. (2014). Dalla popolazione residente a quella abitualmente dimorante: modelli di previsione a confronto sui dati del censimento 2011. ISTAT working papers (in Italian).

Meijer, E., Rohwedder, S., and Wansbeek, T.J. (2012). Measurement error in earnings data: using a mixture model approach to combine survey and register data. *Journal of Business & Economic Statistics* 30: 191–201.

Mushkudiani, N., Daalmans, J., and Pannekoek, J. (2014). Macro-integration for solving large data reconciliation problems. *Austrian Journal of Statistics* 43: 29–48.

Mushkudiani, N., Daalmans, J., and Pannekoek, J. (2015). Reconciliation of labour market statistics using macro-integration. *Statistical Journal of the IAOS* 31: 257–262.

Mushkudiani, N., Pannekoek, J., and Zhang, L.-C. (2016). Uncertainty measurement for economic accounts. Statistics Netherlands: Discussion paper.

Myrskyla, P. (1991). Census by questionnaire census by registers and administrative records: the experience of Finland. *Journal of Official Statistics* 7: 457–474.

Nirel, R. and Glickman, H. (2009). Sample surveys and censuses. In: *Sample Surveys: Design, Methods and Applications* (Chapter 21), vol. 29A (eds. D. Pfeffermann and C.R. Rao), 539–565. North Holland: Elsevier.

ONS – Office for National Statistics (2013). Beyond 2011: Producing Population Estimates Using Administrative Data: In Practice. ONS Internal Report. http://www.ons.gov.uk/ons/about-ons/who-ons-are/programmes-and-projects/beyond-2011/reports-and-publications/index.html (accessed 04 August 2020).

Pannekoek, J. and Zhang, L.-C. (2015). Optimal adjustments for inconsistency in the presence missing data. *Survey Methodology* 41: 127–144.

Pannekoek, J., Shlomo, N., and DeWaal, T. (2013). Calibrated imputation of numerical data under linear edit restrictions. *The Annals of Applied Statistics* 7: 1983–2006.

Pavlopoulos, D. and Vermunt, J.K. (2015). Measuring temporary employment. Do survey or register data tell the truth? *Survey Methodology* 41: 197–214.

Purcell, N. and Kish, L. (1980). Postcensal estimates for local areas (or domains). *International Statistical Review* 48: 3–18.

Rao, J.N.K. and Molina, I. (2015). *Small Area Estimation*, 2e. New York: Wiley.

Renaud, A. (2007). Estimation of the coverage of the 2000 census of population in Switzerland: methods and results. *Survey Methodology* 33: 199–210.

Renssen, R.H. and Nieuwenbroek, N.J. (1997). Aligning estimates for common variables in two or more sample surveys. *Journal of the American Statistical Association* 92: 369–374.

Särndal, C.-E., Swensson, B., and Wretman, J. (1992). *Model Assisted Survey Sampling*. New York: Springer-Verlag.

Shlomo, N., de Waal, T., and Pannekoek, J. (2009). Mass imputation for building a numerical statistical database. Presented at the UNECE Statistical Data Editing Workshop, Neuchatel (October 2009).

Statistics Canada (2015). Statistics Canada Quality Guidelines. http://www.statcan.gc .ca/pub/12-539-x/12-539-x2009001-eng.htm (accessed 4 August 2020).

Stoerts, R., Hall, R., and Fienberg, S. (2015). A Bayesian approach to graphical record linkage and de-duplication. https://arxiv.org/abs/1312.4645 (accessed 04 August 2020).

Thomsen, I. and Villund, O. (2011). Using register data to evaluate the effects of proxy interviews in the Norwegian Labour Force Survey. *Journal of Official Statistics* 27: 87–98.

Thomsen, I. and Zhang, L.-C. (2001). The effects of using administrative registers in economic short term statistics: the Norwegian Labour Force Survey as a case study. *Journal of Official Statistics* 17: 285–294.

UNECE (2011). *Using Administrative and Secondary Sources for Official Statistics: A Handbook of Principles and Practices*. United Nations Economic Commission for Europe. Available at: http://www.unece.org.

UNECE (2014). *Measuring Population and Housing. Practices of UNECE Countries in the 2010 Round of Censuses*. United Nations Economic Commission for Europe. Available at: http://www.unece.org.

Upton, G. and Cook, I. (2008). *A Dictionary of Statistics*. Oxford University Press.

Valliant, R., Dorfman, A.H., and Royall, R.M. (2000). *Finite Population Sampling and Inference: A Prediction Approach*. New York: Wiley.

Van Delden, A., Scholtus, S., and Burger, J. (2016). Accuracy of mixed-source statistics as affected by classification errors. *Journal of Official Statistics* 32 (3): 619–642.

de Waal, T. (2016). Obtaining numerically consistent estimates from a mix of administrative data and surveys. *Statistical Journal of the IAOS* 32: 231–243.

de Waal, T., Pannekoek, J., and Scholtus, S. (2011). *Handbook of Statistical Data Editing and Imputation*. Hoboken, NJ: Wiley.

Wallgren, A. and Wallgren, B. (2014). *Register-Based Statistics: Statistical Methods for Administrative Data*, 2e. Wiley.

Wolter, K. (1986). Some coverage error models for census data. *Journal of the American Statistical Association* 81: 338–346.

Zhang, L.-C. (2009a). A triple-goal imputation method for statistical registers. Presented at the UNECE Statistical Data Editing Workshop, Neuchatel (October 2009).

Zhang, L.-C. (2009b). Estimates for small area compositions subjected to informative missing data. *Survey Methodology* 35: 191–201.

Zhang, L.-C. (2011). A unit-error theory for register-based household statistics. *Journal of Official Statistics* 27: 415–432.

Zhang, L.-C. (2012). Topics of statistical theory for register-based statistics and data integration. *Statistica Neerlandica* 66: 41–63.

Zhang, L.-C. (2015a). On proxy variables and categorical data fusion. *Journal of Official Statistics* 31: 783–807.

Zhang, L.-C. (2015b). On modelling register coverage errors. *Journal of Official Statistics* 31: 381–396.

Zhang, L.-C. and Chambers, R.L. (2004). Small area estimates for cross-classifications. *Journal of the Royal Statistical Society, Series B* 66: 479–496.

Zhang, L.-C. and Dunne, J. (2017). Trimmed dual system estimation. In: *Capture–Recapture Methods for the Social and Medical Sciences* (Chapter 17) (eds. D. Böhning, J. Bunge and P. van der Heijden), 239–259. Chapman & Hall/CRC.

Zhang, L.-C. and Fosen, J. (2012). A modeling approach for uncertainty assessment of register-based small area statistics. *Journal of the Indian Society of Agricultural Statistics* 66: 91–104.

Zhang, L.-C. and Giusti, C. (2016). Small area methods and administrative data integration. In: *Analysis of Poverty Data by Small Area Estimation* (ed. M. Pratesi), 61–82. Wiley.

Zhang, L.-C. and Pritchard, A. (2013). Short-term turnover statistics based on VAT and Monthly Business Survey data sources. European Establishment Statistics Workshop, Nuremberg, Germany (9–11 September 2013).

2

Disclosure Limitation and Confidentiality Protection in Linked Data

John M. Abowd[1], Ian M. Schmutte[2] and Lars Vilhuber[3]

[1]*U.S. Census Bureau and Cornell University, Suitland, MD, USA*
[2]*University of Georgia, Athens, GA, USA*
[3]*Department of Economics and Executive Director of Labor Dynamics Institute (LDI) at Cornell University, Ithaca, NY, USA*

2.1 Introduction

The use of administrative data has long been a part of the procedures at national statistical offices (NSOs), as evidenced in the various chapters in this book. The censuses and surveys conducted by NSOs may use sampling frames built at least partially from administrative data. For instance, the U.S. Census Bureau has used a business register – a list of all domestic businesses – derived from administrative tax filings since at least 1968. This register is the frame for its quinquennial censuses and annual surveys of business activity (DeSalvo, Limehouse, and Klimek 2016). It is also used to link businesses across surveys, to link surveyed businesses to other administrative record data, and as a direct source of statistical information on the levels and growth of business activity, published as the County Business Patterns (CBP) and Business Dynamics Statistics (BDS).[1] Similar examples can be found in most countries that maintain some kind of registry for their businesses. In many countries, similar centrally maintained registers are used as frames for censuses and surveys of a country's inhabitants and workers. Chapter 17 illustrates the Swedish approach to this problem for a national population census.[2] The Institute for Employment Research (IAB), the research institute of the German Employment Agency, uses social security notifications filed by firms, and data generated from the administration of its mandated programs, to sample firms and workers.

1 See www.census.gov/programs-surveys/cbp.html and https://www.census.gov/programs-surveys/ces.html.
2 In the United States, a combination of multiple lists and input by regional and subject matter experts is used to compile the frame for the Census of Population and Housing.

Administrative Records for Survey Methodology, First Edition.
Edited by Asaph Young Chun, Michael D. Larsen, Gabriele Durrant, and Jerome P. Reiter.

McMaster University and later Statistics Canada used administrative job termination notifications ("record of employment") filed by employers to survey departing employees for the Canadian Out-of-Employment Panel (COEP) (Browning, Jones, and Kuhn 1995). Other uses of administrative data in NSOs include linkage for quality purposes (Chapters 8, 14, and 15), and data augmentation (Chapter 12 for the National Center for Health Statistics [NCHS] approach).

In addition, the increasing computerization of administrative records, has facilitated more extensive linking of previously disconnected administrative databases, to create more comprehensive and extensive information. Methods to link databases within administrative units based on common identifiers are easy to implement (see Chapter 9 for more details). In the United States, which does not have a legal national identifier or ID document, the increased use of the Social Security Number (SSN) has facilitated linkage of government databases and among commercial data providers. In many European countries, individuals have national identifiers, and efforts are underway to allow for cross-border linkages within the European Union, in order to improve statistics on the workforce and the businesses of the common economic area created by what is now called the European Union. However, even when common identifiers are not available, linkage is possible (see Chapter 15).

The result has been that data on individuals, households, and business have become richer, collected from an increasing variety of sources, both as designed surveys and censuses, as well as organically created "administrative" data. The desire to allow policy makers and researchers to leverage the rich linked data has been held back, however, by the concerns of citizens and businesses about privacy. In the 1960s in the United States, researchers had proposed a "National Data Bank" with the goal of combining survey and administrative data for use by researchers. Congress held hearings on the matter, and ultimately the project did not go forward (Kraus 2013). Instead, and partially as a consequence, privacy laws were formalized in the 1970s. The U.S. "Privacy Act" (Public Law 93-579, 5 U.S.C. § 552a), passed in 1974, specifically prohibited "matching" programs, linking data from different agencies. More recently, the 2016 Australian Census elicited substantial controversy when the Australian Bureau of Statistics (ABS) decided to keep identifiable data collected through the census for a substantially longer time period, with the explicit goal of enabling linkages between the census and administrative data, as well as linkages across historical censuses (Australian Bureau of Statistics 2015; Karp 2016).

Subsequent decades saw a decline in public availability of highly detailed microdata on people, households, and firms, and the emergence of new access mechanisms and data protection algorithms. This chapter will provide an overview of the methods that have been developed and implemented to safeguard privacy, while providing researchers the means to draw valid conclusions from protected data.

The protection mechanisms we will describe are both physical and statistical (or algorithmic), but exist because of the need to balance the privacy of the respondents, including the confidentiality protection their data receive, with society's need and desire for ever more detailed, timely, and accurate statistics.

2.2 Paradigms of Protection

There are no methods for disclosure limitation and confidentiality protection specifically designed for linked data. Protecting data constructed by linking administrative records, survey responses, and "found" transaction records relies on the same methods as might be applied to each source individually. It is the richness inherent in the linkages, and in the administrative information available to some potential intruders, that pose novel challenges.

Statistical confidentiality can be viewed as "a body of principles, concepts, and procedures that permit confidentiality to be afforded to data, while still permitting its use of for statistical purposes" (Duncan, Elliot, and Salazar-González 2011, p. 2). In order to protect the confidentiality of the data they collect, NSOs and survey organizations (henceforth referred to generically as data custodians) employ many methods. Very often, data are released to the public as tabular summaries. Many of the protection mechanisms in use today evolved to protect published *tables* against disclosure. Generically, the idea is to limit the publication of cells with "too few" respondents, where the notion of "too few" is assessed heuristically.

We will not provide a detailed history or taxonomy of statistical disclosure limitation (SDL) and formal privacy models, instead will refer the reader to other publications on the topic (Duncan, Elliot, and Salazar-González 2011; Dwork and Roth 2014; FCSM 2005). We do need to set up the problem, which we will do by reviewing suppression, coarsening, swapping, and noise infusion (input and output). These are widely used techniques and the main issues that arise in applications to linked data can be understood with reference to these methods.

Suppression is widely used to protect published tables against statistical disclosure. Suppression describes the removal of sub-tables, cells, or items in a cell from a published collection of tables if the item's publication would pose a high risk of disclosure. This method attempts to forge a middle ground between the users of tabular summaries, who want increasingly detailed disaggregation, and publication rules based on cell count thresholds. The Bureau of Labor Statistics (BLS) uses suppression as its primary SDL technique for data releases based on business establishment censuses and surveys. From the outset, it was understood that primary suppression – not publishing easily identified data items – did not protect anything if the agency published the rest of the data, including summary statistics. Users could infer the missing items from what was published (Fellegi 1972).

The BLS, and other agencies that rely on suppression, make "complementary suppressions" to reduce the probability that a user can infer the sensitive items from the published data (Holan et al. 2010). But there is no optimal complementary suppression technology – there are usually multiple complementary suppression strategies that achieve the same protection.

Researchers, however, are not indifferent to these strategies. A researcher who needs detailed geographic variation will benefit from data in which the complementary suppressions are based on removing detailed industries. A researcher who needs detailed industry variation will prefer data with complementary suppression based on geography. Ultimately, the committee that chooses the complementary suppression strategy will determine which research uses are possible and which are ruled out.

But the problem is deeper than this: suppression is a very ineffective SDL technique. Researchers working with the cooperation of the BLS have shown that the suppression strategy used in major BLS business data publications provides almost no protection if it is applied, as is currently the case, to each data release separately (Holan et al. 2010). Some agencies may use cumulative suppression strategies in their sequential data releases. In this case, once an item has been designated for either primary or complementary suppression, it would disappear from the release tables until the entire product is redesigned.

Many social scientists believe that suppression can be complemented by restricted access agreements that allow the researcher to use all of the confidential data but limit what can be published from the analysis. Such a strategy is not a complete solution because SDL must still be applied to the output of the analysis, which quickly brings the problem of which output to suppress back to the forefront.

Custom tabulations and data enclaves. Another traditional response by data custodians to the demand by researchers for more extensive and detailed summaries of confidential data, was to create a custom tabulation, a table not previously published, but generated by data custodian staff with access rights to the confidential data, and typically subject to the same suppression rules. As these requests increased, the tabulation and analysis work was offloaded onto researchers by providing them with access to protected microdata. This approach has expanded rapidly in the last two decades, and is widely used around the world. We discuss it in detail later in this chapter.

Coarsening is a method for protecting data that involves mapping confidential values into broader categories. The simplest method is a histogram, which maps values into (fixed) intervals. Intuitively, the broader the interval, the more protection is provided.

Sampling is a protection mechanism that can be applied either at the collection stage or at the data publication stage. At the collection stage, it is a natural part of conducting surveys. In combination with coarsening and the use of statistical weights, the basic idea is simple: if a table cell is based on only a few sampled individuals which collectively represent the underlying population, then statistical inference will not reveal the attributes of any particular individual with any precision, as long as the identity of the sampled individuals is not revealed. Both coarsening and sampling underlie the release of public use microdata samples.

2.2.1 Input Noise Infusion

Protection mechanisms for microdata are often similar in spirit, though not in their details, to the methods employed for tabular data. Consider coarsening, in which the more detailed response to a question (say, about income), is classified into a much smaller set of bins (for instance, income categories such as "[10 000; 25 000]"). In fact, many tables can be viewed as a coarsening of the underlying microdata, with a subsequent count of the coarsened cases.

Many microdata methods are based on **input noise infusion**: distorting the value of some or all of the inputs before any publication data are built. The Census Bureau uses this technique before building publication tables for many of its business establishment products and in the American Community Survey (ACS) publications, and we will discuss it in more detail for one of those data products later in this chapter. The noise infusion parameters can be set such that all of the published statistics are formally unbiased – the expected value of the published statistic equals the value of the confidential statistic with respect to the probability distribution of the infused noise – or nearly so. Hence, the disclosure risk and data quality can be conveniently summarized by two parameters: one measuring the absolute distortion in the data inputs and the other measuring the mean squared error of publication statistics (either overall for censuses or relative to the undistorted survey estimates).

From the viewpoint of empirical social sciences, however, all input distortion systems with the same risk-quality parameters are not equivalent. In a regression discontinuity design, for example, there will now be a window around the break point in the running variable that reflects the uncertainty associated with the noise infusion. If the effect is not large enough, it will be swamped by noise even though all the inputs to the analysis are unbiased, or nearly so. Once again, using the unmodified confidential data via a restricted access agreement does not completely solve the problem because once the noisy data have been published, the agency has to consider the consequences of allowing the publication of a clean regression discontinuity design estimate where the plot of the unprotected outcomes versus the running variable can be compared to the similar plot produced from the public noisy data.

An even more invasive input noise technique is **data swapping**. Sensitive data records (usually households) are identified based on *a priori* criteria. Then, sensitive records are compared to "nearby" records on the basis of a few variables. If there is a match, the values of some or all of the other variables are swapped (usually the geographic identifiers, thus effectively relocating the records in each other's location). The formal theory of data swapping was developed shortly after the theory of primary/complementary suppression (Dalenius and Reiss 1982, first presented at American Statistical Association (ASA) Meetings in 1978). Basically, the marginal distribution of the variables used to match the records is preserved at the cost of all joint and conditional distributions involving the swapped variables. In general, very little is published about the swapping rates, the matching variables, or the definition of "nearby," making analysis of the effects of this protection method very difficult. Furthermore, even arrangements that permit restricted access to the confidential files still require the use of the swapped data. Some providers destroy the unswapped data. Data swapping is used by the Census Bureau, NCHS, and many other agencies (FCSM 2005). The Census Bureau does not allow analysis of the unswapped decennial and ACS data except under extraordinary circumstances that usually involve the preparation of linked data from outside sources then reimposition of the original swap (so the records acquire the correct linked information, but the geographies are swapped according to the original algorithm before any analysis is performed). NCHS allows the use of unswapped data in its restricted access environment but prohibits publication of most subnational geographies when the research is published.

The basic problem for empirical social scientists is that agencies must have a general purpose data publication strategy in order to provide the public good that is the reason for incurring the cost of data collection in the first place. But this publication strategy inherently advantages certain analyses over others. Statisticians and computer scientists have developed two related ways to address this problem: synthetic data combined with validation servers and privacy-protected query systems. Statisticians define "synthetic data" as samples from the joint probability distribution of the confidential data that are released for analysis. After the researcher analyzes the synthetic data, the validation server is used to repeat some or all of the analyses on the underlying confidential data. Conventional SDL methods are used to protect the statistics released from the validation server.

2.2.2 Formal Privacy Models

Computer scientists define a privacy-protected query system as one in which all analyses of the confidential data are passed through a noise-infusion filter before they are published. Some of these systems use input noise infusion – the confidential data are permanently altered at the record level, and then all analyses are done on the protected data. Other formally private systems apply output noise infusion to the results of statistical analyses before they are released.

All formal privacy models define a cumulative, global privacy loss associated with all of the publications released from a given confidential database. This is called the total privacy-loss budget. The budget can then be allocated to each of the released queries. Once the budget is exhausted, no more analysis can be conducted. The researcher must decide how much of the privacy-loss budget to spend on each query – producing noisy answers to many queries or sharp answers to a few. The agency must decide the total privacy-loss budget for all queries and how to allocate it among competing potential users.

An increasing number of modern SDL and formal privacy procedures replace methods like deterministic suppression and targeted random swapping with some form of noisy query system. Over the last decade these approaches have moved to the forefront because they provide the agency with a formal method of quantifying the global disclosure risk in the output and of evaluating the data quality along dimensions that are broadly relevant.

Relatively recently, formal privacy models have emerged from the literature on database security and cryptography. In formal privacy models, the data are distorted by a randomized mechanism prior to publication. The goal is to explicitly characterize, given a particular mechanism, how much private information is leaked to data users.

Differential privacy is a particularly prominent and useful approach to characterizing formal privacy guarantees. Briefly, a formal privacy mechanism that grants ε-differential privacy places an upper bound, parameterized by ε, on the ability of a user to infer from the published output whether any specific data item, or response, was in the original, confidential data (see Dwork and Roth 2014 for an in-depth discussion).

Formal privacy models are very intriguing because they solve two key challenges for disclosure limitation. First, formal privacy models by definition provide provable guarantees on how much privacy is lost, in a probabilistic sense, in any given data publication. Second, the privacy guarantee does not require that the implementation details, specifically the parameter ε, be kept secret. This allows researchers using data published under formal privacy models to conduct fully *SDL-aware* analysis. This is not the case with many traditional disclosure limitation methods which require that key parameters, such as the swap rate, suppression rate, or variance of noise, not be made available to data users (Abowd and Schmutte 2015).

2.3 Confidentiality Protection in Linked Data: Examples

To illustrate the application of new disclosure avoidance techniques, we describe three examples of linked data and the means by which confidentiality protection is applied to each. First, the **Health and Retirement Study (HRS)** links extensive survey information to respondents' administrative data from the Social Security Administration (SSA) and the Center for Medicare and Medicaid Services (CMS). To protect confidentiality in the linked HRS–SSA data, its data custodians use a combination of restrictive licensing agreements, physical security, and restrictions on model output. Our second example is the Census Bureau's Survey of Income and Program Participation (SIPP), which has also been linked to earnings data from the Internal Revenue Service (IRS) and benefit data from the SSA. Census makes the linked data available to researchers as the **SIPP Synthetic Beta File (SSB)**. Researchers can directly access synthetic data via a restricted server and, once their analysis is ready, request output based on the original harmonized confidential data via a validation server. Finally, the **Longitudinal Employer-Household Dynamics Program** (LEHD) at the Census Bureau links data provided by 51 state administrations to data from federal agencies and surveys and censuses on businesses, households, and people conducted by the Census Bureau. Tabular summaries of LEHD are published with greater detail than most business and demographic data. The LEHD is accessible in restricted enclaves, but there are also restrictions on the output researchers can release. There are many other linked data sources. These three are each innovative in some fashion, and allow us to illustrate the issues faced when devising disclosure avoidance methods for linked data.

2.3.1 HRS–SSA

2.3.1.1 Data Description
The HRS is conducted by the Institute for Social Research at the University of Michigan. Data collection was launched in 1992 and has reinterviewed the original sample of respondents every two years since then. New cohorts and sample refreshment have made the HRS one of the largest representative longitudinal samples of Americans over 50, with over 26 000 respondents in a given wave (Sonnega and Weir 2014). In 2006, the HRS started collecting measures of physical function, biomarkers, and DNA samples. The collection of these additional sensitive attributes reinforces confidentiality concerns.

2.3.1.2 Linkages to Other Data
The HRS team requests permission from respondents to link their survey responses to other data resources, as described below. For consenting respondents,

HRS data are linked at the individual level to administrative records from Social Security and Medicare claims, thus allowing for detailed characterizations of income and wealth over time. Through this, the HRS is linked to at least a dozen other datasets. See Abowd, Schmutte, and Vilhuber (2018) or http://hrsonline.isr .umich.edu/index.php?p=reslis for a more complete list of linked datasets.

The CMS maintain claims records for the medical services received by essentially all Americans age 65 and older and those less than 65 years who receive Medicare benefits. These records include comprehensive information about hospital stays, outpatient services, physician services, home health care, and hospice care. When linked to the HRS interview data, this supplementary information provides far more detail on the health circumstances and medical treatments received by HRS participants than would otherwise be available.

Data from HRS interviews are also linked to information about respondents' employers. This improves information on employer-provided benefits, including pensions. While most pension-eligible workers have some idea of the benefits available through their pension plans, they generally are not knowledgeable about detailed provisions of the plans. By linking HRS interview data with detailed information on pension plans, researchers can better understand the contribution of the pension to economic circumstances and the effects of the pension structure on work and retirement decisions.

HRS data are also linked at the individual level to administrative records from Social Security and Medicare, Veteran's Administration, the National Death Index, and employer-provided pension plan information (Sonnega and Weir 2014).

2.3.1.3 Disclosure Avoidance Methods

To ensure privacy and confidentiality, all study participants' names, addresses, and contact information are maintained in a secure control file (National Institute on Aging and the National Institutes of Health 2017). Anyone with access to identifying information must sign a pledge of confidentiality. The survey data are only released to the research community after undergoing a rigorous process to remove or mask any identifying information. First a set of sensitive variables (such as state of residence or specific occupation) are suppressed or masked. Next, the remaining variables are tested for any possible identifying content. When testing is complete, the data files are subject to final review and approval by the HRS Data Release Protocol Committee. Data ready for public use are made available to qualified researchers via a secure website. Registration is required of all researchers before downloading files for analyses. In addition, use of linked data from other sources, such as Social Security or Medicare records, is strictly controlled under special agreements with specially approved researchers operating in secure computing environments that are periodically audited for compliance.

Additional protections involve distortion of the microdata prior to dissemination to researchers. Earnings and benefits variables such as those from SSA in the HRS are rounded or top coded (Deang and Davies 2009). Similarly, geographic classifications are limited to broad levels of aggregation (e.g. census divisions instead of states or states instead of counties).

The HRS uses licensing as its primary method of giving access to restricted files. A license can be secured only after meeting a stringent set of criteria that leads to a contractual agreement between the HRS, the researcher, and the researcher's employer. The license enables the user to receive restricted files and use them at the researcher's own institutional facility.

2.3.2 SIPP–SSA–IRS (SSB)

2.3.2.1 Data Description

The SIPP/SSA/IRS Public Use File, known as the SIPP Synthetic Beta File or SSB, combines variables from the Census Bureau's SIPP, the IRS individual lifetime earnings data, and the SSA individual benefit data. Aimed at a user community that was primarily interested in national retirement and disability programs, the selection of variables for the proposed SIPP/SSA/IRS-PUF focused on the critical demographic data to be supplied from the SIPP, earnings histories going back to 1937 from the IRS data maintained at SSA, and benefit data from SSA's master beneficiary records, linked using respondents' SSNs. After attempting to determine the feasibility of adding a limited number of variables from the SIPP directly to the linked earnings and benefit data, it was decided that the set of variables that could be added without compromising the confidentiality protection of the existing SIPP public use files was so limited that alternative methods had to be used to create a useful new file.

The technique adopted is called *partially synthetic data* with *multiple imputation of missing items*. The term "partially synthetic data" means that the person-level records are released containing some variables from the actual responses and other variables where the actual responses have been replaced by values sampled from the posterior predictive distribution (PPD) for that record, conditional on all of the confidential data. From 2003 until 2015, seven preliminary versions of the SSB were produced. In this chapter, we will focus on the protections that pertain to the linked nature of the data. The interested reader is referred to Abowd, Stinson, and Benedetto (2006) for details on data sources, imputation, and linkage. The analysis here is for the SSB version 4. Since version 4, two additional versions have been released with slightly different structure.[3] Subsequent versions are well-illustrated by the extensive analysis described here.

---

3 The latest version as of this writing is version 6.0.2 (U.S. Census Bureau 2015).

2.3.2.2 Disclosure Avoidance Methods

The existence of SIPP public use files poses a key challenge for disclosure avoidance. To protect the confidentiality of survey respondents, it was deemed necessary to prevent reidentification of a record that appears in the synthetic data against the existing SIPP public use files. Hence, all information regarding the dating of variables whose source was a SIPP response, and not administrative data, has to be made consistent across individuals regardless of the panel and wave from which the response was taken. The public use file contains several variables that were never missing and are not synthesized. These variables are: gender, marital status, spouse's gender, initial type of Social Security benefits, type of Social Security benefits in 2000, and the same benefit type variables for the spouse. All other variables in the SSB v4 were synthesized.

The model first imputes any missing data, then synthesizes the completed data (Reiter 2004). For each iteration of the missing data imputation phase and again during the synthesis phase, a joint PPD for all of the required variables is estimated according to the following protocol. At each node of the parent/child tree, a statistical model is estimated for each of the variables at the same level. The statistical model is a Bayesian bootstrap, logistic regression, or linear regression (possibly with transformed inputs). The missing data phase included nine iterations of estimation. The synthetic data phase occurred on the 10th iteration. Four missing data implicates were created. These constitute the completed data files that are the inputs to the synthesis phase. Four synthetic implicates were created for each missing data implicate, for a total of 16 synthetic implicates on the released file. Because copying the final weight to each implicate of the synthetic data would have provided an additional unsynthesized variable with 55 552 distinct values, the disclosure risk associated with the weight variable had to be addressed. A synthetic weight using a PPD based on the Multinomial/Dirichlet natural conjugate likelihood and prior was created.

2.3.2.3 Disclosure Avoidance Assessment

The link of administrative earnings, benefits and SIPP data adds a significant amount of information to an already very detailed survey and could pose potential disclosure risks beyond those originally managed as part of the regular SIPP public use file disclosure avoidance process. The synthesis of the earnings data meets the IRS disclosure officer's criteria for properly protecting the federal tax information found in the summary and detailed earnings histories used to create the longitudinal earnings variables.

The Census Bureau Disclosure Review Board at the time of release used two standards for disclosure avoidance in partially synthetic data. First, using the best available matching technology, the percentage of true matches relative to the size of the files should not be excessively large. Second, the ratio of true matches to the total number of matches (true and false) should be close to one-half.

The disclosure avoidance analysis (Abowd, Stinson, and Benedetto 2006) uses the principle that a potential intruder would first try to reidentify the source record for a given synthetic data observation in the existing SIPP public use files. Two distinct matching exercises – one *probabilistic* (Fellegi and Sunter 1969), one *distance-based* (Torra, Abowd, and Domingo-Ferrer 2006) – between the synthetic data and the harmonized confidential data were conducted.[4] The harmonized confidential data – actual values of the data items as released in the original SIPP public use files – are the equivalent of the best available information for an intruder attempting to reidentify a record in the synthetic data. Successful matches between the harmonized confidential data and the synthetic data represent potential disclosure risks. In practice, the intruder would also need to make another successful link to exogenous data files that contain direct identifiers such as names, addresses, telephone numbers, etc. The results from the experiments are conservative estimates of reidentification risk. For the probabilistic matching, the assessment matched synthetic and confidential files exactly on the unsynthesized variables of gender and marital status, and success of the matching exercise is assessed using a person identifier which is not, in fact, available in the released version of the synthetic data. Without the personid, an intruder would have to compare many more record pairs to find true matches, would not find any more true matches (the true match is guaranteed to be in the blocks being compared), and would almost certainly find more false matches. In fact, the records that can be reidentified represent only a very small proportion (less than 3%) of candidate records, and correct reidentifications are swamped by a sea of false reidentifications (Abowd, Stinson, and Benedetto 2006, p. 6).

In distance-based matching, records between the harmonized confidential and synthetic data are blocked in a similar way, and distances (or similarity scores) are computed for a given confidential record and every synthetic record within a block. The three closest records are declared matches, and the personid again checked to verify how often a true match is obtained. A putative intruder who treated the closest record as a match would correctly link about 1% of all synthetic records, and less than 3% in the worst-case subgroup (Abowd, Stinson, and Benedetto 2006, p. 8).

---

4 In much of the documentation for the SSB, the internal confidential files, harmonized across the SIPP panels and waves, and completed using the multiple imputation procedures that produced the four implicates at the root of the synthesis for confidentiality protection, are called the "Gold Standard" files. This nomenclature means that these are the files that would be provided to a researcher in the Census Bureau's restricted access environment (FSRDC). Chapter 9 in this volume discusses linking methodologies.

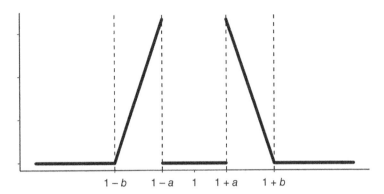

Figure 2.1 Probability density function of the ramp distribution used in LEHD disclosure avoidance system.

2.3.2.4 Analytical Validity Assessment

Although synthetic data are designed to solve a confidentiality protection problem, the success of this solution is measured by both the degree of protection provided and the user's ability to reliably estimate scientifically interesting quantities. The latter property of the synthetic data is known as analytical (or statistical) validity. Analytical validity exists when, at a minimum, estimands can be estimated without bias and their confidence intervals (or the nominal level of significance for hypothesis tests) can be stated accurately (Rubin 1987). To verify analytical validity, the confidence intervals surrounding the point estimates obtained from confidential and synthetic data should completely overlap (Reiter, Oganian, and Karr 2009), presumably with the synthetic confidence interval being slightly larger because of the increased variation arising from the synthesis. When these results are obtained, inferences drawn about the coefficients will be consistent whether one uses synthetic or completed data. The reader interested in detailed examples that show how analytic validity is assessed in the SSB should consult Figures 2.1 and 2.2 and associated discussion in Abowd, Schmutte, and Vilhuber (2018).

Box 2.1 Sidebox: Practical Synthetic Data Use

The SIPP–SSA–IRS Synthetic Beta File is accessible to users in its current form since 2010. Interested users can request an account by following links at https://www.vrdc .cornell.edu/sds/. Applications are judged solely on feasibility (i.e. the necessary variables are on the SSB). After projects are approved by the Census Bureau, researchers will be given accounts on the Synthetic Data Server. Users can submit validation requests, following certain rules, outlined on the **Census Bureau's website**. Deviations

---

(Continued)

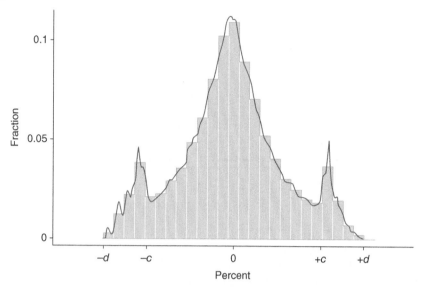

Figure 2.2 Distribution of ΔB in Maryland. For details, see text.

Box 2.1 (Continued)

from the guidelines may be possible with prior approval of the Census Bureau, but are typically only granted if specialized software is needed (other than SAS or Stata), and only if said software also exists already on Census Bureau computing systems. Between 2010 and 2016, over one hundred users requested access to the server, using a succession of continuously improved datasets.

2.3.3 LEHD: Linked Establishment and Employee Records

2.3.3.1 Data Description

The LEHD data links employee wage records extracted from Unemployment Insurance (UI) administrative files from 51 states with establishment-level records from the Quarterly Census of Employment and Wages (QCEW, also provided by the partner states), the SSA-sourced record of applications for SSNs ("Numident"), residential addresses derived from IRS-provided individual tax filings, and data from surveys and censuses conducted by the U.S. Census Bureau (2000 and 2010 decennial censuses, as well as microdata from the ACS). Additional information is linked in from the Census Bureau's Employer Business Register and its derivative files. The merged data are subject both to United States Code (U.S.C.) Title 13 and Title 26 protections. For more details, see Abowd, Haltiwanger, and Lane (2004) and Abowd et al. (2009).

From the data, multiple output products are generated. The Quarterly Workforce Indicators (QWI) provide local estimates of a variety of employment and earnings indicators, such as job creation, job destruction, new hires, separations, worker turnover, and monthly earnings, for detailed person and establishment characteristics, such as age, gender, firm age, and firm size (Abowd et al. 2009). The first QWI were released in 2003. The data are used for a variety of analyses and research, emphasizing detailed local data on demographic labor market variables (Gittings and Schmutte 2016; Abowd and Vilhuber 2012). Based on the same input data, the LEHD Origin-Destination Employment Statistics (LODES) describe the geographic distribution of jobs according to the place of employment and the place of worker residence (Center for Economic Studies 2016). New job-to-job flow statistics measure the movement of jobs and workers across industries and regional labor markets (Hyatt et al. 2014). The microdata underlying these products is heavily used in research, since it provides nearly universal coverage of U.S. workers observed at quarterly frequencies. Snapshots of the statistical production database are made available to researchers regularly (McKinney and Vilhuber 2011a,2011b; Vilhuber and McKinney 2014).

2.3.3.2 Disclosure Avoidance Methods

We describe in detail the disclosure avoidance method used for workplace tabulations in QWI and LODES (Abowd et al. 2012). Not discussed here are the additional disclosure avoidance methods applied in advance of publishing data on job flows (Abowd and McKinney 2016). Focusing on QWI and LODES is sufficient to highlight the types of confidentiality concerns that arise from working with these linked data, and the kinds of strategies the Census Bureau uses to address them.

In the QWI confidentiality protection scheme, confidential micro-data are considered protected by noise infusion if one of the following conditions holds: (1) any inference regarding the magnitude of a particular respondent's data must differ from the confidential quantity by at least c% even if that inference is made by a coalition of respondents with exact knowledge of their own answers (FCSM 2005, p. 72), or (2) any inference regarding the magnitude of an item is incorrect with probability not less than y%, where c and y are confidential but generally "large." Condition (1) is intended to prevent, say, a group of firms from "backing out" the total payroll of a specific competitor by combining their private information with the published total. Condition (2) prevents inference of counts of the number of workers or firms that satisfy some condition (say, the number of teenage workers employed in the fast food industry in Hull, GA) assuming item suppression or some additional protection, like synthetic data, when the count is too small.

Complying with these conditions involves the application of SDL throughout the data production process. It starts with the job-level data that record characteristics of the employment match between a specific individual and a specific workplace, or establishment, at a specific point in time. When the job-level data are aggregated to the establishment level, the QWI system adds statistical noise. This noise is designed to have three important properties. First, every job-level data point is distorted by some minimum amount. Second, for a given workplace, the data are always distorted in the same direction (increased or decreased) and by the same percentage magnitude in every period. Third, when the estimates are aggregated, the distortions added to individual data points tend to cancel out in a manner that preserves the cross-sectional and time-series properties of the data. The chosen distribution is a ramp distribution centered on unity, with a distortion of at least $a\%$ and at most $b\%$ (Figure 2.1).

All published data from QWI use the same noise-distorted data, and any special tabulations released from the QWI must follow the same procedures. The QWI system extends the idea of multiplicative noise infusion as a cross-sectional confidentiality protection mechanism first proposed by Evans, Zayatz, and Slanta (1998). A similar noise-infusion process has been used since 2007 to protect the confidentiality of data underlying the Census Bureau's CBP (Massell and Funk 2007) and was tested for application to the Commodity Flow Survey (Massell, Zayatz, and Funk 2006).

In addition to noise infusion, the QWI confidentiality protection system uses weighing, which introduces an additional difference between the confidential data item and the released data item. Finally, when a statistic meant to be published turns out to be based on data from fewer than three persons or establishments, it is suppressed. Suppression is only used when the combination of noise infusion and weighing may not distort the publication data with a high enough probability to meet the criteria laid out above; however the suppression rate is much lower than in comparable tabular publications, such as the QCEW.[5] An alternative to suppression (proposed by Gittings 2009; Abowd et al. 2012) uses a synthetic data model that replaces suppressed values with samples drawn from an appropriate PPD. The hybrid system incorporating both noise-infused and synthetic data allows the release of data without suppressions. The confidentiality protection provided by the hybrid system without suppressions is comparable to the protection afforded by the system using the noise infusion system with suppressions, but the analytical validity of the data produced by the hybrid system is improved because the synthetic data are better than the best inference an external user can make regarding the suppressions (Gittings 2009).

The LODES provides aggregated information on where workers are employed (Destinations) and where they live (Origins), along with the characteristics of those places. As the name implies, the data are intended for use in understanding

---

5 Not all estimates are subject to suppression. Estimates such as employment are subject to suppression. Continuous dollar measures like payroll are not (Abowd et al. 2009, 2012).

commuting patterns and the nature of local labor markets. The fundamental geographic unit in LODES is a Census block, and thus much more detailed than QWI for which data are published as county-level aggregates. LODES is tabulated from the same microdata as the QWI, and for workplaces (the destination), uses a variation of the QWI noise infusion technique. Cells that do not meet the publication criteria of the QWI continue to be suppressed in LODES, but are replaced using synthetic data.[6] For residences (the origin), the protection system relies on a provably-private synthetic data model (Machanavajjhala et al. 2008). A statistical model is built from the data, as the PPD of release data X' given the confidential data X: $\Pr[X'|X]$. Synthetic data points are sampled from the model X', *and released*. In general, to satisfy *differential privacy* (Dwork 2006; Dwork et al. 2006, 2017), the amount of noise that must be injected into the synthetic data model is quite large, typically rendering the releasable data of low utility. The novelty of the LODES protection system was to introduce the concept of "probabilistic differential privacy," and early variant of what are now called approximate differential privacy systems. By allowing the differential privacy guarantee (parametrized by ε) to fail in certain rare cases (which occur with probability δ), (ε, δ)-probabilistic differential privacy (Machanavajjhala et al. 2008) improves the analytical validity of the data greatly. LODES uses Census tract-to-tract relations to estimate the PPD for the block-to-block model. A unique model is estimated for each block, recovering the likelihood of a place of residence conditional on place of work and characteristics of the workers and the workplaces. Several additional measures further improve the privacy and analytical validity of the model (see Machanavajjhala et al. 2008 for further details). The resulting privacy-preserving algorithm guarantees ε-differential privacy of 8.99 with 99.999 999% confidence ($\delta = 10^{-6}$).

2.3.3.3 Disclosure Avoidance Assessment for QWI

The extent of the protection of the QWI micro-data can be measured in two ways: showing the percentage deviation as a measure of the uncertainty about the true value that one can infer from the released value, and the amount of reallocation of small cells (less than five entities in a tabulation cell).[7] Each cell underlying the tabulation is for a statistic X_{kt} where k is a cell defined by a combination of age, gender, industry, and county, and for all released time periods for the states at the time of these experiments.[8] The interested reader may find an example assessment in table 1 of Abowd, Schmutte, and Vilhuber (2018) undistorted, unweighted data.

---

6 Similar methods have been discussed for the QWI (Abowd et al. 2012; Gittings 2009), but not yet implemented.

7 The comparisons were computed using custom internal tabulations as well as published numbers, for two states (Illinois and Maryland). Only Maryland is reported here.

8 The disclosure avoidance assessment was run when first releasing the QWI, in 2003, and are reproduced here as they were presented to the Disclosure Review Board then. At the time, QWI

2.3.3.4 Analytical Validity Assessment for QWI

The noise infusion algorithm for QWI is designed to preserve validity of the data for particular analysis tasks. We demonstrate analytical validity using two statistics: time-series properties of the distorted data relative to the confidential data of several estimates, and the cross-sectional unbiasedness of the published data for beginning-of-quarter employment B. The unit of analysis is an interior substate geography $\times$ industry $\times$ age $\times$ sex cell kt.[9] Analytical validity is obtained when the data display no bias and the additional dispersion due to the confidentiality protection system can be quantified so that statistical inferences can be adjusted to accommodate it.

Time-Series Properties of Distorted Data

We estimate an AR(1) for the time series associated with each cell kt. For each cell, the error $\Delta r = r - r^*$ is computed, where r and r^* are the first-order serial correlation coefficient computing using confidential data and protected data, respectively. Table 2.1 shows the distribution of the errors Δr across SIC-division $\times$ county cells, for accessions A, beginning-of-quarter employment B, full-quarter employment F, net job flows JF, and separations S (for additional tables, see Abowd et al. 2012). Table 2.1 shows that the time series properties of the QWI remain largely unaffected by the distortion. The central tendency of the bias (as measured by the median of the Δr distribution) is never greater than 0.001, and the error distribution is tight: the semi-interquartile range of the distortion for B in Table 2.1 is 0.022, which is less than the precision with which estimated serial correlation coefficients are normally displayed.[10] The overall spread of the distribution is slightly higher when considering two-digit SIC $\times$ county and three-digit SIC $\times$ county cells (not reported here), due to the greater sparsity. The time series properties of the QWI data are unbiased. The small amount additional noise in the time series statistics is, in general, economically meaningless.

Cross-sectional Unbiasedness of the Distorted Data

The distribution of the infused noise is symmetric, and allocation of the noise factors is random. The data distribution resulting from the noise infusion should thus be unbiased. We compute the bias ΔX in each cell kt, expressed in percentage

---

were available for industry classifications according to the U.S. Department of Labor (DOL), 1987 definitions. Modern QWI are available for North American Industry Classification System (NAICS), 2012 definitions. The basic conclusion does not change.

9 Substate geography in all cases is a county, whereas the industry classification is the U.S. Department of Labor (DOL 1987).

10 The maximum semi-interquartile range for any SIC2-based variables is 0.0241, and for SIC3-based variables is 0.0244.

Table 2.1 Distribution of errors Δr in first-order serial correlation, QWI.

Variable	Median	Semi-interquartile range
Accessions	−0.000 542	0.026 314
Beginning-of-quarter employment	0.000 230	0.021 775
Full-quarter employment	0.000 279	0.018 830
Net job flows	−0.000 025	0.002 288
Separations	0.000 797	0.025 539

terms:

$$\Delta X_{kt} = \frac{X_{kt}^* - X_{kt}}{X_{kt}} \times 100$$

Evidence of unbiasedness is provided by Figure 2.2, which shows the distribution of the bias for $X = B$.[11] The distribution of ΔB has most mass around the mode at 0%. Also, as is to be expected, secondary spikes are present around $\pm c$, the inner bound of the noise distribution.

Box 2.2 Sidebox: Do-It-Yourself Noise Infusion

The interested user might consult a simple example (with fake data) at https://github .com/labordynamicsinstitute/rampnoise (Vilhuber 2017) that illustrates this mechanism.

2.4 Physical and Legal Protections

The provision of very detailed micro-tabulations or public-use microdata may not be sufficient to inform certain types of research questions. In particular, for business data the thresholds that trigger SDL suppression methods are met far more often than for individuals or households. In those cases, the research community needs controlled access to confidential microdata. Three key reasons why access to microdata may be beneficial are:

(i) microdata permit policy makers to pose and analyze complex questions. In economics, for example, analysis of aggregate statistics does not give a sufficiently accurate view of the functioning of the economy to allow analysis of the components of productivity growth;

---

11 Data for Maryland. For additional variables and states, see Abowd et al. (2012). All histograms are weighted by B_{kt}. Industry classification is three-digit SIC (industry groups).

(ii) access to microdata permits analysts to calculate marginal rather than just average effects. For example, microdata enable analysts to do multivariate regressions whereby the marginal impact of specific variables can be isolated;

(iii) broadly speaking, widely available access to microdata enables replication of important research

(United Nations 2007, p. 4)

As we've outlined above, many of the concerns about confidentiality have either removed or prevented creation of public-use microdata versions of linked files, exacerbating the necessity of providing alternate access to the confidential microdata.

NSOs and survey organizations usually provide access to confidential linked data within restricted-access data centers. In the United States, this means either using 1 of 30 secure sites managed by the Census Bureau as part of the Federal Statistical Research Data Center System (FSRDC),[12] or going to the headquarters of the statistical agency. Similarly, in other countries, access is usually restricted to headquarters of NSOs. Secure enclaves managed by NSOs used to be rare. In the 1990s and early 2000s, an expansion of existing networks and the creation of new, alternate methods of accessing data housed in secure enclaves occurred in several countries. Access methods may be through physical travel, remote submission, or remote processing. However, all methods rely on two fundamental elements. First, the researchers accessing the data are mostly free to choose the modeling strategy of their choice, and is not restricted to the tables or queries that the data curator has used for published statistics. Second, the output from such models is then analyzed to avoid unauthorized disclosure, and subsequently released to the researcher for publication.

Several methods are currently used by NSOs and other data collecting agencies to provide access to confidential data. Sections 2.4.1–2.4.5 will describe each of them in turn.[13]

2.4.1 Statistical Data Enclaves

Statistical data enclaves, or Research Data Centers, are secure computing facilities that provide researchers with access to confidential microdata, while putting restrictions on the content that can be removed from the facility. The different advisory committees of the two largest professional association (ASA, and the American Economic Association, AEA), pushed for easier and broader access for researchers as far back as the 1960s, though the emphasis then was on the avoiding the cost of making special tabulations. The AEA suggested creating

---

12 See https://www.census.gov/fsrdc (accessed on 15 December 2017).
13 The section draws on Weinberg et al. (2007) and Vilhuber (2013).

Census data centers at selected universities (Kraus 2013). In the 1990s and early 2000s, similar networks started in other countries. In Canada, the Canadian Foundation for Innovation (CFI) awarded a number of grants to open research data centers, with the first opening at McMaster University (Hamilton, Ontario) in 2000.[14] The creation of the RDCs was specifically motivated by the inability to ensure confidentiality while providing usability of longitudinally linked survey data (Currie and Fortin 2015).

In the United States, a 2004 grant by the National Science Foundation laid the groundwork for subsequent expansion of the (then Census) Research Data Center network from 8 locations, open since the mid-1990s, to over 30 locations in 2017. One of the key motivations was to make the newly available linked administrative data at LEHD accessible to researchers. The network operates under physical security constraints managed by the Census Bureau and the IRS, in locations that are considered part of the Census Bureau itself, and staffed by Census Bureau employees.

Statistical data enclaves can be central locations, in which a single location at the statistical agency is made available to approved researchers. In the United States, NCHS and BLS follow this model, in addition to using the FSRDC network. In Canada, business data can be accessed at Statistics Canada headquarters, while other data may be accessed both there and at the geographically dispersed RDCs, which obtain physical copies of the confidential data.

Some facilities are hybrid facilities. The statistical processing occurs at a central location, but the secure remote *access* facilities are distributed geographically. The U.S. FSRDCs have worked this way since the early 2000s. A central computing facility is housed in the Census Bureau's primary data center. Secure remote access is provided to approved researchers at designated sites throughout the county, namely the FSRDCs. Each of the FSRDC sites is a secure Census Bureau facility that is physically located on controlled premises provided by the partner organization, often a university or Federal Reserve Bank. The German IAB locates certified thin clients in dedicated rooms at partner institutions. Secure spaces are costly to build and certify. Recently, institutions in the United Kingdom have attempted to reduce the cost by commoditizing such secure spaces (Raab, Dibben, and Burton 2015). In France, the Centre d'accès sécurisé distant aux données (CASD) has a secure central computing facility, and allows for remote access through custom secure devices from designated but otherwise ordinary university offices, which satisfy certain physical requirements, but are not dedicated facilities. Similar arrangements are used by Scandinavian NSOs, as well as by survey organizations such as the HRS. Remote access to full

---

14 For an extensive history of the Canadian Research Data Center Network, see Currie and Fortin (2015).

desktop environments within the secure data enclave, commonly referred to as "virtual desktop infrastructure" (VDI), from regular laptops or workstations, is increasingly common.

The location of remote access points is often limited to the country of the data provider (United States, Canada), or to countries with reciprocal or common enforcement mechanisms (within the European Union, for European NSOs). Cross-border access, even within the European Union, remains exceedingly rare, with only a handful of cross-border secure remote access points open in the European Union. The most prolific user of cross-border secure remote access points, as of this writing, is the German IAB, with multiple data access points in the United States and a recently opened one in the United Kingdom.

2.4.2   Remote Processing

Two other alternative remote access mechanisms are often used: manual and automatic remote processing. Manual remote processing occurs when the remote "processor" is a staff member of the data provider. This can be as simple as sending programs in by email, or finding a co-author who is an employee of the data provider. The U.S. NCHS, German IAB, and Statistics Canada provide this type of access. Generally, the costs of manual remote processing are paid by the users.

More sophisticated mechanisms automate some or all of the data flow. For instance, programs may be executed automatically based on email or web submission, but disclosure review is performed manually. This method is used by the IAB's JoSuA (Institute for Employment Research 2016). Fully automated mechanisms, such as LISSY (Luxembourg), ANDRE (U.S. NCHS), DAS (U.S. NCES), Australia's Remote Access Data Laboratory (RADL), Canada's Real Time Remote Access (RTRA), generally restrict the command set from the allowed statistical programming languages (SAS, Stata, and SPSS) and limit what the users can do to certain statistical procedures and languages for which known automated disclosure limitation procedures have been implemented.

Most of these systems only provide access to household and person surveys. Of the known systems surveyed above, only Australia's RADL systems and the Bank of Italy's implementation of LISSY (Bruno, D'Aurizio, and Tartaglia-Polcini 2009, 2014) seem to provide access to *business* microdata through automated remote processing facilities.

2.4.3   Licensing

Users of secure research data centers always sign some form of legally binding user or licensing agreement. These agreements describe acceptable user behavior, such as not copying or photographing screen contents. However, licensing alone may also be used to provide access to restricted-use microdata outside of formal restricted access data centers. In general, the detail in licensed microdata files is greater than in the equivalent (or related) public-use file, and may allow for disclosure of confidential data if inappropriately exploited. For this reason, licensed microdata files tend to have several additional levels of disclosure avoidance methods applied, including output review in some cases. For instance, even without linkages, the HRS licensed files have more detailed geography on respondents (county, say, rather than Census region), but do not have the most detailed geography (GPS coordinates or exact address). Generally, the legally enforceable license imposes restrictions on what can be published by the researchers, and restricts who can access the data, and for what purpose. The contracting organization is the researcher's university, which is subject to penalties such as loss of eligibility status for research grants if the license is violated.

In the United States, some surveys (NCES, NLSY, and HRS) use licensing to distribute portions of the data they collect on their respondents. Commercial data providers (COMPUSTAT, etc.) also license the data distributed to researchers. Penalties for license infractions range from restricting future research grant funding, for example in HRS, to monetary penalties, for example in commercial data licenses. We are not aware of any studies that quantify the violation rates or financial penalties actually incurred due to license violations. Licensing may be limited by the enforceability of laws or contracts, and thus may be limited to residents of the same jurisdiction in which the data provider is housed. Often, some licensing is combined with the creation of ad-hoc data enclaves, the simplest of these being stand-alone, nonnetworked computer workstations.

2.4.4 Disclosure Avoidance Methods

Data enclaves exist to allow researchers to perform analyses within the restricted environment, and then extract or publish some form of statistical summary that can be released from the secure environment. Generally, these summaries are estimates from a statistical model. In general, model-based output is evaluated in accordance with the same criteria traditionally used for tabular output (minimum number of units within a reporting cell, minimum percentage of global activity within a reporting cell). In contrast to licensing arrangements, which allow researchers to self-monitor, statistical data enclaves have regimented output monitoring, typically by staff of the data provider. Generally, released statistical outputs are registered in some fashion, but documentation of the full provenance chain may be limited.

No systematic attempt has been made, to our knowledge, to measure formally the cumulative privacy impact of model-based releases because the science and technology for doing so are rudimentary. Remote processing facilities, on the other hand, when using automated mechanisms, rely on several practices to reduce the risk of disclosure. First, they limit the scope of possible analyses to those for which the agency has developed safe procedures. The number of times a researcher may request releases may also be limited. Nevertheless, most agencies recognize that this review system does not scale because the infeasibility of a full accounting of all possible query combinations over time. In general, they apply basic disclosure avoidance techniques such as suppression, perturbation, masking, recoding, and bootstrap sampling of the input data to each project separately. Some systems apply automated analysis of log and output files (Schouten and Cigrang 2003), although often a manual review is also included (O'Keefe et al. 2013). Some systems provide for self-monitored release of model results, either under licensing or remote access. There are also limitations on quantity and frequency of self-released results, combined with sampling by human reviewers. More sophisticated tools, such as perturbation or synthesizing of estimated model parameters, have been proposed (Reiter 2003). Finally, such systems require review of the draft research paper before submission to any publication medium including online preprint repositories like ArXiv.org.

All three of the examples of linked data provided in this paper rely on some version of secure data enclaves to provide microdata access to approved researchers. HRS data are made available to tenure-track researchers who sign a data use agreement and provide documentation of a secure local computing environment. An additional option for HRS data is to visit to the Michigan Center on the Demography of Aging data enclave, which makes data accessible to researchers in a physical data enclave at "headquarters," like many NSOs. More recently, HRS has started to offer secure VDI access to researchers. The confidential data underlying the SSB, and against which validation requests are run, are also available either within the FSRDC network, or by sending validation requests by email to staff at Census headquarters (a form of "remote processing"). LEHD microdata are only available through the FSRDC.

An open question is whether the disclosure risks addressed through physical security measures are greater for linked data. Enabling researchers to measure some of the heuristic disclosure risk such as n cell count or p-percent rule (O'Keefe et al. 2013) becomes more important when any possible combination of k variables (k large) leads to small cells or dominated cells. Even subject matter experts cannot assess these situations *a priori*.

2.4.5 Data Silos

One concern with the increasing move to multiple distinct access points for confidential data is the "**siloing**" of data. The critical symptom is a physical separation of files in distinct secure data enclaves. The underlying causes are the incompatible legal restrictions on different data. Typically, these restrictions impose administrative barriers to combining data sources for which linking is technically possible.

Such administrative barriers may also be driven by ethical or confidentiality concerns. The question of consent by survey or census respondents may explicitly prevent the linkage of their survey responses or of their biological specimen with other data. For example, the Canadian Census long form of 2006 offered respondents the option to either answer survey questions on earnings, or consent to linking in their tax data on earnings. In the 2016 census, the question was no longer asked, and users were simply notified that linkage would happen.

In the case of the LEHD data, as of December 2015, all 50 states as well as the District of Columbia had signed agreements with the Census Bureau to share data and produce public-use statistics. It would thus seem possible for researchers to access a comprehensive LEHD jobs database through the FSRDC network, by linking together the job databases from 51 administrative entities. However, all but 12 of the States had declined to automatically extend the right to use the data to external researchers within the FSRDC network. Nevertheless, some of the same states that declined to provide such permission in the FSRDC give access to researchers through their state data centers or other means. The UI state-level data is thus siloed, and researchers may be faced with nonrepresentative data on the American job market. Several European projects, such as Data without Boundaries (DwB), have investigated cross-national access with elevated expectations but relatively limited success (Schiller and Welpton 2014; Bender and Heining 2011). Increasingly, the U.S. Census Bureau and CASD also host data from other data providers, through collaborative agreements, moving toward a reduction of the siloing of data.

Secure multiparty computing may be one solution to this problem (Sanil et al. 2004; Karr et al. 2005, 2006, 2009). However, implementation of such methods, at least in the domain of the social and medical sciences cooperating with NSOs, is in its infancy (Raab, Dibben, and Burton 2015). The typical limitations are the throughput of the secure interconnection between the sources and the requirement of manual model output checking. These limitations drastically slow down any iterative procedure.

2.5 Conclusions

The goal of this chapter has been to illustrate how confidentiality protection methods can be and have been applied to linked administrative data. Our examples provide a guide to best-practices for data custodians endeavoring to walk the fine line between making data accessible and protecting individual privacy and confidentiality. Our examples also illustrate different paradigms of protection ranging from the more traditional approach of physical security to more modern formal privacy systems and the provision of synthetic data.

In concluding, we note that from a theoretical perspective, there does not appear to be a clear distinction between the threats to confidentiality in linked data relative to unlinked data, or in survey data relative to administrative data. Richly detailed data pose disclosure risks, irrespective of whether that richness is inherent in the data design, or comes from linkages of variables from multiple sources. Likewise, there are no special methods to protect confidentiality in linked versus unlinked data. Any data with a network, relational, panel or hierarchical structure poses special challenges to data providers to protect confidentiality while preserving analytical validity. Our example of the QWI shows one way this challenge has been successfully managed in a linked data setting, but the same tools could be effective in application to the QCEW, which uses the same frame, but does not involve worker-firm linkages.

However, from a legal perspective, linking two datasets can change the nature of confidentiality protection in a more practical manner. Any output must conform to the strongest privacy protections required across each of the linked datasets. For example, when the LEHD program links SSA data on individuals to IRS data on firms, any downstream research must comply with the confidentiality demands of all three agencies. Likewise, the data must conform to the U.S. Census Bureau publication thresholds for data involving individuals and firms. Hence, linking data can produce a maze of confidentiality requirements that are difficult to articulate, comply with, and monitor. Harmonizing or standardizing such requirements and practices across data providers, both public and private, and across jurisdictions would be helpful. Privacy and confidentiality issues also invite updated and continuing research on the demand for privacy from citizens and businesses, as well as the social benefit that arises from the dissemination of data.

2.A Appendix: Technical Terms and Acronyms

- ACS – American Community Survey, a large survey conducted continuously by the U.S. Census Bureau, on topics such as jobs and occupations, educational attainment, veterans, housing characteristics, and several other topics (https://www.census.gov/programs-surveys/acs/)
- BDS – Business Dynamics Statistics, produced by the U.S. Census Bureau, see https://www.census.gov/programs-surveys/bds.html for more details.
- CBP – County Business Patterns, produced by the U.S. Census Bureau, see www.census.gov/programs-surveys/cbp.html for more details.

- COEP – Canadian Out-of-Employment Panel, a survey initially conducted by McMaster University in Canada, subsequently taken over by the Statistics Canada (Browning, Jones, and Kuhn 1995)
- COMPUSTAT – a commercial database maintained by Standard and Poor's, with information on companies in the United States and around the world (http://www.compustat.com/).
- HRS – Health and Retirement Study, a long-running survey run by the Institute for Social Research at the University of Michigan in the United States on aging in the United States population (http://hrsonline.isr.umich.edu/)
- LEHD – Longitudinal Employer-Household Dynamics Program at the U.S. Census Bureau, which links data provided by 51 state administrations to data from federal agencies and surveys (https://lehd.ces.census.gov/)
- LODES – LEHD Origin-Destination Employment Statistics describe the geographic distribution of jobs according to the place of employment and the place of worker residence, in part through the flagship webapp OnTheMap (https://onthemap.ces.census.gov/)
- QWI – Quarterly Workforce Indicators, a set of local statistics of employment and earnings, produced by the Census Bureau's LEHD program (https://lehd.ces.census.gov/data/)
- SIPP – Survey of Income and Program Participation is conducted by the U.S. Census Bureau on topics such as economic well-being, health insurance, and food security (https://www.census.gov/sipp/).
- SSB – the SIPP Synthetic Beta File, also known as "SIPP/SSA/IRS Public Use File"

2.A.1 Other Abbreviations

- ABS – Australian Bureau of Statistics, the Australian NSO (http://abs.gov.au/)
- AEA – American Economic Association (https://www.aeaweb.org)
- ASA – American Statistical Association (https://www.amstat.org)
- BLS – Bureau of Labor Statistics, the NSO in the United States providing data on "labor market activity, working conditions, and price changes in the economy." (https://bls.gov)
- CASD – Centre d'accès sécurisé distant aux données, the French remote access system to most administrative data files (https://casd.eu)
- Census Bureau – the largest statistical agency in the United States (https://census.gov)
- CMS – Center for Medicare and Medicaid Services administers US government health programs such as Medicare, Medicaid, and others (https://cms.gov/)

- EIA – Energy Information Agency, collecting and disseminating information on energy generation and consumption in the United States (https://eia.gov).
- FICA – Federal Insurance Contribution Act, the law regulating the system of social security benefits in the United States
- IAB – Institute for Employment Research at the German Ministry of Labor (http://iab.de/en/iab-aktuell.aspx)
- FSRDC – Federal Statistical Research Data Centers were originally created as the U.S. Census Bureau Research Data Centers. They provide secure facilities for authorized remote access government restricted-use microdata, and are structured as partnerships between federal statistical agencies and research institutions (https://www.census.gov/fsrdc)
- IRS – Internal Revenue Service handles tax collection for the US government (https://irs.gov)
- NCHS – National Center for Health Statistics, the US NSO charged with collecting and disseminating information on health and well-being (https://www.cdc.gov/nchs/)
- NSO – National statistical offices. Most countries have a single national statistical agency, but some countries (USA, Germany) have multiple statistical agencies
- OASDI – Old Age, Survivors and Disability Insurance program, the official name for Social Security in the United States
- QCEW – Quarterly Census of Employment and Wages is a program run by the BLS, collecting firm-level reports of employment and wages, and publishing quarterly estimates for about 95% of US jobs (https://www.bls.gov/cew/)
- SER – Summary Earnings Records on SSA data
- SSA – Social Security Administration, administers government-provided retirement, disability, and survivors benefits in the United States (https://ssa.gov)
- SSN – Social Security Number, an identification number in the United States, originally used for management of benefits administered by the SSA, but since expanded and serving as a quasi-national identifier number
- UI – Unemployment Insurance, which in the United States are administered by each of the states (and District of Columbia)
- U.S.C – United States Code is the official compilation of laws and regulations in the United States

2.A.2 Concepts

- *Analytical validity*: It exists when, at a minimum, estimands can be estimated without bias and their confidence intervals (or the nominal level of significance for hypothesis tests) can be stated accurately (Rubin 1987). The estimands can be summaries of the univariate distributions of the variables, bivariate measures of association, or multivariate relationships among all variables.

- *Coarsening*: A method for protecting data that involves mapping confidential values into broader categories, e.g. a histogram.
- *Confidentiality*: A "quality or condition accorded to information as an obligation not to transmit [...] to unauthorized parties" (Fienberg 2005, as quoted in Duncan, Elliot, and Salazar-González 2011). Confidentiality addresses data already collected, whereas privacy (see below) addresses the right of an individual to consent to the collection of data.
- *Data swapping*: Sensitive data records (usually households) are identified based on *a priori* criteria, and matched to "nearby records." The values of some or all of the other variables are swapped, usually the geographic identifiers, thus effectively relocating the records in each other's location.
- *Differential privacy*: A class of formal privacy mechanisms. For instance, ε-differential privacy places an upper bound, parameterized by ε, on the ability of a user to infer from the published output whether any specific data item, or response, was in the original, confidential data (Dwork and Roth 2014).
- *Dirichlet-multinomial distribution*: A family of discrete multivariate probability distributions on a finite support of nonnegative integers. The probability vector p of the better-known multinomial distribution is obtained by drawing from a Dirichlet distribution with parameter α.
- *Input noise infusion*: Distorting the value of some or all of the inputs before any publication data are built or released.
- *Posterior predictive distribution (PPD)*: In Bayesian statistics, the distribution of all possible values conditional on the observed values.
- *Privacy*: "An individual's freedom from excessive intrusion in the quest for information and [...] ability to choose [... what ...] will be shared or withheld from others" (Duncan, Jabine, and de Wolf 1993, quoted in Duncan, Elliot, and Salazar-González 2011). See also confidentiality, above.
- *Sampling*: As part of SDL, works by only publishing a fractional part of the data.
- *Statistical confidentiality or SDL – Statistical disclosure limitation*: Can be viewed as "a body of principles, concepts, and procedures that permit confidentiality to be afforded to data, while still permitting its use for statistical purposes" (Duncan, Elliot, and Salazar-González 2011, p. 2).
- *Suppression*: Describes the removal of cells from a published table if its publication would pose a high risk of disclosure.

Acknowledgments

John M. Abowd is the Associate Director for Research and Methodology and Chief Scientist, U.S. Census Bureau, the Edmund Ezra Day Professor of Economics, Professor of Statistics and Information Science, and the Director of the Labor Dynamics Institute (LDI) at Cornell University, Ithaca, NY, USA. https://johnabowd.com. Ian M. Schmutte is Associate Professor of Economics at the University of Georgia, Athens, GA, USA. http://ianschmutte.org. Lars Vilhuber is Senior Research Associate in the Department of Economics and Executive Director of Labor Dynamics Institute (LDI) at Cornell University, Ithaca, NY, USA. https://lars.vilhuber.com. The authors acknowledge the support of a grant from the Alfred P. Sloan Foundation (G-2015-13903), NSF Grants SES-1131848, BCS-0941226, TC-1012593. Any opinions and conclusions expressed herein are those of the authors and do not necessarily represent the views of the U.S. Census Bureau, the National Science Foundation, or the Sloan Foundation. All results presented in this work stem from previously released work, were used by permission, and were previously reviewed to ensure that no confidential information is disclosed.

References

Abowd, J.M. and McKinney, K.L. (2016). Noise infusion as a confidentiality protection measure for graph-based statistics. *Statistical Journal of the IAOS* 32 (1): 127–135. https://doi.org/10.3233/SJI-160958.

Abowd, J.M. and Schmutte, I.M. (2015). Economic analysis and statistical disclosure limitation. *Brookings Papers on Economic Activity* 50 (1): 221–267.

Abowd, J.M. and Vilhuber, L. (2012). Did the housing price bubble clobber local labor market job and worker flows when it burst? *The American Economic Review* 102 (3): 589–593. https://doi.org/10.1257/aer.102.3.589.

Abowd, J.M., Haltiwanger, J., and Lane, J. (2004). Integrated longitudinal employer–employee data for the United States. *The American Economic Review* 94 (2): 224–229.

Abowd, J.M., Stinson, M., and Benedetto, G. (2006). Final Report to the Social Security Administration on the SIPP/SSA/IRS Public Use File Project. 1813/43929. U.S. Census Bureau. http://hdl.handle.net/1813/43929.

Abowd, J.M., Stephens, B.E., Vilhuber, L. et al. (2009). The LEHD infrastructure files and the creation of the quarterly workforce indicators. In: *Producer Dynamics: New Evidence from Micro Data* (eds. T. Dunne, J.B. Jensen and M.J. Roberts). University of Chicago Press.

Abowd, J.M., Kaj Gittings, R., McKinney, K.L., et al. (2012). Dynamically consistent noise infusion and partially synthetic data as confidentiality protection measures for related time series. US Census Bureau Center for Economic Studies Paper No. CES-WP-12-13. http://dx.doi.org/10.2139/ssrn.2159800.

Abowd, J.M., Schmutte, I.M., and Vilhuber, L. (2018). Disclosure avoidance and confidentiality protection in linked data. U.S. Census Bureau Center for Economic Studies Working Paper CES-WP-18-07.

Australian Bureau of Statistics (2015). Media release – ABS response to privacy impact assessment. Australian Bureau of Statistics. http://abs.gov.au/AUSSTATS/abs@ .nsf/mediareleasesbyReleaseDate/C9FBD077C2C948AECA257F1E00205BBE? OpenDocument (accessed 05 August 2020).

Bender, S. and Heining, J. (2011). The research-data-centre in research-data-centre approach: a first step towards decentralised international data sharing. *IASSIST Quarterly/International Association for Social Science Information Service and Technology* 35 (3) https://www.iassistquarterly.com/index.php/iassist/article/ view/119.

Browning, M., Jones, S., and Kuhn, P.J. (1995). *Studies of the Interaction of UI and Welfare Using the COEP Dataset*. LU2-153/224-1995E, Unemployment Insurance Evaluation Series. Ottawa: Human Resources Development Canada. http:// publications.gc.ca/collections/collection_2015/rhdcc-hrsdc/LU2-153-224-1995- eng.pdf.

Bruno, G., D'Aurizio, L., and Tartaglia-Polcini, R. (2009). Remote processing of firm microdata at the Bank of Italy. No. 36, Bank of Italy. http://dx.doi.org/10.2139/ssrn .1396224 (accessed 05 August 2020).

Bruno, G., D'Aurizio, L., and Tartaglia-Polcini, R. (2014). Remote processing of business microdata at the Bank of Italy. In: *Statistical Methods and Applications from a Historical Perspective*, Studies in Theoretical and Applied Statistics (eds. F. Crescenzi and S. Mignani), 239–249. Springer International Publishing. http://link .springer.com/chapter/10.1007/978-3-319-05552-7_21.

Center for Economic Studies (2016). LODES Version 7. OTM20160223. U.S. Census Bureau. http://lehd.ces.census.gov/doc/help/onthemap/OnTheMapDataOverview .pdf (accessed 05 August 2020).

Currie, R. and Fortin, S. (2015). Social statistics matter: history of the Canadian Research Data Center Network. Canadian Research Data Centre Network. http:// rdc-cdr.ca/sites/default/files/social-statistics-matter-crdcn-history.pdf (accessed 05 August 2020).

Dalenius, T. and Reiss, S.P. (1982). Data-swapping: a technique for disclosure control. *Journal of Statistical Planning and Inference* 6 (1): 73–85. https://doi.org/10.1016/ 0378-3758(82)90058-1.

Deang, L.P. and Davies, P.S. (2009). Access restrictions and confidentiality protections in the Health and Retirement Study. No. 2009–01, U.S. Social Security Administration. https://www.ssa.gov/policy/docs/rsnotes/rsn2009-01.html.

DeSalvo, B., Limehouse, F.F., and Klimek, S.D. (2016). Documenting the business register and related economic business data. Working Papers 16–17. Center for Economic Studies. U.S. Census Bureau. https://ideas.repec.org/p/cen/wpaper/16-17.html.

Duncan, G.T., Jabine, T.B., and de Wolf, V.A. (eds.); Panel on Confidentiality and Data Access, Committee on National Statistics, Commission on Behavioral and Social Sciences and Education, National Research Council and the Social Science Research Council (1993). *Private Lives and Public Policies: Confidentiality and Accessibility of Government Statistics*. Washington, DC: National Academy of Sciences.

Duncan, G.T., Elliot, M., and Salazar-González, J.J. (2011). *Statistical Confidentiality: Principles and Practice*, Statistics for Social and Behavioral Sciences. New York: Springer-Verlag.

Dwork, C. (2006). Differential privacy. In: *Automata, Languages and Programming*, Lecture Notes in Computer Science, vol. 4052 (eds. M. Bugliesi, B. Preneel, V. Sassone and I. Wegener), 1–12. Berlin, Heidelberg: Springer Berlin Heidelberg. http://link.springer.com/10.1007/11787006_1.

Dwork, C. and Roth, A. (2014). The algorithmic foundations of differential privacy. *Foundations and Trends® in Theoretical Computer Science* 9 (3–4): 211–407. https://doi.org/10.1561/0400000042.

Dwork, C., McSherry, F., Nissim, K., Smith, A. (2006). Calibrating noise to sensitivity in private data analysis. In: *Proceedings of the 3rd Theory of Cryptography Conference*, pp. 265–284.

Dwork, C., Smith, A., Steinke, T., Ullman, T. (2017). Exposed! A Survey of Attacks on Private Data. *Annual Review of Statistics and Its Application*, 4 (1): 61–84.

Evans, T., Zayatz, L., and Slanta, J. (1998). Using noise for disclosure limitation of establishment tabular data. *Journal of Official Statistics* 14 (4): 537–551.

FCSM (2005). Report on statistical disclosure limitation methodology. Working Paper 22 (second version, 2005). Federal Committee on Statistical Methodology. https://s3.amazonaws.com/sitesusa/wp-content/uploads/sites/242/2014/04/spwp22.pdf.

Fellegi, I.P. (1972). On the question of statistical confidentiality. *Journal of the American Statistical Association* 67 (337): 7–18.

Fellegi, I.P. and Sunter, A.B. (1969). A theory for record linkage. *Journal of the American Statistical Association* 64 (328): 1183–1210. https://doi.org/10.1080/01621459.1969.10501049.

Fienberg, S.E. (2005). Confidentiality and disclosure limitation. In: *Encyclopedia of Social Measurement* (ed. K. Kempf-Leonard), 463–469. New York, NY: Elsevier.

Gittings, R. (2009). Essays in labor economics and synthetic data methods. PhD thesis. Cornell University, Ithaca, NY, USA. https://ecommons.cornell.edu/handle/1813/14039.

Gittings, R.K. and Schmutte, I.M. (2016). Getting handcuffs on an octopus: minimum wages, employment, and turnover. *ILR Review* 69 (5): 1133–1170. https://doi.org/10.1177/0019793915623519.

Holan, S.H., Toth, D., Ferreira, M.A.R., and Karr, A.F. (2010). Bayesian multiscale multiple imputation with implications for data confidentiality. *Journal of the American Statistical Association* 105 (490): 564–577. https://doi.org/10.1198/jasa.2009.ap08629.

Hyatt, H., McEntarfer, E., McKinney, K., et al. (2014). JOB-TO-JOB (J2J) flows: new labor market statistics from linked employer-employee data. Working Papers 14–34. Center for Economic Studies. U.S. Census Bureau. https://ideas.repec.org/p/cen/wpaper/14-34.html.

Institute for Employment Research (2016). Job submission application (JoSuA) at the Research Data Centre of the Federal Employment Agency: user manual. https://josua.iab.de/gui/manual.pdf (accessed 05 August 2020).

Karp, P. (2016). Census controversy shows ABS 'needs to do better', says Statistical Society. *The Guardian* (9 August 2016). http://www.theguardian.com/australia-news/2016/aug/09/census-controversy-shows-abs-needs-to-do-better-says-statistical-society.

Karr, A.F., Lin, X., Sanil, A.P., and Reiter, J.P. (2005). Secure regression on distributed databases. *Journal of Computational and Graphical Statistics* 14 (2): 263–279. https://doi.org/10.1198/106186005X47714.

Karr, A.F., Lin, X., Sanil, A.P., and Reiter, J.P. (2006). Secure statistical analysis of distributed databases. In: *Statistical Methods in Counterterrorism* (eds. A.G. Wilson, G.D. Wilson and D.H. Olwell), 237–261. New York: Springer. http://link.springer.com/chapter/10.1007/0-387-35209-0_14.

Karr, A.F., Lin, X., Sanil, A.P., and Reiter, J.P. (2009). Privacy-preserving analysis of vertically partitioned data using secure matrix products. *Journal of Official Statistics* 25 (1): 125–138.

Kraus, R. (2013). Statistical Déjà vu: the national data center proposal of 1965 and its descendants. *Journal of Privacy and Confidentiality* 5 (1) : 1–37. https://doi.org/10.29012/jpc.v5i1.624 Accessed online (01/19/2021) at https://journalprivacyconfidentiality.org/index.php/jpc/article/view/624.

Machanavajjhala, A., Kifer, D., Abowd, J.M. et al. (2008). Privacy: theory meets practice on the map. In: *Proceedings of the International Conference on Data Engineering IEE*, 277–286. https://doi.org/10.1109/ICDE.2008.4497436.

Massell, P.B. and Funk, J.M. (2007). Recent developments in the use of noise for protecting magnitude data tables: balancing to improve data quality and rounding that preserves protection. *Proceedings of the 2007 FCSM Research Conference*. Council of Professional Associations on Federal Statistics. Washington, DC, USA (5–7 November, 2007). https://nces.ed.gov/FCSM/pdf/2007FCSM_Massell-IX-B.pdf

Massell, P., Zayatz, L., and Funk, J. (2006). Protecting the confidentiality of survey tabular data by adding noise to the underlying microdata: application to the commodity flow survey. In: *Privacy in Statistical Databases*, Lecture Notes in Computer Science (eds. J. Domingo-Ferrer and L. Franconi), 304–317. Springer Berlin Heidelberg. https://doi.org/10.1007/11930242_26.

McKinney, K.L. and Vilhuber, L. (2011a). LEHD data documentation LEHD-Overview-S2008-rev1. Working Papers 11–43. Center for Economic Studies. U.S. Census Bureau. https://ideas.repec.org/p/cen/wpaper/11-43.html.

McKinney, K.L. and Vilhuber, L. (2011b). LEHD infrastructure files in the Census RDC: overview of S2004 snapshot. Working Papers 11–13. Center for Economic Studies. U.S. Census Bureau. https://ideas.repec.org/p/cen/wpaper/11-13.html.

National Institute on Aging and the National Institutes of Health (2017). *Growing Older in America: The Health and Retirement Study*. University of Michigan. http://hrsonline.isr.umich.edu/index.php?p=dbook.

O'Keefe, C.M., Westcott, M., Ickowicz, A., et al. (2013). Protecting confidentiality in statistical analysis outputs from a virtual data centre. Joint UNECE/Eurostat work session on statistical data confidentiality.

Raab, G.M., Dibben, C., and Burton, P. (2015). Running an analysis of combined data when the individual records cannot be combined: practical issues in secure computation. Joint UNECE/Eurostat work session on statistical data confidentiality. http://www1.unece.org/stat/platform/display/SDCWS15/Statistical+Data+Confidentiality+Work+Session+Oct+2015+Home.

Reiter, J.P. (2003). Model diagnostics for remote-access regression servers. *Statistics and Computing* 13: 371–380.

Reiter, J.P. (2004). Simultaneous use of multiple imputation for missing data and disclosure limitation. *Survey Methodology* 30: 235–242.

Reiter, J.P., Oganian, A., and Karr, A.F. (2009). Verification servers: enabling analysts to assess the quality of inferences from public use data. *Computational Statistics & Data Analysis* 53 (4): 1475–1482. https://doi.org/10.1016/j.csda.2008.10.006.

Rubin, D.B. (1987). The calculation of posterior distributions by data augmentation: comment: a noniterative sampling/importance resampling alternative to the data augmentation algorithm for creating a few imputations when fractions of missing information are modest: the SIR algorithm. *Journal of the American Statistical Association* 82 (398): 543–546.

Sanil, A.P., Karr, A.F., Lin, X., and Reiter, J.P. (2004). Privacy preserving regression modelling via distributed computation. In: *Proceedings of the Tenth ACM SIGKDD International Conference on Knowledge Discovery and Data Mining*, 677–682. ACM. https://doi.org/10.1145/1014052.1014139.

Schiller, D. and Welpton, R. (2014). Distributing access to data, not data – providing remote access to European microdata. *IASSIST Quarterly* 38 (3). https://www.iassistquarterly.com/index.php/iassist/article/view/122.

Schouten, B. and Cigrang, M. (2003). Remote access systems for statistical analysis of microdata. Discussion Paper 03004. Statistics Netherlands. https://www.oecd.org/std/37502934.pdf.

Sonnega, A. and Weir, D.R. (2014). The Health and Retirement Study: a public data resource for research on aging. *Open Health Data* 2 (1): 576. https://doi.org/10.5334/ohd.am.

Torra, V., Abowd, J.M., and Domingo-Ferrer, J. (2006). Using Mahalanobis distance-based record linkage for disclosure risk assessment. In: *Privacy in Statistical Databases*, vol. 4302 (eds. J. Domingo-Ferrer and L. Franconi), 233–242. Berlin, Heidelberg: Springer Berlin Heidelberg. http://link.springer.com/10.1007/11930242_20.

U.S. Census Bureau (2015). SIPP Synthetic Beta Version 6.0.2. Washington, DC and Ithaca, NY, USA. http://www2.vrdc.cornell.edu/news/data/sipp-synthetic-beta-file/ (accessed 05 August 2020).

U.S. Department of Labor (DOL). (1987). Standard Industrial Classification (SIC) Manual. Occupational Safety and Health Administration (OSHA): https://www.osha.gov/data/sic-manual (accessed 01/19/2021).

United Nations (2007). *Managing Statistical Confidentiality and Microdata Access – Principles and Guidelines of Good Practice.* United Nations Economic Commission for Europe – Conference of European Statisticians. https://www.unece.org/fileadmin/DAM/stats/publications/Managing.statistical.confidentiality.and.microdata.access.pdf.

Vilhuber, L. (2013). Methods for Protecting the Confidentiality of Firm-Level Data: Issues and Solutions. 19. Labor Dynamics Institute. http://digitalcommons.ilr.cornell.edu/ldi/19/.

Vilhuber, L. (2017). Labordynamicsinstitute/rampnoise: code for multiplicative noise infusion. https://doi.org/10.5281/zenodo.1116352 (accessed 05 August 2020).

Vilhuber, L. and McKinney, K. (2014). LEHD infrastructure files in the Census RDC – overview. Working Papers 14–26. Center for Economic Studies, U.S. Census Bureau. http://ideas.repec.org/p/cen/wpaper/14-26.html.

Weinberg, D.H., Abowd, J.M., Steel, P.M., et al. (2007). Access methods for United States microdata. Working Papers 07–25. Center for Economic Studies, U.S. Census Bureau. https://ideas.repec.org/p/cen/wpaper/07-25.html.

Part II

Data Quality of Administrative Records and Linking Methodology

3

Evaluation of the Quality of Administrative Data Used in the Dutch Virtual Census

Piet Daas, Eric S. Nordholt, Martijn Tennekes and Saskia Ossen

Statistics Netherlands

3.1 Introduction

All European Union countries conducted population and housing censuses in 2011. The way the census was conducted was up to the countries. In the Netherlands, so-called virtual censuses have been used ever since the last traditional Census was held in 1971. This means that census forms are no longer used to collect data and that the required information has to be derived from data included in already existing administrative registers and available survey data (Schulte Nordholt 2014). In this way, the Virtual Censuses of 1981, 1991, 2001, and 2011 were conducted. The censuses of 1981 and 1991 were of a limited character; the data were much less detailed than the set of tables of the 2001 Census. In 2001, Statistics Netherlands published census information on the municipal level. For the 2011 Census, even more registers were combined. The population register forms the backbone for the integration activities that eventually resulted in coherent and detailed demographic and socio-economic statistical information on persons and households.

A generic problem in using administrative registers for statistical purposes is that the data in these sources are collected and maintained by other organizations for nonstatistical purposes. The underlying process is beyond the control of statistical agencies. This not only makes the agencies highly dependent on other organizations, but also affects the quality of the output. Statistical agencies, particularly those of the Netherlands and the Nordic countries, are expected to use more and more registers in the future in order to lower the administrative burden of data collection (United Nations 2007). In order to determine the potential use

Administrative Records for Survey Methodology, First Edition.
Edited by Asaph Young Chun, Michael D. Larsen, Gabriele Durrant, and Jerome P. Reiter.
© 2021 John Wiley & Sons, Inc. Published 2021 by John Wiley & Sons, Inc.

of externally collected registers, a framework has been developed to determine the quality of the data provided prior to use (Daas et al. 2009); this is also referred to as input quality. This framework was used to study the input quality of the most important registers used in the Dutch Virtual Census 2011. The results of these studies are described in this chapter.

In Section 3.2, the data sources and variables of the 2011 Dutch Census considered in this chapter are introduced. The focus of this chapter is on the quality of a number of registers used to obtain information on *the highest level of educational attainment* and *current activity status*. In addition, a source for *housing information* is included. These were the most important variables for which quality was a serious concern in the virtual census of 2011 and – as a consequence – also for other future virtual censuses. In Section 3.3, the quality framework is described in detail. Next, the results of applying the framework are discussed. Finally, some conclusions are drawn. This chapter is closely tied to the work described in Chapters 5, 7, and 8 in this book. In contrast to Chapters 5, 7, and 8, however, this chapter focuses on evaluating the potential use of administrative sources, at various levels, at the start of the statistical production process. The question dominating the work described here is: What is the quality of the registers when used as input for the virtual census?

3.2 Data Sources and Variables

The Dutch Population register is the backbone of the Census. Information from other registers and surveys is added to eventually derive all 2011 Census variables. All information on persons can be easily combined using the unique Citizen Service Number, in Dutch: "Burgerservicenummer" (BSN 2017), that was introduced in 2007. In this context, it is important to realize that the composition of registers may change over time and so does their quality. So, although all available registers can be joined to the population register in order to derive the variables for a particular census, the part of the population for which the registers provide information, as well as the quality of that information, may differ from previous censuses.

The decisions about which data sources were used to produce the different variables in the 2011 Census were predominantly based on the quality of the sources containing information about the variables. In this chapter, a number of registers will be compared for a limited set of variables. The most important are the *highest level of educational attainment* and *current activity status*. In addition, *housing information* is compared.

The *highest level of educational attainment* is an important variable for comparison to past time periods and across society. Information regarding this variable can be found in the Dutch Labor Force Survey (LFS) which contains only a small fraction (approximately 1%) of the population aged 15–64 per calendar year. Information about many more people can be found in the education register. All graduates of secondary education and higher are registered here, except for a small part that graduates at some special-private schools. However, the coverage of the education register gradually increased over the years meaning that, for people of increasing age, less information is available. Ideally, information from both sources would be combined. For the Census, information from one of these sources might be enough to produce reliable and consistent tables.

Current activity status is a variable that includes many different categories such as employed, unemployed, and homemakers. Information about employed people comes from a number of combined registers not discussed here. Information about unemployment according to the International Labor Organization definition can be obtained from LFS survey data. However, another option is to derive unemployment from registers containing information on benefits. The unemployment benefit register and the social security register are the most important sources of information for this. By combining all information available in registers, most categories of the variable *current activity status* can be derived. All persons for which no employment data is available and are included in the unemployment benefit register or the social security register are considered to be unemployed. However, the information in these combined registers is nearly integral but does not have the exact definition of unemployment needed for the census. The research question here was what information was best for the 2011 Census: sample information from the LFS with the correct definition or the nearly integral information from registers with an approximation of the official definition?

The new housing register is an example of a register that is introduced in the 2011 Census to derive the census variable *housing information*. For the 2001 Census, this variable has been derived from two other data sources; viz. the old housing register and the survey on housing conditions. A disadvantage of the new housing register is that it lacks some information. Since some of the variables in the housing register are also available in other sources (e.g. in the land use register), the question is which of the sources should be used to derive specific census variables?

The brief overview given above clearly reveals that the education, unemployment benefit, social security, housing, and population registers may provide useful information for deriving one of the variables under concern. In this chapter, the current state and quality of the information about *highest level of education attainment*, *current activity status*, and *housing information* available in the registers (and in the LFS) will be studied using the quality framework for registers.

3.3 Quality Framework

The quality framework for registers was developed to standardize the quality assessment of administrative registers (Daas et al. 2009). The aim of this framework was to provide a complete overview of the quality by means of three high level views (Daas, Ossen, and Tennekes 2010), also referred to as hyper dimensions (Karr, Sanil, and Banks 2006). They are called the source, metadata, and data hyper dimension. Each hyper dimension is composed of several dimensions of quality and each dimension contains a number of quality indicators. A quality indicator is measured or estimated by one or more methods which can be qualitative or quantitative (Daas et al. 2009).

In Section 3.3.1, an overview is provided of the quality aspects in the Source and Metadata hyper dimension and the methods developed to determine them. Next, the quality aspects in the data hyper dimension are described (Daas et al. 2011).

3.3.1 Source and Metadata Hyper Dimensions

A statistical agency that plans to use an administrative register should start by exploring the quality of the information that enables the use of the data source on a regular basis. This quality aspect is located in the source hyper dimension of the quality framework. In Table 3.1, the dimensions, quality indicators, and method descriptions for this hyper dimension are listed (Daas et al. 2009). The dimensions are labeled (1) supplier, (2) relevance, (3) privacy and security, (4) delivery, and (5) procedures.

The second hyper dimension in the framework, the metadata hyper dimension, focuses on the conceptual and process-related quality aspects of the metadata of the source. Prior to use, it is essential that a statistical agency fully understands the metadata-related quality aspects because any misunderstanding highly affects the quality of the output based on the data in the source. In Table 3.2, the dimensions, quality indicators, and method descriptions are listed for the Metadata hyper dimension (Daas et al. 2009). The dimensions are called (1) clarity, (2) comparability, (3) unique keys, and (4) data treatment.

For the evaluation of the quality indicators in the source and metadata hyper dimensions, a checklist has been developed. It is included in Daas et al. (2009) and its application by various European statistical agencies is described in Daas, Ossen, and Tennekes (2010), and Daas et al. (2011, 2013). The checklist guides the user through the measurement methods for each of the quality indicators in both hyper dimensions. By answering the questions in the checklist, the "value" of every method for each indicator in Tables 3.1 and 3.2 is determined, ranging from good to poor. Evaluation of the metadata part requires that the user has a

Table 3.1 Quality framework for secondary data sources, source hyper dimension.

Dimensions	Quality indicators	Methods
1. Suplier	1.1 Contact	– Name of the data source – DSH contact information – NSI contact person
	1.2 Purpose	– Reason for use of the data source by DSH
2. Relevance	2.1 Usefulness	– Importance of data source for NSI
	2.2 Envisaged use	– Potential statistical use of data source
	2.3 Information demand	– Does the data source satisfy information demand?
	2.4 Response burden	– Effect of data source on response burden
3. Privacy and security	3.1 Legal provision	– Basis for existence of data source
	3.2 Confidentiality	– Does the Personal Data Protection Act apply? – Has use of data source been reported by NSI?
	3.3 Security	– Manner in which the data source is send to NSI – Are security measures required? (hard-/software)
4. Delivery	4.1 Costs	– Costs of using the data source
	4.2 Arrangements	– Are the terms of delivery documented? – Frequency of deliveries
	4.3 Punctuality	– How punctual can the data source be delivered? – Rate at which exceptions are reported – Rate at which data is stored by DSH
	4.4 Format	– Formats in which the data can be delivered
	4.5 Selection	– What data can be delivered? – Does this comply with the requirements of NSI?
5. Procedures	5.1 Data collection	– Familiarity with the way the data is collected
	5.2 Planned changes	– Familiarity with planned changes of data source – Ways to communicate changes to NSI
	5.3 Feedback	– Contact DSH in case of trouble? – In which cases and why?
	5.4 Fall-back scenario	– Dependency risk of NSI – Emergency measures when data source is not delivered according to arrangements made

DSH, Data Source Holder; NSI, National Statistical Institute.
Source: Daas et al. 2009. http://pietdaas.nl/beta/pubs/pubs/Paper_with_checklist.pdf

Table 3.2 Quality framework for secondary data sources, metadata hyper dimension.

Dimensions	Quality indicators	Methods
1. Clarity	1.1 Population unit definition	– Clarity score of the definition
	1.2 Classification variable definition	– Clarity score of the definition
	1.3 Count variable definition	– Clarity score of the definition
	1.4 Time dimensions	– Clarity score of the definition
	1.5 Definition changes	– Familiarity with occurred changes
2. Comparability	2.1 Population unit definition comparability	– Comparability with NSI definition
	2.2 Classification variable definition comparability	– Comparability with NSI definition
	2.3 Count variable definition comparability	– Comparability with NSI definition
	2.4 Time differences	– Comparability with NSI reporting periods
3. Unique keys	3.1 Identification keys	– Presence of unique keys – Comparability with unique keys used by NSI
	3.2 Unique combinations	– Presence of useful combinations of variables
4. Data treatment (by DSH)	4.1 Checks	– Population unit checks performed – Variable checks performed – Combinations of variables checked – Extreme value checks
	4.2 Modifications	– Familiarity with data modifications – Are modified values marked and how? – Familiarity with default values used

Source: Daas et al. 2009. http://pietdaas.nl/beta/pubs/pubs/Paper_with_checklist.pdf

particular use in mind, such as the inclusion of certain variables in the 2011. The next step is the determination of the quality of the data.

3.3.2 Data Hyper Dimension

Indicators for the evaluation of the quality of the data in a register are part of the data hyper dimension. The focus of the indicators in this dimension is the quality of the data in the registers used as input in the statistical process (Daas et al. 2011). For each indicator measurement methods have been described, which were implemented in R, and a quality-indicator report card has been developed (Daas et al. 2013). The indicators and dimensions identified are listed in Table 3.3. The dimensions are named (1) technical checks, (2) accuracy, (3) completeness, (4) a time-related dimension, and (5) integrability.

Application of the report card by various European National Statistical Agencies demonstrated the advantage of noting quality findings in a structured way (Daas et al. 2013). The general observation was that the indicators in the first dimension, i.e. technical checks, always needed to be determined first. Subsequently, indicators in the other dimensions become relevant but without a clear sequence of importance. For the variables *highest level of education attainment* and *current activity status* the third dimension was found to be the most relevant. Certainly in relation to age, this revealed some important quality consideration. For *housing information* no major issues were identified by the indicators in the data hyper dimension.

3.4 Quality Evaluation Results for the Dutch 2011 Census

The checklist referring to the source and metadata hyper dimensions has been applied to the aforementioned registers. A first step has been made in applying the indicators corresponding to the data hyper dimension. This study focused on the variables *highest level of educational attainment*, *current activity status*, and *housing information*.

3.4.1 Source and Metadata: Application of Checklist

The checklist was applied to the education, unemployment benefit, social security, housing, and population registers. The evaluation results obtained for the source and metadata hyper dimensions are shown in Table 3.4. Evaluation scores are indicated at the dimensional level. The dimensional scores were obtained by selecting the most commonly observed score for every measurement method in each dimension. The symbols for the scores used are: good (+), reasonable (0), poor (−), and unclear (?); intermediary scores are created by combining symbols with a slash (/) as a separator.

The overall scores for the majority of the data sources were quite good in source. The education register was an exception, because a poor score was observed for *delivery*. This is the result of the low frequency of delivery (not more often than once a year). The education register also has a reasonable (o) score for *relevance* because this source did not satisfy all information demands of the census. This register suffers severely from selective undercoverage (see Section 3.4.2). The

Table 3.3 Quality framework for secondary data sources, data hyper dimension.

Dimensions	Quality indicators	Methods
1. Technical checks	1.1 Readability	– Accessibility of the file and data in the file
	1.2 File declaration	– Compliance of the data to the metadata agreements
	1.3 Convertibility	– Conversion of the file to the NSI-standard format
2. Accuracy	*Objects*	
	2.1 Authenticity	– Legitimacy of objects
	2.2 Inconsistent objects	– Extent of erroneous objects in source
	2.3 Dubious objects	– Presence of untrustworthy objects
	Variables	
	2.4 Measurement error	– Deviation of actual data value from ideal error-free measurements
	2.5 Inconsistent values	– Extent of inconsistent combinations of variable values
	2.6 Dubious values	– Presence of implausible values or combinations of values for variables
3. Completeness	*Objects*	
	3.1 Undercoverage	– Absence of target object in the source
	3.2 Overcoverage	– Presence of non-target objects in the source
	3.3 Selectivity	– Statistical coverage and representativeness of objects
	3.4 Redundancy	– Presence of multiple registrations of objects
	Variables	
	3.5 Missing values	– Absence of values for (key) variables
	3.6 Imputed values	– Presence of values resulting from imputation
4. Time-related dimension	4.1 Timeliness	– Time between end of reference period and receipt of source
	4.2 Punctuality	– Time lag between the actual and agreed delivery date
	4.3 Overall time lag	– Overall time difference between end of reference period and moment NSI concluded source can be used
	4.4 Delay	– Extent of delays in registration
	Objects	

Table 3.3 (Continued)

Dimensions	Quality indicators	Methods
	4.5 Dynamics of objects	– Changes in the population of objects over time
	Variables	
	4.6 Stability of variables	– Changes of variables or values over time
5. Integrability	*Objects*	
	5.1 Comparability of objects	– Similarity of objects in source with the NSI-objects
	5.2 Alignment of objects	– Linking-ability of objects in source with NSI-objects
	Variables	
	5.3 Linking variable	– Usefulness of linking variables (keys) in source
	5.4 Comparability of variables	– Proximity (closeness) of variables

Source: Modified from Daas et al. 2009.
http://pietdaas.nl/beta/pubs/pubs/Paper_with_checklist.pdf

unemployment benefit and social security registers scored reasonable for *supplier* and *procedures* because of the sometimes problematic and not very clear purpose of the data provider and the high dependency risk of Statistics Netherlands. The housing register has a reasonable score for *relevance* because this source did not satisfy all information demands; it was missing some variables, e.g. whether a dwelling is owned or rented. The population register only had good scores.

The overall scores for the data sources were also quite good for most dimensions in the metadata hyper dimension. The *clarity* and *data treatment* dimensions showed only good results. Again, the education register was the only data source with a poor score. This data source scored poor on *comparability* because the time period variables could not be transformed easily to the time points used by Statistics Netherlands. The housing register only had a reasonable score for *unique keys* because of the difficult comparability of the unique keys used in this source. This considerably hindered combining this data source with the other sources of information. The unemployment benefit and the social security registers had reasonable scores for *comparability* because of time differences in the reporting periods. Positive exception to all of this was, again, the population register which only had good scores.

The overall evaluation results for the five data sources revealed that attention should be paid to the *supplier, relevance, procedures,* and *comparability*-related quality aspects. For the housing register attention should be paid to *unique keys.*

Table 3.4 Evaluation results for the source hyper dimension.

Source dimensions	Data sources				
	ER	UR	SR	HR	PR
1. Supplier	+	0	0	+	+
2. Relevance	0	+	+	0	+
3. Privacy and security	+	+	+	+	+
4. Delivery	−	+	+	+	+
5. Procedures	0	0	0	+	+

Metadata dimensions	Data sources				
	ER	UR	SR	HR	PR
1. Clarity	+	+	+	+	+
2. Comparability	−	0	0	+	+
3. Unique keys	+	+	+	0	+
4. Data treatment	+	+	+	+	+

The symbols for the scores used are: good (+), reasonable (0), and poor (−). The data sources are the education (ER), unemployment benefit (UR), social security (SR), housing (HR), and population (PR) registers.
Source: Modified from Daas et al. 2009.
http://pietdaas.nl/beta/pubs/pubs/Paper_with_checklist.pdf

The results for the population register demonstrate that it is possible to have every quality aspect in the source and metadata hyper dimensions under control. For the other data sources, it can be argued that the results suggest that one or more of the quality aspects in both hyper dimensions require attention. It was concluded that not many problems were observed in the source and metadata hyper dimensions and that the data should be studied for these registers.

3.4.2 Data Hyper Dimension: Completeness and Accuracy Results

In this section, results of applying the indicators referring to the data hyper dimension are discussed. In the available dataset, raw data were already preprocessed to a limited extent and linked to the population register. All data furthermore referred to the same date: 1 January 2008. This implied that the indicators referring to the dimensions: *technical checks*, *time-related*, and *integrability* should not be considered. For more information on these topics the reader is referred to Daas et al. (2013). The analysis therefore focuses on the *completeness* and the *accuracy* dimension.

3.4.2.1 Completeness Dimension

The analysis on completeness concentrates on the variables *highest level of educational attainment* (derived from the education register) and *current activity status* (derived from the unemployment benefit register, social security register, and LFS). The housing variables for the 2011 Census all came from registers that were complete, and are therefore not discussed further in this chapter.

To get the first impression of the level of *undercoverage* of the information available regarding these variables, it was assumed that the population considered consisted of all persons included in the population register. As the data used contained one row for every person in the population register, the number of rows for which a value was missing was determined. This can be misleading of course when variables are studied that are only applicable to a part of the population. However, this did not apply to the variables considered here, as the categories of these variables were such that every person belongs to a category. The highest level of educational attainment was missing for 9 238 212 individuals, which was 56.3% of the population. The current activity status was missing for 2 140 266 (13%).

The undercoverage regarding the variable *current activity status* was not surprising, as for three categories of this variable (i.e. unemployed, homemakers, and others) not all information for the entire populations was available in the registers. In that case, only information from the LFS (based on samples) can be used. Advantage of the register-based data was that much more data on unemployed data were available but the comparability of the definition of unemployment with LFS data was an issue; in the metadata evaluation this comparability was scored reasonable. At the data level this may result in some measurement error. More serious, undercoverage existed for the variable *highest level of educational attainment*. This was also not unexpected, as the registers used for deriving this variable contained mainly information about people under the age of 45. This knowledge suggests that the undercoverage was seriously selective regarding age.

The selectivity of the information available regarding the *highest level of educational attainment* was further examined as shown in Figure 3.1. The dark gray bars in Figure 3.1 refer to the part of the population having a known value for this variable. The light gray bars refer to the population as a whole. In the histogram on the left, people are grouped into age groups (five-year classes), whereas the bar chart on the right groups people based on the variable "place of usual residence" (provinces).

Figure 3.1 clearly shows that the available information regarding the *highest level of educational attainment* was seriously selective regarding age, i.e. much more information is available about the lower age groups. On the other hand, the shares of people living in the 12 provinces of the Netherlands show a strong resemblance

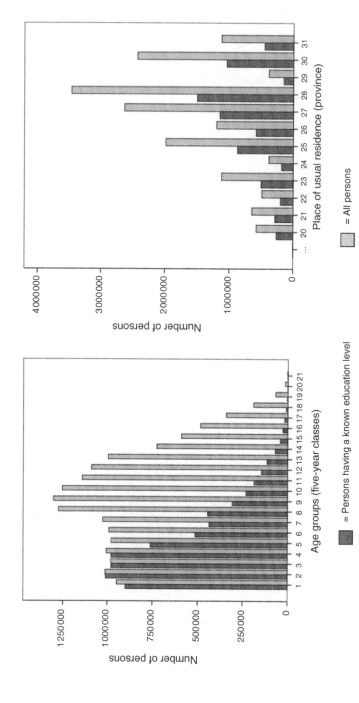

Figure 3.1 **Educational attainment and place of residence by per five-year age group.** Light gray bars are for the population as a whole (according to the population register). Dark gray bars are based on the education register.

between the whole population and the part of the population for which information about the *highest level of educational attainment* was available. This suggests that the information was not selective regarding the place where people live.

Another indicator regarding *completeness* is *redundancy*, i.e. the presence of multiple registrations of objects. To investigate whether data suffered from redundancy, rows in the dataset were compared that showed equal values for a certain combination of variables, such as *address, age group, gender, marital status, position in household*, and *household size* (including the *highest level of educational attainment* and *current activity status*). There turned out to be 67 644 "duplicates" in the test data, corresponding to 0.4% of the data. A further analysis of the duplicates revealed that most duplicated records corresponded to people living in institutes. People living in homes for the elderly, for example, do all have the same address, are all in the same age category, and so on. Given that it is possible that people in institutions do have the same values for the limited set of variables available in our test database, it was decided to focus future research on the selective part of duplicates that do not correspond to people living in institutions.

3.4.2.2 Accuracy Dimension

Regarding the *accuracy* dimension, the focus was on whether there were any dubious values in the data. Again the variables *highest level of educational attainment* and *current activity status* were looked at. For these variables, especially the relation with age was interesting. For illustration, reaching the highest levels of education takes time. Therefore, for example, a person of 18 years old can (normally) not yet have a PhD degree. Furthermore, it was expected that (almost) only elderly people will have a value for the variable *current activity status* equal to 3 (pension or capital income recipients).

To be able to analyze whether these expected relations between the variables of interest and age hold, cross tabulations were created. The results are shown in Tables 3.5 and 3.6. In interpreting these results, care has to be taken of the fact that especially for the variable *highest level of educational attainment*, a lot of values are missing and that the "amount of missingness" depends on age (see the first column of Table 3.5). Because of this, not much can be concluded from, for example, the counts per cell of the table. Despite this, for the variable *highest level of educational attainment*, it is valid to conclude that for the youngest part of the Dutch population either no value was present, or that it was "not applicable." The youngest people that have reached education level 6 (second stage of tertiary education) were in the age group 20–24. Furthermore, there turned out to be some people who reached educational level 5 (first stage of tertiary education) already within the age group 15–19. Most people reached this level however at a higher age. It can thus be cautiously concluded that most young people continued

Table 3.5 Cross tabulation of the variable "highest level of educational attainment" versus age group.

Age group	Highest level of educational attainment								
	Missing	No level	Level 1	Level 2	Level 3	Level 4	Level 5	Level 6	Not appl.
1: [0,5)	47 674	0	0	0	0	0	0	0	898 187
2: [5,10)	1 414	0	0	0	0	0	0	0	1 009 745
3: [10,15)	1 275	0	0	0	0	0	0	0	977 689
4: [15,20)	27 192	147	258 459	539 275	163 161	43	16 684	0	0
5: [20,25)	216 918	1 830	14 448	126 841	405 943	2 987	209 128	4	0
6: [25,30)	476 523	3 800	10 723	36 161	147 537	2 779	312 380	79	0
7: [30,35)	589 228	4 661	11 692	22 976	104 633	3 041	288 873	217	0
8: [35,40)	830 625	5 136	12 811	27 478	112 666	5 832	280 059	399	0
9: [40,45)	996 017	4 888	12 890	31 923	100 996	7 906	148 114	485	0
10: [45,50)	1 022 110	4 596	13 128	33 829	79 051	8 578	91 443	502	0
11: [50,55)	955 668	4 245	14 695	32 938	58 110	7 787	68 153	409	0
12: [55,60)	944 786	4 018	16 745	30 682	41 449	6 790	45 224	392	0
13: [60,65)	881 054	3 536	16 656	28 929	30 744	5 720	28 315	335	0
14: [65,70)	656 135	2 962	11 408	19 367	18 166	3 353	13 914	190	0
15: [70,75)	547 283	2 059	8 049	12 352	10 552	1 818	6 867	101	0
16: [75,80)	457 713	1 254	5 361	8 994	6 520	1 094	3 900	51	0
17: [80,85)	324 613	493	3 356	6 650	3 922	617	2 321	35	0
18: [85,90)	181 535	201	1 911	3 531	1 811	262	909	13	0
19: [90,95)	64 749	85	852	888	496	61	268	7	0
20: [95,100)	14 174	22	201	157	78	10	40	2	0
21: [100,∞)	1 526	2	7	16	8	1	5	0	0

Highest level of educational attainment: (Missing) No data available, (No level) No formal education, (Level 1) ISCED level 1 Primary education, (Level 2) ISCED level 2 Lower Secondary Education, (Level 3) ISCED level 3 Upper Secondary Education, (Level 4) ISCED level 4 Post Secondary non-tertiary study, (Level 5) ISCED level 5 First stage of tertiary education, (Level 6) ISCED level 6 Second stage of tertiary education, (Not appl.) Not applicable (persons < 15 year).

studying after they have reached level 1. This is in line with the expectations as youngsters are obliged to go to school till the age of 16 in the Netherlands.

In Table 3.6, the numbers of unemployed people, homemakers, and others were solely based on the LFS meaning that for these categories only sample information was used. The results for these categories were not weighted to the population totals. Table 3.6 is in line with the fact that the pensionable age in the Netherlands

Table 3.6 Cross tabulation of the variable "current activity status" versus age group.

Age group	Current activity status							
	Missing	Empl.	Unempl.	Pension	Student	Home	Other	Not appl.
1: [0,5)	0	0	0	0	0	0	0	945 861
2: [5,10)	0	0	0	0	0	0	0	1 011 159
3: [10,15)	0	0	0	0	0	0	0	978 964
4: [15,20)	34 911	482 180	33	0	487 533	11	293	0
5: [20,25)	113 286	716 411	106	0	147 395	190	711	0
6: [25,30)	142 149	818 167	107	0	28 396	486	677	0
7: [30,35)	163 141	856 030	129	0	4 506	744	771	0
8: [35,40)	216 807	1 053 407	180	0	2 418	1 138	1 056	0
9: [40,45)	228 634	1 070 204	228	0	1 853	1 076	1 224	0
10: [45,50)	236 102	1 013 249	242	0	1 134	1 076	1 434	0
11: [50,55)	262 473	875 724	253	1	504	1 261	1 789	0
12: [55,60)	330 898	714 959	263	39 705	232	1 776	2 253	0
13: [60,65)	390 062	343 089	122	256 826	78	2 348	2 764	0
14: [65,70)	8 730	88 209	1	628 490	16	3	46	0
15: [70,75)	5 306	35 690	1	548 059	3	0	22	0
16: [75,80)	3 822	14 705	0	466 339	2	0	19	0
17: [80,85)	2 166	5 897	0	333 936	0	0	8	0
18: [85,90)	1 115	2 360	0	186 690	0	0	8	0
19: [90,95)	405	662	0	66 339	0	0	0	0
20: [95,100)	162	136	0	143 886	0	0	0	0
21: [100, ∞)	97	18	0	1 450	0	0	0	0

Current activity status: (Missing) Data not available, (Empl.) Employed, (Unempl.) Unemployed, (Pension) Pension or capital income recipients, (Student) Students not economically active, (Home) Homemakers, (Other) Others, and (Not appl.) Persons below minimum age for economic activity.

is 65 years in general, i.e. there was a clear peak of records with a value of 3 (pension or capital income recipients) for the variable *current activity status* in the age groups 60–69. Related to this, it can be seen that the part of people having status 1 (employed) significantly decreased once they have reached the age of 65 years. The status 4 (students not economically active) was also in line with the expectations, as this status occurred mostly for people below the age of 25 years. It would be an interesting future extension to compare these survey distributions with those solely based on register data.

3.4.2.3 Visualizing with a Tableplot

In addition to the more traditional ways to study the quality of the data, visualization methods can be used to display quality relevant information in a single plot. Here, the tableplot was used as a method to visualize large multivariate datasets (Tennekes, De Jonge, and Daas 2013). It is particularly useful for displaying aggregated distribution patterns of up to a dozen variables in a single figure providing information on data quality and the presence and selectivity of missing data. A tool to create tableplots has been implemented as the package "tabplot" for the open source statistical software environment R (Tennekes and de Jonge 2016). This package was used to visualize the 2011 Census.

Because of the findings described above, there was a particular interest in the variables *highest level of educational attainment* and *current activity status*. These variables together with *age* were selected. Next, all records were sorted according to age and the rows were subsequently divided into 100 equally sized bins. For numeric variables, the mean values per bin were plotted as bars, while for categorical variables, the category fractions were plotted as stacked bars in distinguishable gray tones, with black used for missing values. The resulting tableplot is shown in Figure 3.2; the use of colors revealed more details (see Tennekes, De Jonge, and Daas 2013). *Age* was plotted both as a numeric and as a categorical variable.

In the third column in Figure 3.2, the various categories for the *highest level of educational attainment* were displayed according to age. It very clearly confirmed the occurrence and distribution of missing values (in black) according to age as stated in Section 3.4.2.2. Advantage of the tableplot was that it immediately revealed two quality issues. The first was related to people with an age below 15. These have, by definition, no formal level of education in the Netherlands as they do not have an educational level yet. These people should therefore all be categorized as "not applicable." However, the lowest two rows in the third column clearly contained a considerable number of missing values (in black). This was an obvious error that should be corrected. The second issue was related to people above 15. With increasing age, the dominating levels of education were: primary, lower secondary, upper secondary, and bachelor/master, respectively. All other categories were only present in small amounts. The tableplot demonstrated that with increasing age the amount of missing information increased dramatically. The latter was caused by the fact that the official registration of the level of education of Dutch graduates gradually increased over the years. As a result, only the level of education of people registered was stored in various public administrations; which was dominated by young people. For all others, the only data sources in which this kind of information was available in the Netherlands were sample surveys. This explained the increasing number of missing values with increasing age.

The last column displayed the various categories of the variable *current activity status*. Similar to the former column, this category also does not apply to people

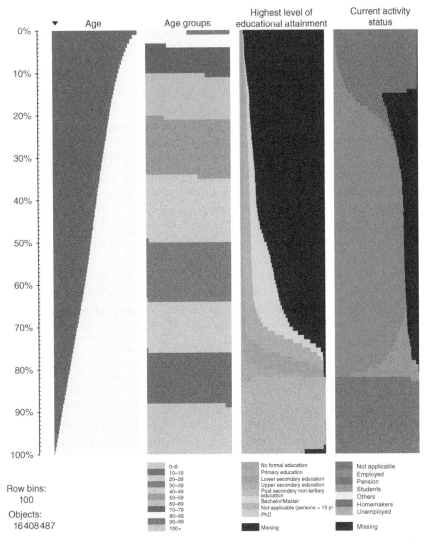

Row bins:
100

Objects:
16 408 487

Figure 3.2 **Tableplot visualization of age, highest level of education, and current activity status of the Dutch 2011 Census file**. Age is used as the sorting variable (from high to low) and is also shown aggregated in groups of 10. Black is used to indicate missing data in the two right columns and gray tones are used to indicate the different categories discerned. Source: Modified from Tennekes, De Jonge, and Daas 2013.

below 15 years of age in the Netherlands. Above 15, people start to get employed, but between 15 and 20 years, this group was mainly composed of students. Above 65, the age at which people become eligible for a pension in the Netherlands, the majority of the people were pensioners. The column, however, also revealed that the case of pensioners started to occur from 55 years onward. The column also showed that a part of the people above 65 were still employed which, according to the definition used in the data source providing this information, meant that these people worked at least one hour a week. However, certainly for people over 80, this finding was remarkable and requires further study (Bakker, Linder, and Van Roon 2008). Between 20 and 65, increasing numbers of missing data were observed. This was again caused by the fact that not all of the activity status data required was available in (public) administrative data sources. For this part of the population, information provided by samples surveys was used.

Visualizing the census data by a tableplot confirmed the observations described in Section 3.4.2.2 of this document. However, it also added additional insights by enabling the comparison of bivariate relations between variables. This proved particularly valuable for missing data. Since this was all done by creating a single figure, it demonstrated that a tableplot was a great way of efficiently inspecting large data sets to identify quality issues. Readers are referred to Tennekes, De Jonge, and Daas (2013) for more details on the tableplot and the findings for some of the other variables in the census data.

3.4.3 Discussion of the Quality Findings

In Section 3.4.2, the quality of all registers studied was evaluated as input for the 2011 Virtual Census of the Netherlands. To enable this, the source, metadata, and data hyper dimension quality indicators of these registers were checked. The results described in this chapter revealed several issues at the source and metadata level which demonstrated that the quality framework developed and the corresponding checklist are valuable tools. This finding was not only relevant for the 2011 Census of Netherlands but also for future censuses in the Netherlands and those in other countries. In principle, it is possible to conduct register-based censuses in many other countries. However, not every country has access to registers from which a considerable number of census variables can be derived. When this is the case, data provided by surveys remain a necessity for at least some of the variables. But for all register-based data, the approach developed is certainly useful.

The focus of the data hyper dimension was on the use of a number of registers to contain information on the variables *highest level of educational attainment, current activity status*, and *housing information*. Clearly the incompleteness of the first two variables was found to be a serious concern. For these variables a decision has been made in 2012 on how the different Dutch Census variables should be derived. The data hyper dimension findings played an important role in this decision making process. For the variable *current activity status* it was decided to derive it from register date, but no distinction was made between the categories homemakers and others. For the variable *highest level of educational attainment* it was decided to base it on the LFS only. This was predominantly caused by the results of more detailed studies, conducted in the Census 2011 project and initiated by the results described in this chapter. In these detailed studies it was confirmed that the register-based value for the *highest level of educational attainment* was indeed very selective. Adding only "half a register" would have led to difficulties in solving estimation problems. See Schulte Nordholt (2014) and Kroese and Renssen (2000) for more details. Currently, projects are foreseen in which information from the education register and LFS and possible other sources are combined, to enable more detailed publications on this variable in future censuses.

Given the detailed information requests for the Census 2011, the available sources for the Dutch Census and our experiences with applying the quality framework, it is clear that the framework discussed forms a sound basis not only for the 2011 Census but also for register-based statistics in general.

3.5 Summary

Since the last census based on a complete enumeration was held in 1971, the willingness of the population in the Netherlands to participate has decreased tremendously. Statistics Netherlands found an alternative in a Virtual Census, by using available registers and surveys as alternative data sources. The increased use of registers in composing a census, however, also lead to several quality considerations. The most important one was the result of the increased dependency on the input quality of the source. For registers, for instance, the collection and maintenance are beyond the control of the statistical institute. It is therefore important that these institutes are able to evaluate all considerations at the source, metadata, and data level. The overall framework developed for these purposes is described in this chapter.

3.6 Practical Implications for Implementation with Surveys and Censuses

One practical implication of this work is that projects that make use of administrative and survey data routinely should evaluate the quality of the data sources used. Evaluations should involve many dimensions and presentations. If standardized tools are used, then experience will be gained with their implementation and interpretation.

A second practical implication of this work is that administrative registers should proactively evaluate their own variables and processes for quality. If quality of some variables is too poor to use, then perhaps they should be dropped from the registry or flagged as such. Steps might be considered to improve the quality of such variables.

Third, reports based on register data could be accompanied by documentation of the quality of the data going into analyses.

3.7 Exercises

Download the census income files from the UC Irvine Machine Learning Repository archive located at: https://archive.ics.uci.edu/ml/machine-learning-databases/census-income-mld/.

Extract the income data file (census-income.data.gz) and make sure the variable names (census-income.names) are used as a header.

1   Create a tableplot for Age (sorting variable), Gender, MartialStatus, Work-Class, Education, EduEnrolled, HourlyWage, EmploymentStatus, and Race.

2   Check the variables HourlyWage and WeeksWorked in the census file. Explain what you observe. What is the best way to display the findings?

References

Bakker, B.F.M., Linder, F., and Van Roon, D. (2008). Could that be true? Methodological issues when deriving educational attainment from different administrative datasources and surveys. *International Association for Official Statistics (IAOS) Conference on Reshaping Official Statistics*, Shanghai, China. http://www.stats.gov.cn/english/specialtopics/IAOSConference/200811/t20081118_65710.html.

BSN (2017). What is a Citizen Service Number? Website of the Dutch Tax Office. https://www.belastingdienst.nl/wps/wcm/connect/bldcontenten/belastingdienst/individuals/other_subjects/citizen_service_number/what_is_a_citizen_service_number_bsn/what_is_a_citizen_service_number_bsn (accessed 05 August 2020).

Daas, P.J.H., Ossen, S.J.L., Vis-Visschers, R.J.W.M., and Arends-Toth, J. (2009). Checklist for the quality evaluation of administrative data sources. Discussion paper 09042, Statistics Netherlands, Heerlen/The Hague, The Netherlands. http:// pietdaas.nl/beta/pubs/pubs/Paper_with_checklist.pdf.

Daas, P.J.H., Ossen, S.J.L., and Tennekes, M. (2010). The determination of administrative data quality: recent results and new developments. *Conference on Quality in Official Statistics*, Helsinki, Finland (4–6 May 2010). Statistics Finland and Eurostat. https://q2010.stat.fi/

Daas, P., Ossen, S., Tennekes, M., et al. (2011). Report on methods preferred for the quality indicators of administrative data sources. Second deliverable of workpackage 4 of the BLUE Enterprise and Trade Statistics project (28 September 2011).

Daas, P., Tennekes, M., Ossen, S., et al. (2013). Report on guidelines on the usage of the prototype of the computerized version of QRCA and Report on the overall evaluation results. Second deliverable of workpackage 8 of the BLUE Enterprise and Trade Statistics project (29 March 2013).

Karr, A.F., Sanil, A.P., and Banks, D.L. (2006). Data quality: a statistical perspective. *Statistical Methodology* 3 (2): 137–173.

Kroese, A.H. and Renssen, R.H. (2000). New applications of old weighting techniques; constructing a consistent set of estimates based on data from different surveys. In: *Proceedings of ICES II*, 831–840. Buffalo, NY: American Statistical Association.

Schulte Nordholt, E. (2014). Introduction to the Dutch Census 2011. In: *Dutch Census 2011, Analysis and Methodology*, 7–18. The Hague/Heerlen: Statistics Netherlands.

Tennekes, M. and De Jonge, E. (2016). tabplot: Tableplot, a visualization of large datasets. R package version 1.3. https://CRAN.R-project.org/package=tabplot (accessed 05 August 2020).

Tennekes, M., De Jonge, E., and Daas, P.J.H. (2013). Visualizing and inspecting large data sets with tableplots. *Journal of Data Science* 11 (1): 43–58.

United Nations (2007). *Register-Based Statistics in the Nordic Countries: Review of Best Practices with Focus on Population and Social Statistics*. New York and Geneva: United Nations.

4

Improving Input Data Quality in Register-Based Statistics: The Norwegian Experience
Coen Hendriks

Statistics Norway, Oslo, Norway

4.1 Introduction

Dating back to the 1960s, Statistics Norway (SN) has used registers in the production of statistics. It has been a major goal for SN to produce the Population and Housing Census from registers (Longva, Thomsen, and Severeide 1998). The Population and Housing Census of 2011 was the first Norwegian purely register-based census. Experiences from the census of 2011 showed there was a need for improved quality on input data. This was supported by reviews of other statistics which were based on Eurostat's Code of Practice.

The particular challenge encountered in 2011 concerns the linkage between households and dwellings. Both Household and Dwelling statistics were already established separately, each based on a different base administrative register. In the census of 2011, not all the households from the data file of the established Household statistics could be directly linked to the dwellings from the data file of the established Dwelling statistics. A technique called Double Nearest Neighbor Imputation (DNNI, Zhang and Hendriks 2012), was developed and used to solve the problem of missing micro-linkage between households and dwellings. It provided an "one-number" input data file (utilizing one-to-one linkage) for the census statistics on households, housing conditions, and dwellings. The method of DNNI provides a general approach to the problem of micro-linkage between different types of units for which direct linking is impossible to start with. To ensure the results of data integration on a more fundamental level, Zhang and Hendriks advise to look for means to improve the quality of the source data.

SN has since strengthened the cooperation on data quality with register owners. The chief aim of this chapter is to describe the systematic approach that has been developed and the related experience, where particular emphasis is given to

Administrative Records for Survey Methodology, First Edition.
Edited by Asaph Young Chun, Michael D. Larsen, Gabriele Durrant, and Jerome P. Reiter.

the so-called base registers (UNECE 2007). The main idea is to avoid errors in the source, rather than repairing errors upon receiving the data. SN has discussed data quality with register owners on numerous occasions and in different settings. After the Census of 2011, it was clear that the cooperation with register owners on data quality should be improved. This was done by means of two types of agreements between SN and register owners. The first type is an agreement on data processing. The second type is an agreement on cooperation on data quality in administrative data systems. In addition, SN is actively taking part in forums for cooperation between SN and register owners. An important aspect of the cooperation is the measurement and documentation of input data quality.

The presentation here rests on the conceptual framework which is widely used in the Nordic countries (UNECE 2007). A register is defined here as a systematic collection of unit-level data organized in such a way that updating is possible. Updating is the processing of identifiable information with the purpose of establishing, bringing up to date, correcting or extending the register, in short keeping track of any changes in the data describing the units and their attributes.

Administrative registers are registers primarily used in administrative information systems. That means that the registers are used in the production of goods and services in public or private institutions or companies, or the information is a result of such production. In the Nordic countries, most administrative registers used for statistical purposes are countrywide registers operated by the state or jointly by local authorities. However, private registers are also used. Some examples are mentioned in the chapter. Statistical registers are created by processing data from administrative registers. Statistical registers could be based on a single administrative register, but they are more frequently based on combined data from several administrative sources.

4.2 The Use of Administrative Sources in Statistics Norway

SN uses many registers, but the following three administrative registers are regarded as *base registers* for the production of statistics.

- The Central Population Register (CPR, owned by the Tax Administration)
- The Central Coordinating Register for Legal Entities (RLE, owned by the Brønnøysund Register Centre)
- The Cadastre (ground properties, addresses, buildings, and dwellings, owned by the Mapping Administration, updated by the municipalities)

The three registers serve primarily administrative purposes and are the authoritative sources for information on the units in the register. The registers are being used to administer rights and duties for the registered units. The CPR administers the rights and duties for registered persons. This may be the right to be a resident person in Norway, to participate in elections, to receive an education, but also the duty to pay taxes or to make sure your children attend school.

Similarly, legal entities have rights and duties which are being administered through the RLE. By registering in the RLE the entity receives an organization number. The organization number is used by public authorities to coordinate information about the entity. If the entity is required to register in one of the other major public registers, such as the Register of Business Enterprises, the Value Added Tax Register, or the Register of Employers and Employees, the entity also needs to register with the RLE. Other legal entities choose to register with the RLE voluntarily. This will get the entity an organization number which is required to open a bank account.

The Cadastre is the official register for legal rights and obligations associated with fixed property and housing cooperatives. The register lists ownership and encumbrances such as mortgages, leasing rights, preemptive purchasing rights, and so on. Details of physical aspects relating to a property, such as borders, areas, buildings, addresses, and coordinates, are registered in the Cadastre. The Cadastre is a public register and is updated by the municipalities.

The expression *basic data* is used for the data which identify and describe the units. Resident persons are registered in the CPR with basic data, e.g. personal identification number (PIN), name, address, sex, and nationality. In the RLE typical basic data are the identification number, the name of the registered legal entity, the address, type of legal entity, the industrial code, date of establishment, and the name of the CEO. For units in the Cadastre basic data are numerical identification codes for ownership, addresses, ground properties, and buildings, in addition to coordinates and alphanumerical information like street names.

It is important that basic data are of high quality, since the data are continuously being distributed and used in the public and private sector (e.g. banking and finances). An example is a notice of change of address which is sent to the CPR. From the CPR, the PIN with the new address is being distributed automatically and daily to the users of the basic data from the CPR, including SN. In this way, other administrative registers are updated with the new address and other basic data.

On a daily base, SN receives information from the registers which are mentioned above. To start off, SN received full copies. From then on, only records with updated information are being sent. To facilitate the statistical use of the registers,

SN has established a database for the statistical versions of the base registers. The statistical database includes:

- The Statistical Population Register (SPR, based on the CPR)
- The Business Register (BR, business enterprises and local kind of activity units, based on the RLE)
- The Statistical Cadastre (SC, based on the Cadastre)

In many cases, the database is extended with information from other sources. Examples of extensions are information on jobs which is linked to persons in the SPR or aggregated information on employment which is added to businesses in the BR. The user can do data inspection at micro level, browse from one register to the other and extract combined data from several registers. The PIN, the business identification number (BIN), and the numerical address code are most often used as keys for linkage.

In addition to the base registers, SN uses a variety of other administrative sources which can be described as *specialized registers* (UNECE 2007). Some of them are well established in the statistical system. An example is employment statistics which is based on The Register on Employers and Employees, the Register of Personal Tax Payers, and several other sources. Another example is income statistics which is based on data from the Register on Tax Returns, the End of the Year Certificate Register, and several other sources. Specialized registers use the basic data from the administrative base registers to identify and describe the units. Information on the subject matter of the specialized register is registered in specific variables (Table 4.1).

SN is constantly adding new administrative sources to the statistical system. Units are identified by identification number. But the identification number is also a key for linkage. By linking information from specialized registers, the base registers can be updated and extended. The additional data can be used to fill out missing values, analyze and correct over and under coverage, reduce inconsistency or simply to add new variables to the base registers.

Resident persons are registered at the "de jure" or legal address in the SPR. It is common knowledge that many persons have a "de facto" or actual address which is different from the legal address. SN has added addresses to the SPR from The Register on Mail Recipients from Norway Post, The National Digital Contact Register from the Agency for Public Management and eGovernment, and from other sources (Hendriks and Åmberg 2011).

Another example of an additional source to the BR is the Euro Group Register (EGR). EGR is the statistical Business Register of multinational enterprise groups having at least one legal entity in the territory of the European Union (EU) or European Free Trade Association (EFTA)-countries. The EGR is created for statistical purposes only, to facilitate the coordination of survey frames in the European Statistical System (ESS) for producing high quality statistics on global business

Table 4.1 Administrative and statistical registers, base registers, and specialized registers.

	Administrative register	**Statistical register/use**
Base registers	Central Population Register – CPR	Statistical Population Register – SPR
	Central coordinating Register for Legal Entities – RLE	Business Register – BR
	Cadastre	Statistical Cadastre – SC
Some examples of specialized registers	Register for personal tax payers	The tax statistics for personal tax payers
	National certificate database State education loan fund	The register of the population's level of education
	The health personnel register	
	Register of business enterprises	Sampling for business statistics
	Register for employers and employees	Employment statistics
	Register of personal tax payers	

activities, like Foreign Affiliates Statistics (FATS) and Foreign Direct Investment statistics (FDI).[1]

The SC has missing values for important variables on buildings, like year of construction or floorspace in square meters. This information is available in the The Taxation Property Register (SERG) from the Tax Administration and in "Homes for sale" in Finn.no (a Norwegian commercial advertising service on the internet). The population manager for the SC has linked the additional sources to the SC to fill out missing values for year of construction or floorspace.

4.3 Managing Statistical Populations

Maintaining the database for the statistical versions of the base registers is called *statistical population management* in SN. Managing statistical populations is a well-known activity in SN. A statistical population database aims to cover a statistical population for a specific subject matter, with regards to units and variables. The database must be updated and quality assured, and the documentation must meet the needs of the users in SN. When a statistical population database is used for sampling or calculating weights, the user must have information on coverage, the quality of stratification variables, and the timestamp of the latest updates.

---

1 http://ec.europa.eu/eurostat/web/structural-business-statistics/structural-business-statistics/eurogroups-register.

A statistical database can be used as a direct source for statistics, e.g. demographic statistics, and as a sampling frame. It is also used during data collection. Contact information from the statistical database is used to contact respondents in sample surveys. In business surveys, the database is used to register and follow up the obligation to report, e.g. when was the questionnaire sent out, has the respondent answered, or when is the deadline.

Ideally, a statistical register should reflect the "real world." This is a constant challenge for the managers of statistical populations in SN. Occasionally units, which are active in the administrative source, might be classified as non-active in the statistical base register. This may be a resident person of extreme high age who has emigrated and may have died abroad. In such a case the person's vital events were not registered in the Norwegian CPR. Legally speaking, the person is still alive. But the person is excluded from the statistical population for demographic statistics because the statistician assumes that the probability for the person being alive is very small. Another example is a business enterprise without any recorded activity during a period of years (no turn-over in the value-added tax [VAT]-register or in the annual income statement to the Tax Authorities, no employees). It might also be a business enterprise which temporarily was put on hold. Such business enterprises are classified as active in the administrative RLE because there are still rights or duties to be administered. E.g. limited companies are required to send annual accounts to the Register of Company Accounts, even if the company has been inactive. However, an active business enterprise which has not contributed to the gross domestic product (GDP) over a period of years (no turn-over, no employees), can be excluded from the statistical population.

Since the turn of the century, SN developed the cooperation on registers within SN, and between the register owners and SN. The approach was put into system. Some of the measures are:

- Appointing a coordinator in SN for all the contact with the register owners
- Established common guidelines for the follow up of the register owners during and after data collection
- Developed a system for receiving the data and facilitating the dataflow to the users of the data in SN
- Used in a better way the common information in the population registers
- Included the statistical divisions in the management and quality assurance of statistical basic data
- Developed ways to measure, document, and report on data quality to the register owners

Two important measures are coordinated contact with the register owners and reporting on quality. These measures will be highlighted later in the chapter.

4.4 Experiences from the First Norwegian Purely Register-Based Population and Housing Census of 2011

SN has been planning a register-based population and housing census for decades (e.g. Longva, Thomsen, and Severeide 1998). At the time of the last Norwegian questionnaire-based census in 2001, every dwelling in Norway was given a unique dwelling identification number and the dwelling number was introduced as a part of addresses in the CPR (Severeide and Hendriks 1998). In principle, the first Norwegian fully register-based census of 2011 was done by linking information from relevant registers and compiling the statistics. However, this turned out to be complicated when it came to the production of register-based census statistics on dwelling households, occupied and non-occupied dwellings and on housing conditions.

When linking the resident population from the CPR to dwellings from the Cadastre using the dwelling identification number, one should expect to get a reasonable number of dwelling households. This was done to produce data for the census of 2011 and compared to the regular statistics on households. In doing so, some remarkable differences became clear. The number of private households from the matched registers was reduced by 6% compared to the established household statistics. The number of persons living alone was reduced by 17%, dropping to the level of the early 1990s. Overall the linked registers resulted in fewer, but larger households.

Errors occur in the register households whenever people who do not live together are grouped into the same household, and/or when people in the same household are divided into different households. Such errors are unit errors. A unit error problem involves more than just a mismatch between two sets of fixed units. There are two reasons why the number of households is unknown. Firstly, the register on dwellings is not perfect. There are both missing and wrongly registered dwellings identification numbers in the dwelling register (which is a part of the Cadastre). There might also be delays in updating. Therefore, the number of dwelling units at a given address cannot be known for sure. Secondly, even when the errors in the dwelling register are disregarded, it is not true that the number of dwelling households will always be the same as the number of dwelling units. Moreover, unit errors will almost certainly arise in a longitudinal perspective, because the update of the dwelling identification number in the population register is not perfect (Zhang 2011).

After much discussion on the methodology it was decided to base the 2011-census file for households on the datafile from the established household statistics and on the datafile from the established dwelling statistics. The technique which was used was called DNNI (Zhang and Hendriks 2012). For the census application of the DNNI, there should be fewer households than dwellings in the respective input data files. Since the publication for the census was planned for each municipality and statistical tract, these variables were used as blocking variables. First, 1 962 000 households were linked directly to dwellings from the Cadastre using the dwelling identification number for linkage. This linked 85% of the households to dwellings. Second, for each household (A) from the first step, which was not linked to dwellings, one finds the closest resembling (nearest neighbor) household (B) which has been linked to a dwelling (C) at the first step. This was done by comparing household type and size, average ages for adults and children in the household and, when available, building type for linkage. Third, for each donor household (B) with dwelling (C) identified in the second step, one finds the closest resembling (nearest neighbor) unoccupied dwelling (D), which was then linked to household (A) which could not be linked to a dwelling in the first step. The second round of matching was done by using the address code, type of building and areal of floorspace for linkage. Unoccupied dwellings which were linked during the second round of imputation, were marked as occupied dwellings in the housing census. The DNNI resulted in 99.6% of the households with dwelling characteristics.

The DNNI ensured that the number of households, their distribution according to the number of persons per household and other statistical properties from the established statistics on households, and the number of dwellings were maintained in the census. Dwellings which were not included in the DNNI, were counted as unoccupied dwellings in the census. The method provides a general approach to the problem of micro-linkage between different types of units for which direct linking is impossible to start with. To ensure the results of data integration on a more fundamental level, however, one must also look for means to improve the quality of the source data (Zhang and Hendriks 2012).

Zhang (2012) has pointed out that administrative registers certainly do not provide perfect statistical data. Zhang describes a two-phase model of integrated statistical microdata. To exemplify several types of errors he refers to a statistical register of examination results which can be used to produce statistics on the highest completed educational attainment or as a source for background variables on education in other statistics like the Population and Housing Census. Such a statistical register is established in SN and updated with information on examination results for students from Norwegian educational institutions. The target population of the register is the resident population of Norway. The target concept in measuring educational attainment is examination results regardless of the country where the examination was taken. Since the register is updated with information on examination results for students from Norwegian educational institutions, there occurs a validity error. Examinations passed abroad are not measured. To compensate for this error SN did a postal survey in October 2011 on the level of education among 217 000 immigrants aged 20 years and older. The immigrants could be resident foreigners or resident Norwegians born abroad. The number of returned envelopes uncovered a frame error in the CPR. Persons who were registered as residents, were in fact not a part of the resident population.

The postal services returned 22 000 envelopes because they could not be delivered correctly to the addressee. An analysis of the address information in the CPR four months after the survey gives an indication on frame error due to measurement errors in the CPR. Among the 22 000 persons with returned envelopes, 7% of the addressees had emigrated. Six percent had residence or work permits which were overdue, indicating that they possibly had left the country. Eleven percent had sent a notification for change of address in Norway. The rest was unaccounted for and they were assumed to be without a permanent place of residence. This confirmed that there can be a considerable time lag between the actual emigration or change of address and the registration of the event in the CPR.

4.5 The Contact with the Owners of Administrative Registers Was Put into System

Improvements can be achieved by several types of cooperation between SN and register owners. Three approaches are described: two types of agreements, on data processing and cooperation, and thirdly the forums for cooperation. The approaches have a sound legal basis. In addition, they have strong elements of communication and co-ordination within SN, between SN and register owners and among register owners.

4.5.1 Agreements on Data Processing

There was much debate in SN on the methodological approach toward register-based census statistics on households and dwellings for 2011. Everyone agreed that the input quality of the data could be improved. Rather than dealing with errors, they should be avoided in the source. This applies to the sources for the Population and Housing Census, as well as every other register source of administrative data in the public and private sector.

SN uses the Code of Practice (Eurostat 2017) as a framework for systematic quality reviewing of statistical products and processes. Since 2012, every statistical division in SN has had at least one review according to the Code of Practice. The reviews have underlined the importance of improved cooperation with the register owners on data quality. This will reduce the need for editing, imputation, and other forms for adjustments.

SN aims to improve the cooperation with register owners. SN has good skills in integrating data from different sources and on quality assurance of register-based sources in general. The register owners have good skills on the subject matter for the register, but often lack skills in the practical work on registers. Cultivating these skills will eventually lead to better quality.

SN can give feedback on quality issues at micro level to the register owners, if the errors are found within the source. The legal department in SN has agreed that statisticians can make a complaint about data quality to the register owner. Such a complaint can be documented by sending a description of the problem including the individual identification numbers for the units with suspicious values. An example could be missing information on a person's spouse. According to the CPR, person A is married to person B, but person B is not married to person A. When this is the case, both identification numbers are returned to the register owner, with a description of the problem. Such reporting, using a single source approach, is well within the provision of confidentiality from the Norwegian Statistics Act.[2] When errors become apparent after linking information from two or more sources, SN can give feedback by means of aggregated data. Sharing information at micro level, from a dataset which combines information from different sources would be in conflict with the Statistics Act and is not practiced in general. In some cases, SN has received a request from a register owner to assess quality in data from combined sources. Provided the register owner is allowed to use data from different administrative registers, an agreement on data processing can be drawn up between SN and the register owner. An agreement on data processing regulates the linking of the sources for quality improvement purposes and the reporting of errors at micro level (the multiple source approach).

An agreement on data processing was signed by SN and the Norwegian Tax Administration in February 2012. The Tax Administration is in charge of the CPR and has access to the Cadastre from the Norwegian Mapping Authority. The Cadastre is being used by the Tax Administration to improve the quality of information in the CPR during population registration, e.g. by checking a notification of change of address against information from the Cadastre. A first check is to confirm that the new address is actually an official address. If not, the notice of change of address will be declined. Another check would be to compare the number of persons on the address to the type of building from the Cadastre, to avoid overcrowded addresses. Refer to the appendix of this chapter for more examples on quality checks.

Once the agreement on data processing was signed, SN can do these and other checks electronically and on a much larger scale than the Tax Administration. After linking persons from the CPR against the Cadastre, SN has identified groups

---

2 https://www.ssb.no/en/omssb/lover-og-prinsipper/statistikkloven.

of persons with incomplete or non-logical data and transferred these to the Tax Administration for follow-up. The number of errors is extensive and there are various types of them. These errors cause many problems for register-based household statistics, and indeed for other administrative and statistical use of the data. Therefore, SN is eager to cooperate with the Tax Administration.

4.5.2 Agreements of Cooperation on Data Quality in Administrative Data Systems

The Code of Practice reviews showed that there is a potential for improved cooperation on data quality with the owners of the administrative registers. Cooperation on quality with register owners has been SN-policy for many years, but it was based on ad hoc measures. When a statistician finds a data quality problem, he/she would pick up the telephone and discuss the problem with someone at the register owner's office. In some cases, the problem could be solved in the administrative source and an update was sent to SN. In other cases, the problem was solved by editing the data. In 2011, the cooperation between register owners and SN was organized differently. The Director General of SN initiated the process. He decided to establish general agreements on cooperation with external register owners to improve input quality in the registers which SN uses in the production of statistics.

The purpose is to improve the quality of administrative data sources before transferring them to SN. This approach will be beneficial for SN, other users of the registers, and for the register owners.

A working group was established in SN with representatives from the statistical departments, from the Divisions for Statistical Populations and Statistical Methods and Standards. The legal advisor in SN participated along with the person who had conducted the Code of Practice reviews. The working group was chaired by the Director for the Department of Data Collection. The working group made templates for the agreements and for quality reports, and an inventory of all the register-based sources in SN. SN invited the major register owners to enter top level agreements, arguing that the public sector will benefit from improved data quality and data owners will experience improved quality in their products and services. After the agreements are signed, they are being reviewed annually.

The agreement on cooperation describes the background and purpose of the agreement, which registers are covered, the names of the contact persons for the day to day contact and a description of the quality improvements that can be made. In addition, practical arrangements for transferring the register, duration of the agreement and the legal basis for the data collection are described. The duties of both parties are described, as well as a list on the datasets which are to be delivered.

The agreements are supplemented by a quality report for each register. The quality reports are based on the quality indicators from the Blue-Ets Work

Package 4 (Daas and Ossen 2011). The indicators were grouped in the following quality-dimensions: Technical checks, Accuracy, Completeness, Integrability, and Time-related dimensions. Refer to the appendix for an example of a quality report.

By the end of 2016, SN has signed agreements with 29 partners. In all SN produces 99 quality reports per year.

4.5.3 The Forums for Cooperation

To organize the practical cooperation on data exchange between the RLE and the affiliated registers (including SN's Business Register), two bodies were established:

- *The Cooperation Forum*: to facilitate the data production and exchange between the affiliated registers
- *The User Forum*: a broader forum where a variety of users of the data from the RLE meet

SN participates actively in both, as a member and occasionally as the chairperson. The participation from SN is appreciated highly because statisticians tend to have a good overview. They have good skills, methods, and tools for quality assessment and checks.

A cooperation forum and a user forum are proposed for the other national base registers as well. Currently there is a user forum for the CPR which meets twice per year.

4.6 Measuring and Documenting Input Data Quality

4.6.1 Quality Indicators

Based on the quality indicators for administrative sources from Blue-Ets WP4 (Daas and Ossen 2011), SN has developed quality checks for the three administrative base registers which were listed in the introduction of this chapter. This was done for the benefit of the users of the registers in SN and for the owners of the administrative registers. The quality checks are run on a regular basis on the original data from the administrative source. For each check, an indicator is added to the dataset (Table 4.2). A positive value indicates that there might be something wrong with a specific unit in the register, but not necessarily. The indicators are summarized in quality reports. The quality reports give an overall view of the quality of each separate register and how quality develops over time. The quality checks indicate quality issues at micro level.

Table 4.2 Dataset with quality indicators.

	Ind$_1$	Ind$_2$	Ind$_3$	Ind$_4$	Ind$_5$	Ind$_6$	Ind$_7$	Ind$_{...}$	Ind$_N$	Sum
Unit$_1$	1	1	0	1	0	0	1	...	...	4
Unit$_2$	0	0	0	0	0	0	0	...	...	0
Unit$_3$	0	0	0	0	0	0	1	...	...	1
Unit$_4$	0	0	0	0	0	0	0	...	...	0
Unit$_5$	0	1	0	1	0	0	1	...	...	3
Unit$_6$	0	0	0	0	0	0	1	...	...	1
Unit...	...	...	...	...	...	...	...	...	...	...
Unit$_M$	...	...	...	...	...	...	...	...	...	...
Sum	1	2	0	2	0	0	4	...	...	P

Source: Coen Hendriks, Division for Statistical Populations, "Improved input data quality from administrative sources though the use of quality indicators," Oslo, 14–17 October 2014. Statistics Norway

The number of positive indicators P can be calculated from the dataset. Let N be the number of checks and M the number of units. The number of positive indicators P in the dataset can be calculated as follows:

$$P = \sum_{i=1}^{M} \sum_{j=1}^{N} x_{ij}$$

A general quality indicator Q can be calculated by dividing the total number of positive checks by the number of cells in the dataset. To improve readability Q is calculated per thousand:

$$Q = (P / (N * M)) * 1000$$

Q can be calculated for the dataset and for subpopulations. The quality indicator Q can be used to monitor quality in one register and its subpopulations over time and to compare quality between subpopulations (Table 4.3). Units/rows with many positive indicators can be of interest to the data owners. Checks/columns with many positive indicators can also be of interest. Despite standardization one must be careful when comparing Q for different registers.

4.6.2 Operationalizing the Quality Checks

To operationalize the quality checks, a working group was established at the Division for Statistical Populations. Experts from each of the statistical base registers participated. They have a good overview of the subject matter, long

Table 4.3 General quality indicators Q for registered persons in the CPR.

	Municipality	Records checked	Positive indicators	Records without positive indicators	General quality indicator Q (%)
1849	Hamarøy	1 819	1 244	1 154	24
0817	Drangedal	4 132	1 561	3 302	13
1854	Ballangen	2 587	976	1 880	13
1857	Værøy	777	288	613	13
1874	Moskenes	1 108	394	882	12
2018	Måsøy	1 244	450	1 014	12
1514	Sande	2 632	872	2 258	11
1835	Træna	489	163	388	11
1840	Saltdal	4 691	1 458	1 940	11
1850	Tysfjord	2 004	623	1 617	11
1851	Lødingen	2 246	735	1 855	11
2014	Loppa	1 027	318	843	11
0301	Oslo	634 249	135 547	556 138	7
1201	Bergen	271 854	46 889	245 024	6
1103	Stavanger	130 755	17 357	121 071	5
1601	Trondheim	182 122	22 166	169 173	4
	Norway	5 107 477	777 584	4 638 325	5

Twelve municipalities with the highest general quality indicator Q, four largest municipalities and Norway, 1 January 2014.
Source: Coen Hendriks, Division for Statistical Populations, "Improved input data quality from administrative sources though the use of quality indicators," Oslo, 14–17 October 2014. Statistics Norway

experience from working on registers, and have excellent programming skills. Each expert proposed quality checks for the administrative register. The proposed checks were grouped according to the dimensions from Blue-ets WP4 in technical checks, accuracy, completeness, integrability, and time-related dimensions. The checks were compared across the other base registers to ensure a minimum degree of comparability. This resulted in indicator files for each base register. A typical set of indicator files for the CPR is made up from an indicator file for registered persons,[3] one for families and one for dwelling addresses which are being used in

---

3 The expression Registered Persons is used in this chapter to distinguish between persons who are registered in the CPR as being alive and with an address in Norway, and Resident Persons as statistical units in population statistics. Due to statistical inference, there might be a difference between the total number of registered persons and the number of resident persons.

the CPR. This means that in the CPR each registered person, each formal family and each residential address is checked, and indicators are produced. The units which are being checked in the RLE are legal entities and local kind of activity units. The units which are being checked in the Cadastre are addresses, buildings, ground properties, and functional units in buildings (e.g. dwellings).

4.6.3 Quality Reports

The positive indicators are counted for each register and grouped in five quality dimensions from Blue-ets WP4, in a quality report. Refer to Appendix for an example of a quality report for registered persons in the CPR.

The quality reports are made available for users of the registers in SN.

The quality reports are in demand by the register owners and are used as a tool for quality improvement. SN has quarterly meetings with the owners of the CPR and the Cadastre, and the specific statistics divisions. E.g. the division in charge of Construction Statistics might join the quarterly meeting on quality in the Cadastre. Depending on the agenda, other public agencies which might benefit from the meeting are invited.

4.6.4 The Approach Is Being Adopted by the Owners of Administrative Data

As explained above, SN can inform the data owners on quality aspects, either by means of an aggregated quality report or by transferring the units with positive indicators. Table 4.3 shows 12 municipalities with the highest value of the general quality indicator Q for registered persons. The municipalities with the worst scores concentrate mainly in the county of Troms (municipality code starts with 18). This indicates that population registration in Troms needs extra attention. A further inspection of the quality report for registered persons showed that indicators for inconsistent values had the highest score. Invalid links between family members was a major problem. The quality report also indicated measurement errors, more specifically quality issues with the registered addresses in the CPR.

Sharing information from a dataset which combines information from different sources conflicts with the Statistics Act and is not practised in general. In a few cases SN has received requests from register owners to assess data quality in one source by linking it to another source. Provided the register owner is authorized to use the other source and has a copy of the other source available, an agreement on data processing can be drawn up between SN and the register owner. An agreement on data processing regulates the linking of the sources for quality improvement purposes and the reporting of errors at micro level.

Once the indicators were established for the separate registers and agreements on data processing were established, SN started measuring quality across registers. By linking information from two sources, inconsistency can be measured. During the spring of 2016, SN transferred lists of suspicious units in the Cadastre to the Mapping Authority. The lists were based on combined data from the Cadastre and the CPR. This was possible because SN has an agreement on data processing with the Mapping Authority. It was not the first time such a list was produced, but this time the list was broken down by municipality. The Mapping Authority passed the lists on to the municipalities, urging them to make corrections directly in the Cadastre and, if necessary, to contact the population register to solve quality issues. The reactions from the Mapping authorities were positive, especially because the list was broken down by and distributed to every municipality.

Register owners are interested in the results of quality measurements across registers. The Tax Administration will modernize the CPR in the period 2016–2020. SN is an important user of the data from the CPR and is supporting the project. The Tax Administration aims to include the principles for quality checks from SN, in population registration. This will be a major step in the right direction and will be beneficial for all the users of the data from the CPR, not only for SN. SN will participate in the main project as a pilot user of the new data, and will share experiences in using the data with other users. SN will also participate actively in securing data quality in the modernized CPR.

4.7 Summary

Even though SN has a long history of using and linking register-based data, the first Norwegian fully register-based census of 2011 revealed linkage problems. The DNNI-method was developed to solve linkage problems when linking households to dwellings. To ensure the results of data integration on a more fundamental level, SN looked for means to improve the quality of the source data.

Improved contact with register owners on the quality of the data, results in better input data for statistics. SN introduced statistical population management as a new profession in the statistical office. An approach for measuring and reporting on data quality was developed for the three base administrative registers.

The efforts are appreciated by the data owners. The principles of SN's quality assurance are being adopted by the register owners to secure the data quality in the basic administrative registers. Since basic administrative data is distributed and used widely in other data systems in the public and private sector, advantages of the approach are apparent. A simple update or correction in the CPR will spread like ripples in water in other register systems which rely on the CPR for the basic data. This also counts for updates in the RLE and the Cadastre.

4.8 Exercises

1 What are the differences and similarities between a data processing agreement and agreement on cooperation?

2 When working on data from the three base registers, three keys for linkage are often in use. Which are these keys, and which one do you consider the most important one? Explain your answer.

3 Explain the difference between an administrative register and a statistical register.

4.A Example of a Quality Report for Registered Persons in the Central Population Register

Quality report for registered persons in the Central Population Register, 2012–2014.

	01 January 2012	01 January 2013	01 January 2014	Change 2013–2014
Quality indicator, per 1000	5.4	5.7	5.2	0.5
Number of checks	29	29	29	—
Number of positive indicators	784 396	841 040	777 584	(63 456)
Number of records (registered persons)	4 983 756	5 049 958	5 107 477	57 519
Number of records (registered persons) with positive indicators	496 247	523 357	469 152	(54 205)
Number of records (registered persons) with negative indicators	4 487 509	4 526 601	4 638 325	111 724
C2 Accuracy				
C21 Identifiability				
C21A Use of discontinued PIN on the transaction	—	7	—	(7)
C22 Inconsistent units				
C22A Spouses/partners have different civil status	49 015	54 768	58 814	4 046
C22B Spouses/partners have different date for civil status	86 502	93 039	97 359	4 320

Quality report for registered persons in the Central Population Register, 2012–2014.

	01 January 2012	01 January 2013	01 January 2014	Change 2013–2014
C22C Registered as a resident on Svalbard in the CPR, but not in the Svalbard reg.	8	19	5	(14)
C22D Persons in a family have a common family id, but not a common dwelling id	1 614	1 409	1 369	(40)
C24 Measurement errors				
C24A Invalid municipality number according to the Cadastre	—	—	—	—
C24B Invalid street number according to the Cadastre	—	—	—	—
C24C Invalid house number according to the Cadastre	3 701	7 028	2 293	(4 735)
C24D Invalid entrance number according to the Cadastre	1 488	1 643	945	(698)
C24E Invalid sub-number according to the Cadastre	490	689	355	(334)
C24F Invalid dwelling number according to the Cadastre	34 634	48 248	33 800	(14 448)
C24G Use of property id as address while street address is available in the Cadastre	4 181	7 500	2 538	(4 962)
C24H Use of invalid address according to the Cadastre	116 215	128 977	103 067	(25 910)
C25 Inconsistent values				
C25A Mother's PIN is invalid	106 283	102 018	97 808	(4 210)
C25D Father's PIN is invalid	134 902	130 588	126 460	(4 128)
C25C Spouse's/partner's PIN is invalid	36 800	43 590	48 660	5 070
C25D Dnr indicates birth before 1900	—	—	—	—
C25E Illogical sequence for civil status values	2 431	2 986	3 605	619
C26 Dubious objects				
C26A More than 10 registered persons per dwelling	71 204	72 340	68 636	(3 704)

Quality report for registered persons in the Central Population Register, 2012–2014.

	01 January 2012	01 January 2013	01 January 2014	Change 2013–2014
C3 Completeness				
C35 Missing values				
C35A Registered in a multi-dwelling building, but missing dwelling number	55 445	67 203	60 070	(7 133)
C36 Imputed values				
C36A Incomplete dwelling number (H0000)	748	764	816	52
C4 Time-related checks				
C44 Object dynamics				
C44A Birth registered more than 31 d after official date of birth	2 120	2 367	2 630	263
C44B Death registered more than 31 d after official date of death	2 001	1 094	97	(997)
C44C Immigration registered more than 31 d after official date of immigration	8 802	11 065	12 391	1 326
C44D Emigration registered more than 31 d after official date of emigration	11 977	8 467	1 109	(7 358)
C44E Moving registered more than 31 d after official date of moving	53 286	54 294	53 800	(494)
C44F Svalbard: Immigration registered more than 31 d after official date of immigration	199	162	143	(19)
C44G Svalbard: Emigration registered more than 31 d after official date of emigration	—	419	491	72
C44H Svalbard: Moving registered more than 31 d after official date of moving	350	356	323	(33)

References

Daas, P. and Ossen, S. (2011). Report on methods preferred for the quality indicators of administrative data sources. *Deliverable 4.2 BLUE-Enterprise and Trade Statistics. European Commission European Research Area, Seventh framework programme.*

Eurostat (2017). European Statistical Code of Practice for the National and Community Statistical Authorities. *Adopted by the European Statistical System Committee (16 November 2017).*

Hendriks, C. and Åmberg, J. (2011). Building and maintaining quality in register populations. Paper to 58th World Statistics Congress of the International Statistical Institute, Session STS50 Methods and Quality of Administrative Data Used in a Census, Dublin, Ireland (21–26 August 2011).

Longva, S., Thomsen, I., and Severeide, P.I. (1998). Reducing costs of censuses in Norway through use of administrative registers. *International Statistical Review* 66 (2): 223–234.

Severeide, P.I. and Hendriks, C. (1998). Establishing a dwelling register in Norway: the missing link. Paper to the Joint ECE-Eurostat Work Session in Population and Housing Censuses, Dublin, Ireland (9–11 November 1998).

UNECE 2007: *Register Based Statistics in the Nordic Countries. Review on Best Practices with Focus on Population and Social Statistics.* http://www.unece.org/fileadmin/DAM/stats/publications/Register_based_statistics_in_Nordic_countries.pdf.

Zhang, L.-C. (2011). A unit-error theory for register-based household statistics. *Journal of Official Statistics* 27 (3): 415–432.

Zhang, L.-C. (2012). Topics of statistical theory for register-based statistics and data integration. *Statistics Neerlandica* 66 (1): 41–63.

Zhang, L.-C. and Hendriks, C. (2012). Micro integration of register-based census data for dwelling and household. Paper to the Conference of European Statisticians, Work Session on Statistical Data Editing, Oslo, Norway (24–26 September 2012).

5

Cleaning and Using Administrative Lists: Enhanced Practices and Computational Algorithms for Record Linkage and Modeling/Editing/Imputation
William E. Winkler

U.S. Census Bureau

5.1 Introductory Comments

This chapter provides an overview of methods for cleaning administrative lists. The National Statistical Institutes (NSIs) originally developed the methods for processing survey data. The NSIs have developed systematic, efficient generalized methods/software based on the Fellegi–Holt (FH) model of statistical data editing (JASA 1976) and the Fellegi–Sunter (FS) model of record linkage (JASA 1969). The generalized software is suitable for processing files for businesses, survey institutes, and administrative organizations. New computational algorithms yield drastic 100+ fold speed increases. The software is suitable for processing hundreds of millions or billions of records. In Sections 5.1.1–5.1.3, following some introductory examples, we provide background on modeling/edit/imputation, record linkage, and adjusting statistical analyses for linkage error.

5.1.1 Example 1

The NSIs through the 1960s primarily performed data collection manually. Enumerators asked questions from survey forms and did minor corrections to the information provided by respondents. Other reviewers systematically went through the collected forms to make additional "corrections." The data were keyed into the computers. Processing software beginning in the 1960s incorporated hundreds (sometimes thousands) of "if–then–else" rules into computer code so that "corrections" were made more quickly and consistently. The issues with the early systems were that there could be logical errors in the specifications of the edits (i.e. and edit such as "a child under 16 could not be married"), that there were coding errors, and that no systematic methods existed for assuring that joint distributions of variables were preserved in a principled manner.

Administrative Records for Survey Methodology, First Edition.
Edited by Asaph Young Chun, Michael D. Larsen, Gabriele Durrant, and Jerome P. Reiter.

Starting in the 1980s, five NSIs developed "edit" systems based on the Fellegi–Holt (1976, hereafter FH) model of statistical data editing. The advantage of the FH systems were that edits resided in easily modified tables, the logical consistency of the system could be checked prior to the receipt of data, the main logic resides in reusable routines, and, in one pass through the data, a "corrected record" is guaranteed to satisfy edits. Prior to FH theory, an "if–then–else" edit system might run correctly on test decks but fail when non-anticipated edit patterns were encountered during production. Additionally, there might be logic errors in the set of "if–then–else" rules. NSIs hired individuals with expertise in Operations Research (typically PhDs) who did theory and computational algorithms for the integer programming logic needed to implement FH systems. NSIs in Canada, Spain, Italy, the Netherlands, and the United States developed general systems. Each was able to demonstrate substantial computational efficiencies with the new systems that were portable across different types of survey data and that often ran on more than one computer architecture.

The generalized software yielded significant time-savings and dramatically increased accuracy. The 1997 U.S. Economic Censuses (with a total of 8 million records in three files) were each processed in less than 12 hours. Kovar and Winkler (1996) showed that the Census Bureau structured programs for economic editing and referrals (SPEER) system maintained the very high quality of Statistics Canada's Generalized Edit/Imputation System while being 60 times as fast (due to new computational algorithms – Winkler 1995b). Because of substantial hardware-speed increases in the last 20 years, most of the NSIs methods/software should be suitable for smaller administrative lists (millions or tens of millions of records). Census Bureau software (Winkler 2013) should be suitable for hundreds of millions of records or even billions of records.

5.1.2 Example 2

Record linkage is the methodology of finding and correcting duplicates within a single list or across two or more lists. The means of merging/finding duplicates involve non-unique (quasi-) identifiers such as name, address, date-of-birth, and other fields. Statistics Canada, the Italian National Statistical Institute (ISTAT), and the U.S. Census Bureau developed systems for record linkage based on the Fellegi–Sunter (hereafter FS) ideas. University Computer Science (CS) professors who were funded by various health agencies developed other systems.

Howard Hogan and Kirk Wolter (1984) introduced a capture–recapture procedure (called the Post Enumeration Survey, PES) in which a second list was matched against the Decennial Census to evaluate/correct-for under- and over-coverage. Their procedure had seven components. They believed that the record-linkage component to have more error than the other six components combined. Hogan and Wolter estimated that manual matching would require 3000 individuals for six months with a false match error rate of 5%. Their PES estimation/adjustment procedures required at most 0.5% false match rate. Some leading statisticians, including David Freedman from UC Berkeley in a New York Times Op-Ed article, stated that it was impossible to do the PES matching with sufficient accuracy and speed.

To do the matching, we needed to get optimal parameters (unsupervised learning) in ~500 contiguous regions in which we had to do matching. We did the matching in six weeks. We needed approximately 200 individuals for clerical matching (using additional information from paper files) and field follow-up. The clerical review region consisted almost entirely of individuals within the same household who were missing first name and age (the only two fields for distinguishing individuals within households). Expectation maximization (EM) procedures for obtaining optimal parameters (Winkler 1988; Winkler and Thibaudeau 1991) yielded a false match rate of at most 0.2% (based on field follow-up, two rounds of clerical review, and one round of adjudication).

5.1.3   Example 3

Because files (particularly large national files) can yield useful information in statistical, demographic, and economic analyses, various groups are interested in cleaning-up and merging individual files and (possibly) doing additional clean-up of merged files to correct for linkage error.

A conceptual picture would link records in file $\mathbf{A} = (a_i, \ldots, a_n, x_1, \ldots, x_k)$ with records in file $\mathbf{B} = (b_1, \ldots, b_m, x_1, \ldots, x_k)$ using common quasi-identifying information $(x_1, \ldots, x_k)$ to produce the merged file $\mathbf{A} \times \mathbf{B} = (a_i, \ldots, a_n, b_1, \ldots, b_m)$ for analyses. The variables $x_1, \ldots, x_k$ are quasi-identifiers such as names, addresses, dates-of-birth, and even fields such as income (when processed and compared in a suitable manner). Individual quasi-identifiers will not uniquely identify correspondence between pairs of records associated with the same entity; sometimes combinations of the quasi-identifiers may uniquely identify. Each of the files $\mathbf{A}$ and $\mathbf{B}$ may be cleaned for duplicates then cleaned up for missing and contradictory data via edit/imputation.

If there are errors in the linkage, then completely erroneous $(b_1, \ldots, b_m)$ may be linked with a given $(a_i, \ldots, a_n)$ and the joint distribution of $(a_i, \ldots, a_n, b_1, \ldots, b_m)$ in $\mathbf{A} \times \mathbf{B}$ may be very seriously compromised. If there is inadequate cleanup (i.e. noneffective edit/imputation) of $\mathbf{A} = (a_i, \ldots, a_n, x_1, \ldots, x_k)$ and $\mathbf{B} = (b_1, \ldots, b_m, x_1, \ldots, x_k)$, then analyses may have other serious errors in addition to the errors due to the linkage errors. For instance, each of the files $\mathbf{A}$ and $\mathbf{B}$ may have 3% duplicates or edit/imputation errors. If there is 3% matching error, then the joint file $\mathbf{A} \times \mathbf{B}$ could have 9% error. If there is 9% error in a file, then how would an NSI determine if any analysis on $\mathbf{A} \times \mathbf{B}$ could be performed? If there were 5% error in either of the files $\mathbf{A}$ or $\mathbf{B}$, how would the NSI determine that the errors existed or how to correct for them? If there were 5% error in the matching between files $\mathbf{A}$ and $\mathbf{B}$, how would the NSI determine that there was too much error and how would the NSI correct the error?

5.2 Edit/Imputation

In this section, we cover edit/imputation primarily for discrete data based on the statistical data editing model of Fellegi and Holt (1976). The imputation uses ideas of filling-in missing data according to extensions of the missing-at-random (*mar*) model of Little and Rubin (2002, chapter 13) and further extensions by Winkler (2008). It then applies the theory of Winkler (2003) that connects editing with imputation. The computational ideas are set-covering algorithms (Winkler 1997) and three model-building algorithms based on the expectation maximization Haberman (EMH) algorithm (Winkler 1993) that extends the expectation conditional maximization (ECM) algorithm of Meng and Rubin (1993) from linear to convex constraints. Key algorithmic breakthroughs increased speed of these four computational algorithms by factors of 100–1000. The generalized software (Winkler 2008, 2010) is suitable for use in most survey situations with discrete data and is sufficiently fast for hundreds of millions or billions of records.

5.2.1 Background

The intent of classical data collection and clean-up was to provide a data file that was free of logical errors and missing data. For a statistical agency, an interviewer might fill out a survey form during a face-to-face interview with the respondent. The "experienced" interviewer would often be able to "correct" contradictory data or "replace" missing data during the interview. At a later time analysts might make further "corrections" prior to the data on paper forms being placed in computer files. The purpose was to produce a "complete" (i.e. no missing values) data file that had no contradictory values in some variables. The final "cleaned" file would be suitable for various statistical analyses. In particular, the statistical file would allow determination of the proportion of specific values of the multiple variables (i.e. joint inclusion probabilities).

Naively, dealing with edits is straightforward. If a child of less than 16 years old is given a marital status of "married," then either the age associated with the child might be changed (i.e. to older than 16) or the marital status might be changed to "single." The difficulty consistently arose, as a (computerized) record r_0 was changed ("corrected") to a different record r_1 by changing values in fields in which edits failed, the new record r_1 would fail other edits that the original record r_0 had not failed.

In a real-world survey situation, subject matter "experts" may develop hundreds or thousands of if–then–else rules that are used for the editing and hot-deck imputation. Because it is exceptionally difficult to develop the logic for such rules, most edit/imputation systems do not assure that records satisfy edits or preserve joint inclusion probabilities. Further, such systems are exceptionally difficult to implement because of (i) logic errors in specifications, (ii) errors in computer code, and (iii) no effective modeling of hot-deck matching rules. As demonstrated by Winkler (2008), it is effectively impossible with the methods (classical if–then–else and hot-deck) that many agencies use to develop edit/imputation systems that preserve either joint probabilities or to create records that satisfy edit restraints. This is true even in the situations when FH methods are used for the editing and hot-deck is used for imputation.

An edit/imputation system that effectively uses the edit ideas of Fellegi and Holt (1976) and modern imputation ideas (such as in Little and Rubin 2002) has distinct advantages. First, it is far easier to implement (as demonstrated in Winkler 2008). Edit rules are in easily modified tables. The logical consistency of the entire system is tested automatically according the mathematics of the FH model and additional requirements on the preservation of joint inclusion probabilities (Winkler 2003). Second, the optimization that determines the minimum number of fields to change or replace in an edit-failing record is in a fixed mathematical routine that does not need to change. Third, imputation is determined from a model (limiting distribution) and automatically preserves joint distributions. Most modeling is very straightforward. It is based on variants of loglinear modeling and extensions of missing data methods that is contained in easily applied, extremely fast computational algorithms (Winkler 2006b, 2008; also 2010). The methods create records that *always* satisfy edits and preserve joint inclusion probabilities.

The generalized software (Winkler 2010) incorporates ideas from statistical matching software (Winkler 2006b) that can be compared to ideas and results of D'Orazio, Di Zio, and Scanu (2006) and earlier discrete-data editing software (Winkler 2008) that could be used for synthetic-data generation (Winkler 2010). The basic methods are closely related to ideas suggested in Little and Rubin (2002, chapter 13) in that they assume a *mar* assumption that can be slightly weakened in some situations (Winkler 2008, 2010). The original theory for the computational algorithms (Winkler 1993) uses convex constraints (Winkler 1990b) to produce an EMH algorithm that generalizes the MCECM algorithm of Meng and Rubin (1993). Winkler (1993) first applied the EMH algorithm to record linkage. D'Orazio, Di Zio, and Scanu (2006) used the EMH algorithm in statistical matching. The main theorem of Winkler (1990b) generalizes a theorem of Della Pietra, Della Pietra, and Lafferty (1997). Winkler (2000) used the theorem and related computational algorithms in an extension of boosting from improving the parameters in a model to simultaneously improve the model and the parameters in the model.

5.2.2 Fellegi–Holt Model

Fellegi and Holt (1976) were the first to provide an overall model to assure that a changed record r_1 would not fail edits. Their theory required the computation of all implicit edits. Implicit edits are edits that we can logically derive from an originally specified set of "explicit" edits. If the implicit edits were available, then it was always possible to change an edit-failing record r_0 to an edit passing record r_1. The availability of "implicit" edits makes it quite straightforward and fast to determine the minimum number of fields to change in an edit-failing record r_0 to obtain an edit-passing record r_1 (Barcaroli and Venturi 1997). In particular, the implicit edits drastically speed up standard IP algorithms such as branch-and-bound by quickly forcing solutions into more suitable regions. Fellegi and Holt indicated how hot-deck might be used to provide the values for filling in missing values or replacing contradictory values.

5.2.3 Imputation Generalizing Little–Rubin

The intent of filling-in missing or contradictory values in edit-failing records r_0 is to obtain a record r_1 that can be used in computing the joint probabilities in a principled manner. The difficulty observed by many individuals is that a well-implemented hot-deck does not preserve joint probabilities. Rao (1997) provided a theoretical characterization of why hot-deck fails even in two-dimensional situations. The failure occurs even in "nice" situations where individuals had previously assumed that hot-deck would work well. Other authors in the 1980s had empirically shown that hot-deck fails to preserve joint distributions. Andridge and Little (2010) provide a review of hot-deck-related literature.

Little and Rubin (2002, chapter 13) provided a model for filling-in missing data according to the *mar* data. The modeling methods can be considered as a correct extension of "hot-deck." With "hot-deck" a record with missing values is matched against the entire set of *complete* records that have no missing data. In practice, NSIs might have a record with 20 variables of which 5 had missing values. Typically, there would be no *donors* (complete data records) that match on the 15 non-missing values. To find matches among the set of complete data records, NSIs developed collapsing rules that would take subsets of the 15 variables to match on. In extreme (but typical) situations, collapsing would continue until only two to four variables were used for matching. In all situations with collapsing, joint distributions of the final data were substantially distorted. Most NSIs have not developed systematic, efficient methods for collapsing.

Winkler (2008, 2010) provides some theoretical extensions and new exceedingly fast computational algorithms for model-building that generalizes the methods of Little and Rubin (2002). The model-building software is suitable for use by junior analysts and programmers and produces very high quality models that preserve joint distributions. Because the models are of such high quality, Winkler (with special additional software) showed that all large- and medium-size margins were preserved in synthetic data from the model and that 80% of small cells in the synthetic data were actual small cells. To reduce re-identification risk, Winkler (2010) used convex constraints (Winkler 1990b, 1993) that produced models with nearly identical high overall quality but in which only 5% of the cells in the synthetic data were actual small cells.

5.2.4 Connecting Edit with Imputation

With classical edit/imputation in the NSIs, the logic in the software consisted of a number (100s or 1000s) of "if–then–else" rules where values in fields in a record *r* associated with failing edits were replaced with values that would cause the "corrected" record to no longer fail edits. With the development of FH systems, the NSIs, for the first time, could assure that a "corrected" record would no longer fail edits. Because the "substitutions" of values due to editing were not designed to assure that joint distributions were preserved, statisticians later developed imputation models for filling in missing values or replacing "erroneous" values associated with failing edits that assured joint distributions. When the imputation software was used to impute, some "corrected" records did not satisfy edits because the imputation methods did not effectively account for the entire set of edit restraints.

Winkler (2003) provided the first theory that connected imputation with the entire set of edit restraints. Winkler (1997) had earlier provided set-covering algorithms that were at least 100 times as fast as set-covering algorithms that IBM developed for the Italian Labor Force Survey (Barcaroli and Venturi 1997 and private communications with the authors). With suitably fast set-covering algorithms all implicit edits were available prior to building imputation models that systematically and globally assured that all records passing through the system would satisfy all edits and, in aggregate, preserve joint distributions. Lemma 1 (Winkler 1990b) showed how structural zeros (edit restraints) in a contingency table could be consistent with the other restraints in a contingency table such as interactions and certain convex constraints. The theory assured that, if individuals constructed suitable structures using implicit edits in just the right manner, then imputation could be made consistent with the edit restraints.

Although the theory assured that it was possible to develop generalized systems, the existing methods for doing the computation associated with the methods of chapter 13 in Little and Rubin (2002) were 10^3–10^6-times too slow for use on national files or large administrative lists.

5.2.5 Achieving Extreme Computational Speed

The current algorithms do the EM fitting as in Little and Rubin (2002) but with computational enhancements that scale subtotals exceedingly rapidly and with only moderate use of memory. The computational speed for a contingency table of size 600 000 cells is 45 seconds on a standard Intel Core I7 machine and for a table of size 0.5 billion cells in approximately 1000 minutes (each with epsilon (ε) 10^{-12} and 200 iterations). In the larger applications, 16 GB of memory are required. The key to the speed is the combination of effective indexing of cells and suitable data structures for retrieval of information so that each of the respective margins of the M-step of EM-fitting are computed rapidly. In my edit/imputation course, I challenge the students to write an algorithm in SAS or R program that converges to a solution in the 600 000-cell situation in less than one week.

Certain convex constraints can be incorporated in addition to the standard linear constraints of classic loglinear EM fitting (Winkler 1990b, 1993). In statistical matching (Winkler 2006b) was able to incorporate closed form constraints P(Variable $X1 = x11 >$ Variable $X1 = x12$) with the same data as D'Orazio, Di Zio, and Scanu (2006) that needed a much slower version of the iterative fitting EMH algorithm (Winkler 1993) for the same data and constraints. The variable $X1$ took four values and the restraint is that one margin of $X1$ for one value is restricted to be greater than one margin of another value. For general edit/imputation, Winkler (2008) was able to put marginal constraints on one variable to assure that the resultant microdata files and associated margins corresponded much more closely to observed margins from an auxiliary data source. For instance, one variable could be an income range and the produced microdata did not produce population proportions that corresponded closely to published Internal Revenue Service (IRS) data until after appropriate convex constraints were additionally applied.

A nontrivially modified version of the indexing algorithms allows near instantaneous location of cells in the contingency table that match a record having missing data. This is the key logical extension of Little and Rubin (2002, chapter 13). An additional algorithm nearly instantaneously constructs an array that allows binary search to locate the cell for the imputation (for the two algorithms: total < 2.0 ms CPU time). For instance, if a record has 12 variables and 5 are missing, we might need to delineate all 100 000+ cells in a contingency table with 0.5 million or 0.5 billion cells and then draw a cell (donor) with probability-proportional-to-size (pps) to impute missing values in the record with missing values. This type of imputation assures that the resultant "corrected" microdata have joint distributions that are consistent with the model. A naively written SAS search and pps-sample procedure might require as much as a minute CPU time for each record being imputed.

For imputation-variance estimation, other closely related algorithms allow direct variance estimation from the model. This is in contrast to after-the-fact variance approximations using linearization, jackknife, or bootstrap. These latter three methods were developed for after-the-fact variance estimation (typically with possibly poorly implemented hot-deck imputation) that are unable to account effectively for the bias of hot-deck or that lack of model with hot-deck. Most of the methods for the after-the-fact imputation-variance estimation have only been developed for one-variable situations that do not account for the multivariate characteristics of the data and assume that hot-deck matching (when naively applied) is straightforward while most hot-deck matching is never straightforward.

5.3 Record Linkage

In Section 5.3.1, we provide background on the Fellegi–Sunter model of record linkage. In Section 5.3.2, we show how to estimate "optimal" parameters for approximately five hundred regions without training data (unsupervised learning) that was first used in a production system for the 1990 Decennial Census. In Section 5.3.3 we show to estimate false match rates without training data. The methods are not as accurate as the semi-supervised methods of Larsen and Rubin (2001) and Winkler (2002). In Section 5.3.4, we indicate how we achieved extreme computational speed suitable for sets of national files and administrative lists with hundreds of millions or billions of records.

5.3.1 Fellegi–Sunter Model

Fellegi and Sunter (1969) provided a formal mathematical model for ideas that had been introduced by Newcombe et al. (1959), Newcombe and Kennedy (1962). They introduced many ways of estimating key parameters without training data. To begin, notation is needed. Two files A and B are matched. The idea is to classify pairs in a product space $\mathbf{A} \times \mathbf{B}$ from two files A and B into M, the set of true matches, and U, the set of true nonmatches. Fellegi and Sunter, making rigorous concepts introduced by Newcombe et al. (1959), considered ratios of probabilities of the form:

$$R = \mathrm{P}\left(\gamma \in \Gamma \,|\, \mathrm{M}\right) / \mathrm{P}\left(\gamma \in \Gamma \,|\, \mathrm{U}\right) \tag{5.1}$$

where γ is an arbitrary agreement pattern in a comparison space Γ. For instance, Γ might consist of eight patterns representing simple agreement or not on the largest name component, street name, and street number. Alternatively, each $\gamma \in \Gamma$ might additionally account for the relative frequency with which specific values of name components such as "Smith" and "Zabrinsky" occur. Then P(agree "Smith" | M) < P(agree last name | M) < P(agree "Zabrinsky" | M) which typically gives a less frequently occurring name like "Zabrinsky" more distinguishing power than a more frequently occurring name like "Smith" (Fellegi and Sunter 1969; Winkler 1995a). Somewhat different, much smaller, adjustments for relative frequency are given for the probability of agreement on a specific name given U. The probabilities in (5.1) can also be adjusted for partial agreement on two strings because of typographical error (which can approach 50% with scanned data (Winkler 2004)) and for certain dependencies between agreements among sets of fields (Larsen and Rubin 2001; Winkler 2002). The ratio R or any monotonically increasing function of it such as the natural log is referred to as a *matching weight* (or score).

The decision rule is given by:

If $R > T_\mu$, then designate pair as a match.

If $T_\lambda \leq R \leq T_\mu$, then designate pair as a possible match
and hold for clerical review. (5.2)

If $R < T_\lambda$, then designate pair as a nonmatch.

The cutoff thresholds T_μ and T_λ are determined by *a priori* error bounds on false matches and false nonmatches. Rule (5.2) agrees with intuition. If $\gamma \in \Gamma$ consists primarily of agreements, then it is intuitive that $\gamma \in \Gamma$ would be more likely to occur among matches than nonmatches and ratio (5.1) would be large. On the other hand, if $\gamma \in \Gamma$ consists primarily of disagreements, then ratio (5.1) would be small. Rule (5.2) partitions the set $\gamma \in \Gamma$ into three disjoint subregions. We refer to the region $T_\lambda \leq R \leq T_\mu$ as the *no-decision region* or *clerical review* region. In some situations, resources are available to review pairs clerically.

Fellegi and Sunter (1969, Theorem 1) proved the optimality of the classification rule given by (5.2). Their proof is very general in the sense it holds for any representations $\gamma \in \Gamma$ over the set of pairs in the product space $\mathbf{A} \times \mathbf{B}$ from two files. As they observed, the quality of the results from classification rule (5.2) were dependent on the accuracy of the estimates of $P(\gamma \in \Gamma \mid M)$ and $P(\gamma \in \Gamma \mid U)$.

Figure 5.1 provides an illustration of the curves of log frequency versus log weight for matches and nonmatches, respectively. The two vertical lines represent the lower and upper cutoffs thresholds T_λ and T_μ, respectively. The x-axis is the log of the likelihood ratio R given by (5.1). The y-axis is the log of the frequency counts of the pairs associated with the given likelihood ratio. The plot uses pairs of records from a contiguous geographic region that we matched in the 1990 Decennial Census. The clerical review region between the two cutoffs primarily

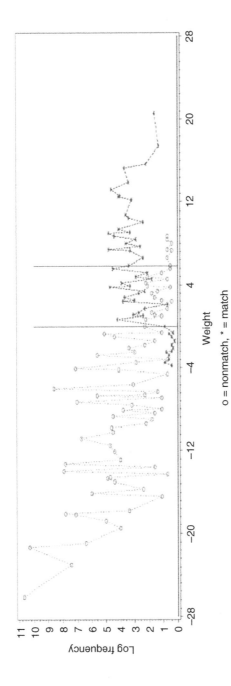

o = nonmatch, * = match
Cutoff "L" = 0 and cutoff "U" = 6

Figure 5.1 Log frequency versus weight matches and nonmatches combined.

consists of pairs within the same household that are missing both first name and age (the only two fields that distinguish individuals within a household).

5.3.2 Estimating Parameters

In this section, we provide a summary of current extensions of the EM procedures for estimating false match rates. With any matching project, we are concerned with false match rates among the set of pairs among designated matches above the cutoff score T_μ in (5.2) and the false nonmatch rates among designated non-matches below the cutoff score T_λ in (5.2). Very few matching projects estimate these rates although valid estimates are crucial to understanding the usefulness of any files obtained via the record linkage procedures. Sometimes we can obtain reasonable upper bounds for the estimated error rates via experienced practitioners and the error rates are validated during follow-up studies (Winkler 1995a). If a moderately large amount of training data is available, then it may be possible to get valid estimates of the error rates.

If a small amount of training data is available, then it may be possible to improve record linkage and good estimates of error rates. Larsen and Rubin (2001) combined small amounts of (labeled) training data with large amounts of unlabeled data to estimate error rates using an Monte Carlo Markov chain (MCMC) procedure. In machine learning (Winkler 2000), the procedures are referred to as *semi-supervised learning*. In ordinary machine learning, the procedures to get parameters are "supervised" by the training data that is labeled with the true classes into which later records (or pairs) will be classified. Winkler (2002) also used semi-supervised learning with a variant of the general EM algorithm.

Both the Larsen and Rubin (2001) and Winkler (2002) methods were effective because they accounted for interactions between the fields and were able to use labeled training data that was concentrated between the lower cutoff T_λ and the upper cutoff T_μ. Winkler (1988) observed that the conditional independence (**CI**) model (naive Bayes) in the paper was an approximation of a general interaction model. Winkler (2014, also Herzog, Scheuren, and Winkler 2010) provided more complete explanations of why the 1988 procedure was a best naïve Bayes approximation of general interaction model. The model-approximation methods were rediscovered by Kim Larsen (2005) who used much-slower-to-compute general-additive models.

Because the EM-based methods of this section serve as a template of other EM-based methods, we provide details of the unsupervised learning methods of Winkler (2006a) that are used for estimating the basic matching parameters and, in some situations, false match rates. The basic model is that of semi-supervised learning in which we combine a small proportion of labeled (true or pseudo-true matching status) pairs of records with a very large amount of unlabeled data. The conditional independence model corresponds to the naïve Bayesian network formulization of Nigam et al. (2000). The more general formulization of Winkler (2000, 2002) allows interactions between agreements (but is not used in this paper).

Our development is similar theoretically to that of Nigam et al. (2000). The notation differs very slightly because it deals more with the representational framework of record linkage. Let γ_i be the agreement pattern associated with pair p_i. Classes C_j are an arbitrary partition of the set of pairs D in $\mathbf{A} \times \mathbf{B}$. Later, we will assume that some of the C_j will be subsets of M and the remaining C_j are subsets of U. Unlike general text classification in which every document may have a unique agreement pattern, in record linkage, some agreement patterns γ_i may have many pairs $p_{i(l)}$ associated with them. Specifically,

$$P\left(\gamma_i \mid \Theta\right) = \sum_i |C| \, P\left(\gamma_i \mid C_j; \Theta\right) P\left(C_j; \Theta\right) \tag{5.3}$$

where i is a specific pair, C_j is a specific class, and the sum is over the set of classes. Under the Naïve Bayes or conditional independence (**CI**), we have

$$P\left(\gamma_i \mid C_j; \Theta\right) = \Pi_k P\left(\gamma_{i,k} \mid C_j; \Theta\right) \tag{5.4}$$

where the product is over the kth individual field agreement γ_{ik} in pair agreement pattern γ_i. In some situations, we use a Dirichlet prior

$$P\left(\Theta\right) = \Pi_j \left(\Theta_{C_j}\right)^{\alpha-1} \Pi_k \left(\Theta_{\gamma_{i,k} \mid C_j}\right)^{\alpha-1} \tag{5.5}$$

where the first product is over the classes C_j and the second product is over the fields. We use Du to denote unlabeled pairs and Dl to denote labeled pairs. Given the set D of all labeled and unlabeled pairs, the log likelihood is given by

$$
\begin{aligned}
l_c\left(\Theta \mid D; z\right) = {} & \log\left(P\left(\Theta\right)\right) \\
& + (1 - \lambda) \sum_{i \in \mathrm{Du}} \sum_j z_{ij} \log\left(P\left(\gamma_i \mid C_j; \Theta\right) P\left(C_j; \Theta\right)\right) \\
& + \lambda \sum_{i \in \mathrm{Du}} \sum_j z_{ij} \log\left(P\left(\gamma_i \mid C_j; \Theta\right) P\left(C_j; \Theta\right)\right)
\end{aligned}
\tag{5.6}
$$

where $0 \leq \lambda \leq 1$. The first sum is over the unlabeled pairs and the second sum is over the labeled pairs. In the third terms of Eq. (5.6), we sum over the observed z_{ij}. In the second term, we put in expected values for the z_{ij} based on the initial estimates $P(\gamma_i \mid C_j; \Theta)$ and $P(C_j; \Theta)$. After re-estimating the parameters $P(\gamma_i \mid C_j; \Theta)$ and $P(C_j; \Theta)$ during the M-step (that is in closed form under condition [**CI**]), we put in new expected values and repeat the M-step. The computer algorithms are easily monitored by checking that the likelihood increases after each combination of E- and M-steps and by checking that the sum of the probabilities add to 1.0. We observe that if λ is 1, then we only use training data and our methods correspond to naïve Bayes methods in which training data are available. If λ is 0, then we are in the unsupervised learning situations of Winkler (1988, 2006a). Winkler (2000, 2002) provides more details of the computational algorithms.

5.3.3 Estimating False Match Rates

In this section, we provide a summary of current extensions of the EM procedures for estimating false match rates. To estimate false-match rates, we need to have reasonably accurate estimates of the right tail of the curve of nonmatches and the left tail of the curve of matches such as given in Figure 5.1. Similar methods, but with less data, were given in Belin and Rubin (1995). With any matching project, we are concerned with false match rates among the set of pairs among designated matches above the cutoff score T_μ in (5.2) and the false nonmatch rates among designated nonmatches below the cutoff score T_λ in (5.2). Very few matching projects estimate these rates although valid estimates are crucial to understanding the usefulness of any files obtained via the record linkage procedures. Sometimes we can obtain reasonable upper bounds for the estimated error rates via experienced practitioners and the error rates are validated during follow-up studies (Winkler 1995a). If a moderately large amount of training data is available, then it may be possible to get valid estimates of the error rates.

Belin and Rubin (1995) were the first to provide an unsupervised method for obtaining estimates of false match rates. The method proceeded by estimating Box–Cox transforms that would cause a mixture of two transformed normal distributions to closely approximate two well-separated curves such as given in Figure 5.1. They cautioned that their methods might not be robust to matching situations with considerably different types of data. Winkler (1995a) observed that their algorithms would typically not work with business lists, agriculture lists, and low quality person lists where the curves of nonmatches were not well separated from the curves of matches. Scheuren and Winkler (1993), who had the Belin–Rubin EM-based fitting software, observed that the Belin–Rubin methods did work reasonably well with a number of well-separated person lists.

5.3.3.1 The Data Files
Three pairs of files were used in the analyses. The files are from 1990 Decennial Census matching data in which the entire set of 1–2% of the matching status codes that we believed to have been in error for these analyses have been corrected. The corrections reflect clerical review and field follow-up that were not incorporated in computer files originally available to us.

A summary of the overall characteristics of the empirical data is in Table 5.1. We only consider pairs that agree on census block id (small geographic area

Table 5.1 Summary of three pairs of files.

	Files		Files		Files	
	A1	A2	B1	B2	C1	C2
Size	15 048	12 072	4 539	4 851	5 022	5 212
No. of pairs	116 305		38 795		37 327	
No. of matches	10 096		3 490		3 623	

representing approximately 50 households) and on the first character of surname. Less than 1–2% of the matches are missed using this set of blocking criteria. They are not considered in the analysis of this chapter.

The matching fields are:

Person characteristics: First Name, Age, Marital Status, Sex
Household characteristics: Last Name, House Number, Street Name, Phone

Typically, everyone in a household will agree on the household characteristics. Person characteristics such as first name and age help distinguish individuals within household. Some pairs (including true matches) have both missing first name and age.

We also consider partial levels of agreement in which the string comparator values are broken out as $[0, 0.66], (0.66, 0.88], (0.88, 0.94]$, and $(0.94, 1.0]$. The intervals were based on knowledge of how string comparators were initially modeled (Winkler 1990a) in terms of their effects of likelihood ratios (5.1). The first interval is what we refer to as disagreement. We combine the disagreement with the three partial agreements and blank to get five value states (base 5). The large base analyses consider five states for all characteristics except sex and marital status for which we consider three (agree/blank/disagree). The total number of agreement patterns is 140 625. In the earlier work (Winkler 2002), the five levels of agreement worked consistently better than two levels (agree/disagree) or three levels (agree/blank/disagree).

The pairs naturally divide into three classes: C_1 – match within household, C_2 – nonmatch within household, C_3 – nonmatch outside household. In the earlier work (Winkler 2002), we considered two dependency models in addition to the conditional independence model. In that work in which small amounts of labeled training data were combined with unlabeled data, the conditional independence model worked well and the dependency models worked slightly better. Newcombe and Smith (1975) and later by Gill (2001) first used the procedures for dividing the matching and estimation procedures into three classes without the formal likelihood models given by Eqs. (5.3)–(5.6).

Table 5.2 "Pseudo-truth" data with actual error rates.

	Matches	Nonmatches	Other
A pairs	8817 (0.008)	98 257 (0.001)	9231 (0.136)
B pairs	2674 (0.010)	27 744 (0.0004)	8377 (0.138)
C pairs	2492 (0.010)	31 266 (0.002)	3569 (0.369)

We create "pseudo-truth" data sets in which matches are those unlabeled pairs above a certain high cutoff and nonmatches are those unlabeled pairs below a certain low cutoff. Figure 5.1 illustrates the situation using actual 1990 Decennial Census data in which we plot log of the probability ratio (5.1) against the log of frequency. With the datasets of this chapter, we choose high and low cutoffs in a similar manner so that we do not include in-between pairs in our designated "pseudo-truth" data sets. We use these "designated" pseudo-truth data sets in a semi-supervised learning procedure that is nearly identical to the semi-supervised procedure where we have actual truth data. A key difference from the corresponding procedure with actual truth data is that the sample of labeled pairs is concentrated in the difficult-to-classify in-between region where, in the "pseudo-truth" situation, we have no way to designate comparable labeled pairs. The sizes of the "pseudo-truth" data is given in Table 5.2. The errors associated with the artificial "pseudo-truth" are given in parentheses following the counts. The *Other* class gives counts of the remaining pairs and proportions of true matches that are not included in the "pseudo-truth" set of pairs of "Matches" and "Nonmatches." In the *Other* class, the proportions of matches vary somewhat and would be difficult to determine without training data.

We determine how accurately we can estimate the lower cumulative distributions of matches and the upper cumulative distribution of nonmatches. This corresponds to the overlap region of the curves of matches and nonmatches. If we can accurately estimate these two tails of distributions, then we can accurately estimate error rates at differing levels. Our comparisons consist of a set of figures in which we compare a plot of the cumulative distribution of estimates of matches versus the true cumulative distribution with the truth represented by the 45° line. As the plots get closer to the 45° lines, the estimates get closer to the truth. Our plotting is only for the bottom 30% of the curves given in Belin and Rubin (1995, figures 2 and 3). Generally, we are only interested in the bottom 10% of the curves for the purpose of estimating false match rates. Because of the different representation with the 45° line, we can compare much better three different methods of estimation for false match rates.

Our primary results are from using the conditional independence model and "pseudo-semi-supervised" methods of this section with the conditional independence model and actual semi-supervised methods of Winkler (2002). With our "pseudo-truth" data, we obtain the best sets of estimates of the bottom 30% tails of the curve of matches with conditional independence and $\lambda = 0.2$ (Eq. (5.6)). Figure 5.2a–c illustrates the set of curves that provide quite accurate fits for the estimated values of matches versus the truth. The 45° line represents the truth whereas the curve represents the cumulative estimates of matches for the right tails of the distribution. The plots are for the estimates of the false match probabilities divided by the true false match probabilities. Although we looked at results for $\lambda = 0.1, 0.5$, and 0.8 and various interactions models, the results under conditional independence (**CI**) were the best with $\lambda = 0.2$ (Eq. (5.6)). We also looked at several different ways of constructing the "pseudo-truth" data. Additionally, we considered other pairs of files in which all of the error-rates estimates were better (closer to the 45° line) than those for the pair of files given in Figure 5.2.

Figures 5.2d–f are the corresponding curves using the methods of Belin and Rubin (1995). The curves are substantially farther from the 45° lines because they are only using the distributions of weights (natural logs of likelihood ratio (5.1)). Using the detailed breakout of the string comparator values, the fact that the three-class EM (Winkler 1993) provides much better estimates, and the breakout available from Eq. (5.6), we use more information that allows the estimated curves in Figure 5.2a–c to be closer to the truth than the corresponding curves in Figure 5.2d–f.

The final sets of curves (Figure 5.2g–i) are similar to the semi-supervised learning of Winkler (2002) that achieved results only very slightly worse than Larsen and Rubin (2001) but for which the EM computational speeds (10 minutes in each of ~500 regions) were at least 100 times as fast as the MCMC methods of Larsen and Rubin. It is difficult for the unsupervised methods to perform as well as the semi-supervised methods because the relatively small sample can be concentrated in the clerical review region between the lower cutoff T_λ and the upper cutoff T_μ. Because we had underlying truth data, we knew that in some regions only 1/40 of the "truth" sample was truly a match whereas in other regions 1/10 of the "truth" sample was truly a match. In the 1990 Decennial Census, the clerical region consisted almost entirely of individuals within the same household who were missing both first name and age (the only two fields for distinguishing within the household). Because we needed to match all 457 regions of the United States in three to six weeks to provide estimates required by law, we could not perform clerical review in each region or use approximations across certain regions because the optimal parameters vary significantly from region to region (Winkler 1989).

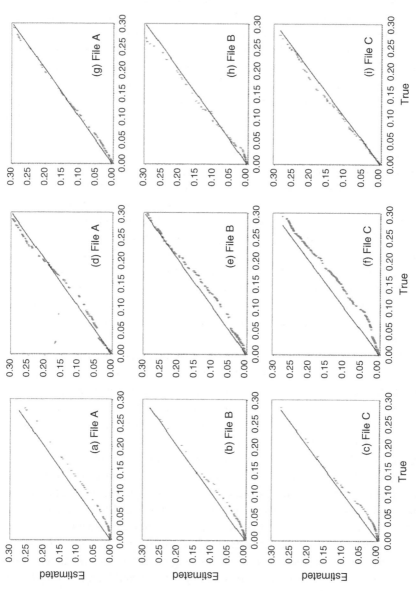

Figure 5.2 Comparison of error rate estimation procedures of Winkler (2006a, first column, unsupervised), Belin and Rubin (1995, second column, unsupervised), and Winkler (2002, third column, semi-supervised). Source: (a) Modified from Winkler (2006a); (b) Modified from Belin and Rubin (1995); (c) Modified from Winkler (2006a).

5.3.4 Achieving Extreme Computational Speed

Early record linkage (Newcombe et al. 1959; Newcombe and Kennedy 1962) had difficulty dealing with the computation associated with files having a few thousand records each. To deal with the computational burden individuals used *blocking*. With blocking, individuals sorted two files on a key such as a postal code and first character of surname. Only pairs agreeing on the sort (blocking) key are considered. To deal with situations where there are typographical error in either the postal code or first character of the surname. A second blocking pass may use part of the postal code and house number.

By the 1990s, NSIs might be working with millions of records to which they might apply 5–10 blocking passes. For the first blocking pass, individuals sorted the two files on the first blocking criterion and then matching performed. Prior to a second pass, individuals created residual files from the two passes after removing the records associated with pairs that were believed to be matches. Individuals sorted the two residual files according to the second blocking criterion and the matching repeated. With five blocking passes, individuals need to read and process each file nine times. Individuals needed to sort each of the two files five times. With a large administrative file with 0.5 billion records, ~3.5 times the size of an original file was needed for each sort. The sort might take 12+ hours. Processing of a pair of files of 0.5 billion records each with five blocking passes might take two to eight weeks. Often the matching could not be performed because too much disk space was needed.

To alleviate the situation and very significantly speed up processing, Yancey and Winkler (2004, 2009) created BigMatch technology. With BigMatch, there is only one pass against each file. We do not need to sort any of the files. We read one file (a smaller file then can be broken into small subparts) into memory with suitable indices created according to the blocking criteria. The suitable indices are in memory. The creation of 10 sets of indices with each corresponding to blocking criteria takes about the same amount of time as a single sort of the file. We read each record in the larger file once. We write all pairs corresponding to all the blocking passes into appropriate files. The U.S. Decennial Census is broken into ~500 contiguous regions. We control the overall matching strategy by Linux scripts. We match each region against ~500 regions. This natural, but crude, parallelization can easily extend to administrative lists.

Winkler (2004) provides some methodological insights into BigMatch and for putting bounds on the false nonmatch rates. Winkler, Yancey, and Porter (2010) provide details related to the 2010 Decennial Census matching system that processed 10^{17} pairs (300 million × 300 million) using four blocking criteria. The four blocking criteria were determined by three individuals who had developed better blocking criteria than Winkler (2004). With the four blocking criteria detailed processing was performed on only 10^{12} pairs. Auxiliary testing in 2005 and 2006 using 2000 Decennial data provided estimates of properly matching 97.5% of all duplicates.

The software ran in 30 hours using 40 CPUs of an SGI Linux machine using 2006 Itanium chips. Each CPU processed 400 000+ pairs per second. Most new High Performance Computer (HPC) machines would be four to eight times as fast due to more memory, more CPUs, and faster CPUs. Based on the number of pairs processed within a given time, BigMatch is 50 times as parallel PSwoosh software (Kawai et al. 2006) developed by computer scientists at Stanford. BigMatch is 10 times as fast as recent Dedoop software (Kolb and Rahm 2013) that uses sophisticated load balancing methods. The later software and two other sets of parallel software (Yan, Xue, and Malin 2013; Karapiperis and Verykios 2014) would not have been available for the 2010 Decennial Census. BigMatch is 500 times as fast as software used in some statistical agencies (Wright 2010).

5.4 Models for Adjusting Statistical Analyses for Linkage Error

5.4.1 Scheuren–Winkler

An early model for adjusting a regression analysis for linkage error is due to Scheuren and Winkler (1993). By making use of the Belin–Rubin predicted false-match rates, Scheuren and Winkler were able to give (somewhat crude) estimates of regressions that had been adjusted for linkage error to correspond more closely with underlying "true" regressions that did not need to account for matching error. Many papers subsequent to Scheuren and Winkler have assumed that accurate values of false match rates (equivalently true match probabilities) are available for all pairs in $\mathbf{A} \times \mathbf{B}$. The difficulty in moving the methods into practical applications is that nobody has developed suitably accurate methods for estimating all false match rates for all pairs in $\mathbf{A} \times \mathbf{B}$ when no training data is available.

Lahiri and Larsen (2005) later extended the model of Scheuren and Winkler with a complete theoretical development. In situations where the true (not estimates) matching probabilities were available for all pairs, the Lahiri–Larsen methods outperformed Scheuren–Winkler methods and were extended to more multivariate situations than the methods of Scheuren and Winkler (1993). Variants of the models for continuous data are due to Chambers (2009) and Kim and Chambers (2012a,b) using estimating equations. The estimating equation approach is highly dependent on the simplifications that Chambers et al. made for the matching process.

Chipperfield, Bishop, and Campbell (2011) provided methods of extending analyses on discrete data. The Chipperfield et al. methods are closely related to Winkler (2002) which contains a full likelihood development. Tancredi and Liseo (2011) applied Bayesian MCMC methods to discrete data. The Tancredi–Liseo methods are impressive because of the number of simultaneous restraints with which they can deal. The Tancredi–Liseo methods are extraordinarily compute intense (possibly requiring as much as one hour computation on each block (approximately 50–100 households). There are 8 million blocks in the United States.

Goldstein, Harron, and Wade (2012) provide MCMC methods for adjusting analyses based on very general methods and software that they developed originally for imputation (Goldstein et al. 2009). They provide methods of estimating the probabilities of pairs based on characteristics of pairs from files that have previously been matched. They are able to leverage relationships between vector $x \in \mathbf{A}$ to $y \in \mathbf{B}$ based on a subset of pairs on which matching error is exceptionally small and then extend the relationships/matching-adjustments to the entire set of pairs in $\mathbf{A} \times \mathbf{B}$. Although we have not encountered situations similar to Goldstein, Harron, and Wade (2012) where estimates of matching probabilities are very highly accurate and where we can obtain highly accurate estimates of relationships for (x, y) pairs on $\mathbf{A} \times \mathbf{B}$ (particularly from previous matching situations), the Goldstein et al. methods are highly promising, possibly in combination with another methods.

The methods for adjusting regression analyses for linkage error have been more successful than the methods for adjusting statistical analyses of discrete data because of various inherent simplifications due to the form of regression models.

5.4.2 Lahiri–Larsen

Unlike the much more mature methods in Sections 5.2 and 5.3, there are substantial research problems. Scheuren and Winkler (1993) extended methods of Neter, Maynes, and Ramanathan (1965) to more realistic record linkage situations in the simple analyses of a regression of the form $y = \beta x$ where y is taken from one file A and x is taken from another file B. Because the notation of Lahiri and Larsen (2005) is more useful in describing extensions and limitations, we use their notation.

Consider the regression model $\mathbf{y} = (y_1, \ldots, y_n)'$:

$$y_i = \mathbf{x}_i' \boldsymbol{\beta} + \varepsilon_i, \quad i = 1, \ldots, n \tag{5.7}$$

where $\mathbf{x}_i = (x_{i1}, \ldots, x_{ip})$ is a column vector of p known covariates $\boldsymbol{\beta} = (\beta_1, \ldots, \beta_p)'$, $E(\varepsilon_i) = 0$, $\text{var}(\varepsilon_i) = \sigma^2$, $\text{covariance}(\varepsilon_i, \varepsilon_j) = 0$ for $i \neq j$, $i, j = 1, \ldots, n$. Scheuren and Winkler (1993) considered the following model for $\mathbf{z} = (z_1, \ldots, z_n)$ given y:

$$z_i = \begin{cases} y_i \text{ with probability } q_{ii} \\ y_j \text{ with probability } q_{ij} \text{ for } i \neq j, i, j = 1, \ldots, n \end{cases} \tag{5.8}$$

where $\sum_{j=1}^{n} q_{ij} = 1$ for $i = 1, ..., n$. Define $\mathbf{q}_i = (q_{i1}, ..., q_{in})', j = 1, ..., n$, and $\mathbf{Q} = (\mathbf{q}_i, ..., \mathbf{q}_n)'$. The naïve least squares estimator of $\boldsymbol{\beta}$, which ignores mismatch errors, is given by

$$\hat{\boldsymbol{\beta}}_N \; (\mathbf{X'X})^{-1}\mathbf{X'z}$$

where $\mathbf{X} = (\mathbf{x}_1, ..., \mathbf{x}_n)'$ is an $n \times p$ matrix.

Under the model described by (5.7) and (5.8)

$$E(z_i) = \mathbf{w'\beta}$$

where $\mathbf{w}_i = \mathbf{q}_i' X = \sum_{j=1}^{n} q_{ij}\mathbf{x}_j', i = 1, ..., n$, is a $p \times 1$ column matrix. The bias of the naïve estimator $\hat{\boldsymbol{\beta}}_N$ is given by

$$\text{bias}(\hat{\boldsymbol{\beta}}_N) = E(\hat{\boldsymbol{\beta}}_N - \boldsymbol{\beta}) \left[(\mathbf{X'X})^{-1}\mathbf{X'W} - I\right]\boldsymbol{\beta} = \left[(\mathbf{X'X})^{-1}\mathbf{X'QX} - I\right]\boldsymbol{\beta}$$

$$(5.9)$$

If an estimator of $\mathbf{B}$ is available where $\mathbf{B} = (B_1, ..., B_n)'$ and $B_i = (q_{ii} - 1)y_i + \sum_{j \neq 1} q_{ij}y_j$. The Scheuren–Winkler estimator is given by

$$\hat{\boldsymbol{\beta}}_{SW} = \hat{\boldsymbol{\beta}}_N - (\mathbf{X'X})^{-1}\mathbf{X'}\hat{\mathbf{B}}$$

$$(5.10)$$

If q_{ij1} and q_{ij2} denote the first and second highest elements of the vector $\mathbf{q}_i$ and z_{j1} and z_{j2} denote the elements of the vector $\mathbf{z}$, then a truncated estimator of $\mathbf{B}$ is given by

$$B_i^{'\text{TR}} = (q_{ij1} - 1)z_{j1} + q_{ij2}z_{j2}$$

$$(5.11)$$

Scheuren and Winkler (1993) used estimates of q_{ij1} and q_{ij2} based on software/methods from Belin and Rubin (1995). Lahiri and Larsen (2000) improve the estimator (5.10) (sometimes significantly) by using the unbiased estimator

$$\hat{\boldsymbol{\beta}}_U = (\mathbf{W'W})^{-1}\mathbf{W'z}$$

$$(5.12)$$

The issues are whether it is possible to obtain reasonable estimates of $\mathbf{q}_i$ or whether the crude approximation given by (5.10) is suitable in a number of situations.

Under a significantly simplified record linkage model where each q_{ij} for $i \neq j$, 1, ..., n, Chambers (2009) (also Kim and Chambers 2012a,b) provides an estimator approximately of the following form

$$\hat{\boldsymbol{\beta}}_U = (\mathbf{W'}\,\mathbf{Cov}_z^{-1}\mathbf{W})^{-1}\mathbf{W'}\,\mathbf{Cov}_z^{-1}\mathbf{z}$$

$$(5.13)$$

that has lower bias than the estimator of Lahiri and Larsen. The matrix $\mathbf{Cov}_z$ is the variance–covariance matrix associated with $\mathbf{z}$. The estimator in (5.12) is the best linear unbiased estimator using standard methods that improve over the unbiased estimator (5.11). Chambers further provides an iterative method for obtaining an empirical best linear unbiased estimator (BLUE) using the observed data.

The issue with the Chambers' estimator is whether the drastically simplified record linkage model is a suitable approximation of the realistic model used by Lahiri and Larsen. The issue with both the models of Chambers (2009) and Lahiri and Larsen (2005) is that they need both a method of estimating q_{ij} for all i, j with all pairs of records and a method of designating which of the q_{ij} is associated with the true match. Scheuren and Winkler (1993) provided a much more ad hoc adjustment with the somewhat crude estimates of the q_{ij} obtained from the model of Belin and Rubin (1995). Lahiri and Larsen demonstrated that the Scheuren–Winkler procedure was inferior for adjustment purposes when the true q_{ij} were known. Winkler and Scheuren (1991), however, were able to determine that their adjustment worked well in a large number of empirical scenarios (approximately one hundred) because the bias of the q_{ij}-estimation procedure and adjustment partially compensated for the increased bias of the Scheuren–Winkler procedure. Further, Winkler (2006a) provided a "generalization" of the Belin–Rubin estimation procedure that provides somewhat more accurate estimates of the q_{ij} and holds in a moderately larger number of situations.

5.4.3 Chambers and Kim

To simplify methods of adjusting Chambers and Kim (Chambers 2009; Kim and Chambers 2012a) provided a model for adjusting statistical analyses under an extreme simplifying assumption for the linkage process originally used by Neter, Maynes, and Ramanathan (1965). Rather than having q_{ij} that differ substantially as they have in most (all?) record linkage applications, they assume that one q_{ij} for the single true match has value λ and the remaining $N - 1$ values of q_{ij} have the same value of $(1 - \lambda)/(N - 1)$. This simplifying assumption makes it much easier to apply an estimating equation approach to get an estimate of $\beta^\wedge$ in the simple regression approach given by Eq. (5.10) and to extend the methods to additional multivariate situations and to situations with logistic regression. They make further assumptions that pairs can be grouped into a moderate number of subsets indexed by G where λ_g, $g = 1, ..., G$, are the appropriate true match probabilities in each group and that each record has a match in their appropriate group. Others have made these secondary assumptions about the grouping in slightly different manners by others.

A key issue is whether the simplifying of the matching process is realistic. Kim and Chambers (2012a) further show in a simulation where the data are generated according to the simplifying-record-linkage model that their method's performance is similar to that of the method of Lahiri and Larsen (2005) when the probabilities q_{ij} are true and satisfy the uniform-λ assumption. They further show that they can compute variances under their model that yields a Best Linear Unbiased Estimate with less variance than the estimator of Lahiri and Larsen.

My concerns is whether the uniform-λ assumption is valid approximation of record linkage processes that have been typically done. I only make a comparison to the Lahiri and Larsen (2005) methods that agree with Chambers (2009) and Kim and Chambers (2012a) under the uniform-λ assumption. The following quantities (weights)

$$\mathbf{w}_i = \mathbf{q}_i' X = \sum_{j=1}^{n} q_{ij} \mathbf{x}_j', \quad i = 1, \ldots, n, \text{ is a } p \times 1 \text{ column matrix}$$

are crucial to the Lahiri–Larsen development. If I replace one q_{ij} by λ and the other q_{ij} with the appropriate quantities under the uniform-λ assumption, then all the estimates in the Lahiri–Larsen development would vary significantly and be wrong. I do not understand how this type of discrepancy can be resolved unless someone is able to come up with situations where the uniform-λ assumption is appropriate and, possibly, the estimating equations need some adjustments.

5.4.4 Chipperfield, Bishop, and Campbell

The natural way of analyzing discrete data is with loglinear models on the pairs of records in $\mathbf{A} \times \mathbf{B}$. Performing analyses for discrete data are more difficult than for situations with certain regression models where the form of the model gives us considerable simplifying information. We like the Chipperfield et al. paper because it gives certain insights that are not available in many of the other papers. For consistency with Chipperfield, Bishop, and Campbell (2011) we follow their notation as consistently as possible. Rather than break out $\mathbf{A}$ and $\mathbf{B}$ as $(a_1, \ldots, a_n)$ and $(b_1, \ldots, b_m)$, we merely enumerate $\mathbf{A}$ with x in $\mathbf{A}$ and $\mathbf{B}$ with y in $\mathbf{B}$. All observed pairs have probabilities p_{xy} where p_{xy} represents all pairs of records in $\mathbf{A} \times \mathbf{B}$ with x in $\mathbf{A}$ and y in $\mathbf{B}$. If we knew the truth, we would know all the p_{xy}. We wish to estimate the p_{xy} in a semi-supervised fashion as with the likelihood equation given in (5.6). Chipperfield, Bishop, and Campbell (2011) take a sample of pairs s_c for which they can determine p_{xy} exactly (no estimation error) for all pairs x and y associated with s_c.

As preliminary notation, we describe contingency tables without the missing data. We assume that x takes G values and y takes C values. Then, the joint distribution of x and y is

$$p(x, y) = p_1(y \mid x, P)\, p_2(x)$$

where $\Pi = (\pi'_1, \ldots, \pi'\_)'$, $(\pi'_g = ((\pi'_1|g, \ldots, \pi'_c|g)'$, $\pi'_c|g$ is the probability the given $x = g$ that $y = c$. The total number of probabilities $p(x, y)$ is equal to C times G. Each $p(x, y)$ is obtained by summing over all pairs $p_i(x, y)$ where x in $\mathbf{A}$ (first component) and y in $\mathbf{B}$ (second component). The standard estimate of $\tilde{n}_{c|x} = n_{c|x}/n_x$), where $n_x = \Sigma_c \Sigma_i w_{ic|x}$, where $w_{ic|x} = 1$ if $y_i = c$ and $x_i = x$ and $w_{ic|x} = 0$ otherwise.

When there is linkage error, we are concerned with methods that adjust for linkage error. If we can observe true matching status, then the underlying truth representation is.

$$w*_{ic|x} = 1 \quad \text{if } y*_i = c, \text{and } x_i = x; \quad \text{else } w*_{ic|x} = 0$$

Ordinarily, we may need to take a (possibly very large) sample to get at the truth and use the following semi-supervised learning procedure.

Take a (likely very large) sample s_c to get (possibly only somewhat) good estimates of $w*_{ic|x}$. Use EM model to get estimates for all p_{xy}. The sample s_c gives

$$\hat{p}_{xy*} = \left(\Sigma_{sc} w*_{ic|x}\, \delta_i \right) / \left(\Sigma_{sc} w*_{ic|x} \right) \tag{5.14}$$

$$\tilde{n}_{c|x} = \tilde{n}_{c|x} / \left(\Sigma_{cc|x} \right)^{-1} \tag{5.15}$$

where

$$\tilde{n}_{c|x} = \Sigma_i \tilde{w}_{ic|x} \tag{5.16}$$

$$\tilde{w}_{ic|x} = w*_{ic|x}\hat{p}_{xy*} + \left(1 - \hat{p}_{xy*} \right) \tilde{n}_{c|x} \quad \text{if } i \in s_c$$

$$= w*_{ic|x} \quad \text{if } i \in s_c$$

$$= \tilde{n}_{c|x} \quad \text{if } i \in s_c \text{ and } \delta_i = 0 \;\; (\delta_i \text{ is indicator that true match}) \tag{5.17}$$

The (semi-supervised) EM procedure is

1. Calculate $\hat{p}_{xy*}$ from (5.7),
2. Initialize $\tilde{\pi}_{c|x}^{(0)}$ and then calculate $\tilde{w}_{c|x}^{(0)}$ from (5.10) and then $\tilde{n}_{c|x}^{(0)}$ from (5.9),
3. Calculate $\tilde{\pi}_{c|x}^{(t)}$ from (5.8) using $\tilde{n}_{c|x}^{(t-1)}$,
4. Calculate $\tilde{w}_{c|x}^{(t)}$ from (5.10) using $\tilde{\pi}_{c|x}^{(t)}$ and then calculate $\tilde{n}_{c|x}^{(t)}$ from (5.9) using $\tilde{w}_{c|x}^{(t)}$,
5. Iterate between 3 and 4 until convergence.

A similar semi-supervised procedure (with full likelihood development) was used in Winkler (2002) and extended to an unsupervised procedure (Winkler 2006a) with a substantial decrease in accuracy for estimating false match rates. With very slight notational changes, the procedure given in steps 1–5 above is the same as the approach using the likelihood given by Eq. (5.6).

The procedure of Chipperfield, Bishop, and Campbell (2011) appears to work well in their simple empirical examples that have substantial similarity to Winkler (2002) but the methods of Chipperfield et al. are more directly generalizable.

Empirical example (Chipperfield, Bishop, and Campbell 2011):

Three values (employed, unemployed, and not in labor force) are compared against same values in another file for a later time period. The total sample size is 1000 which represents ~100 for each combination of cells across time periods. With Chipperfield et al., there is very little variation between the three labor-force values in one time period to another. There are only 3×3 possible patterns. With more realistic data, we might have thousands or millions of patterns. Each false match (x, y) might associate a completely unrelated $y \in \mathbf{B}$ that is chosen approximately randomly from thousands of $\mathbf{B}$ records.

5.4.4.1 Empirical Data

The empirical data consists of 55 926 records from one U.S. State (1% sample) from a public-use file. In the following diagram, we have collapsed a number of the value-states of fields into a smaller number of value-states to make the analysis easier. There are approximately 3.0 million possible data patterns. Even if this relatively straightforward situation, it will be apparent that it is very difficult to extend the existing statistical-adjustment procedures to achieve high or moderate accuracy with complicated real data (Table 5.3).

Matching error (Table 5.4) was induced at the following rate in parts of the file at rates that might correspond to a "good" matching situation with certain types of real data. We only consider the simplest situation where each $x \in \mathbf{A}$ will either

Table 5.3 Data (2000 PUMS data for one state) (number of values for each field).

A data					B data		
Sex	Age	Race	Marital status	Education	Occupation	House	Income
2	16	2	5	16	3	5	40
5120 data patterns					600 data patterns		

Table 5.4 Sampling rates by strata.

	Split records – induce matching error (overall matching error 8–10%)		
1.	8000	0.01	Error
2.	8000	0.02	Error
3.	8000	0.05	Error
4	8000	0.08	Error
5.	8000	0.12	Error
6.	8000	0.15	Error
7.	7926	0.20	Error

be matched with the correct $y \in \mathbf{B}$ or not. With this simplification, each matching error represents a type of permutation of the records with $y \in \mathbf{B}$. Different authors (Scheuren and Winkler 1993; Lahiri and Larsen 2005) have suggested methods for extending methods to the situations where some $x \in \mathbf{A}$ do not have a corresponding $y \in \mathbf{B}$ and vice versa. Chambers (2009) and Kim and Chambers (2012a,b) have given specific extensions along with empirical simulations with continuous data but we will not consider any extensions in this subsection.

The last column in Table 5.5 represents the counts after distortion due to matching error. The numbers in next-to-last column are the counts prior to matching error (i.e. truth). The first eight columns are associated with the values (0, ..., $n_j - 1$) associated with the n_j values associated with the jth field. Higher truth counts are usually reduced in the observed data due to matching error. When an initial value (9th column is blank) followed by 1 it is because a new matching

Table 5.5 Sample data records (counts for true patterns followed by observed patterns).

								True	Observed
1	020	1	5	07	00	00	000	53	50
1	020	1	5	07	00	00	001	3	5
1	020	1	5	07	00	00	004		1
1	020	1	5	07	00	00	006		1
1	020	1	5	07	00	01	001	1	
1	020	1	5	07	00	02	000	1	1
1	020	1	5	07	00	02	001	2	2
1	020	1	5	07	00	04	001		1
1	020	1	5	07	00	04	004		1
1	020	1	5	07	00	04	005	1	1
1	020	1	5	07	00	05	001	1	1
1	020	1	5	07	00	05	005		1
1	020	1	5	07	00	05	006		1
1	020	1	5	07	00	05	007		1
1	020	1	5	07	05	00	000	16	12
1	020	1	5	07	05	00	001	1	1
1	020	1	5	07	05	00	004		1
1	020	1	5	07	05	01	000	18	17
1	020	1	5	07	05	01	001	3	3
1	020	1	5	07	05	02	000	32	29

pattern is created as a result of matching error. Approximately 16 000 (20% of the counts) have 1 in the 10th column of which 1/7 are false matches. There is no way to distinguish these false matches (presently) except via follow-up of a sample. The counts in the 10th column are such that the loglinear models associated with the initial (true) counts are quite different than the models associated with the final counts (that have no correction for matching error).

The Chipperfield, Bishop, and Campbell (2011) approach gives nice insight into some of the difficulties of the general problem. Here are some observations.

1. The empirical counts from the observed data *Obs* with the specified distortions vary significantly from the original data *Orig*. The loglinear models on *Obs* and *Orig* are very different.

2. The empirical example might have 5120 data patterns in one file and 600 data patterns in another file. This would correspond to a relatively small administrative-list example. If the sample size is 0.01 of $\mathbf{A} \times \mathbf{B}$, then most small cells with counts 3 or less will be given $\hat{p}_{xy*} = 0$. It seems unlikely that this will yield suitable estimates of cell counts to improve loglinear modeling. Without correction for matching error, this data does not yield loglinear models that correspond to the loglinear models from the original "truth" data. This means that, due to matching error (8–10%), we cannot reproduce analyses on the original data (even approximately) on the observed data.

3. If the sample size is 0.25–0.50 of the total number of pairs, then too many pairs will need to be reviewed for this procedure to work in practice. Even with a sample size with a proportion on the order of 0.25 pairs, it is unlikely that there will be sufficient information to move the estimates of $\tilde{\pi}_{c|x}^{(t)}$ effectively away from the initial default values of 1/600. If there are informative priors for $\tilde{\pi}_{c|x}^{(0)}$, it is unlikely there is sufficient information to move away from the starting informative priors.

5.4.5 Goldstein, Harron, and Wade

Goldstein et al. (2009) showed how to apply multi-level modeling methods in a number of quite different applications. The freely available software is available as compiled Matlab programs, together with training materials, from http://www.cmm.bristol.ac.uk. In the paper, they demonstrate a large number of ways that the software can be applied. They indicate that their methods can be applied to the problem of adjusting a statistical analysis for record linkage error (in a procedure that they call *Prior Informed Imputation* – PII) that they apply to several other types of problems. Their computational procedures are general MCMC procedures that they indicate are superior to many EM-based procedures as the number of parameters in the models increase significantly.

Goldstein, Harron, and Wade (2012) show more specifics and provide an example of how the multilevel software can be applied to the problem of adjusting a statistical analysis (in their situation to a straightforward regression problem similar to that examined by other previously). At the first level is the observed data; at the second level are data pairs with the matching weights from the record linkage. The record linkage weights inform the imputation by causing the pairs with higher matching weights to be given more emphasis. The results seem quite good (like the other methods individuals have suggested). The advantage of their methods is the seeming great flexibility that the general software provides in working with a number of problems.

5.4.6 Hof and Zwinderman

Hof and Zwinderman (2012, 2015) provide EM-based methods for adjusting a statistical analysis for matching error. Like the other methods, their procedure successively and iteratively make adjustment for linkage error and errors in the statistical models. They provide a considerable number of details related to the likelihoods and how the various terms are approximated. For individuals familiar with EM-based methods, their procedures appear fairly straightforward to apply.

5.4.7 Tancredi and Liseo

Tancredi and Liseo (2011, 2015) provide a full MCMC development for the methods of adjusting a statistical analysis for linkage error. Tancredi and Liseo (2015) give specifics about how the latter methods are a simplification of Tancredi and Liseo (2011). The latter methods should be much faster to compute. They provide considerable additional insight into how the slow improvement in the parameters of the statistical analysis improve the matching parameters and vice-versa.

The methods have some similarity to statistical matching methods of Gutman, Afendulis, and Zaslavsky (2013). Gutman et al. give many additional details of how they approximate the likelihoods in their MCMC development. Although their methods are a special case of Tancredi and Liseo (2015), they provide many details that should be very useful to individuals who wish to do their own development.

5.5 Concluding Remarks

The chapter provides background and an overview of methods of edit/imputation and record linkage that have been developed and used in the NSIs. The generalized systems apply the theory of Fellegi and Holt (1976) and Fellegi and Sunter (1969), respectively. While improving quality, the generalized systems also yield significant cost- and time-savings. The methods of adjusting statistical analyses for linkage error are a large and difficult area of current research.

5.6 Issues and Some Related Questions

a. Record linkage has been subject to considerable error primarily due to idiosyncrasies from different representations of the quasi-identifying fields, missing data, and the ability to perform exceedingly large amounts of computation within a certain amount of time. If there is suitable data preprocessing and matching software, then it may take several individuals two or more years to develop sufficient skill to minimize record linkage error. Are there any training methods that might reduce the training/skill-development times substantially?
b. How do NSIs develop the teams to assure that record linkage error is minimized? Winkler and Hidiroglou (1998) provide information how successful teams that created generalized systems were created and monitored.
c. If an NSI does not have the skill and/or resources to develop their own systems, then how does the NSI create suitable test decks to evaluate the quality of record linkage and edit/imputation software developed by vendors or in shareware?

References

Andridge, R.A. and Little, R.J.A. (2010). A review of hot deck imputation for survey nonresponse. *International Statistical Review* 78 (1): 40–64.

Barcaroli, G. and Venturi, M. (1997). DAISY (design, analysis and imputation system): structure, methodology, and first applications. In: *Statistical Data Editing*, vol. II (eds. J. Kovar and L. Granquist), 40–51. U.N. Economic Commission for Europe.

Belin, T.R. and Rubin, D.B. (1995). A method for calibrating false-match rates in record linkage. *Journal of the American Statistical Association* 90: 694–707.

Chambers, R. (2009), Regression analysis of probability-linked data, Official Statistics Research Series, 4. Available from .
http://www.statisphere.govt.nz/official-statistics-research/series/default.htm.

Chipperfield, J.O., Bishop, G.R., and Campbell, P. (2011). Maximum likelihood estimation for contingency tables and logistic regression with incorrectly linked data. *Survey Methodology* 37 (1): 13–24.

Della Pietra, S., Della Pietra, V., and Lafferty, J. (1997). Inducing features of random fields. *IEEE Transactions on Pattern Analysis and Machine Intelligence* 19: 380–393.

D'Orazio, M., Di Zio, M., and Scanu, M. (2006). Statistical matching for categorical data: displaying uncertainty and using logical constraints. *Journal of Official Statistics* 22 (1): 137–157.

Fellegi, I.P. and Holt, D. (1976). A systematic approach to automatic edit and imputation. *Journal of the American Statistical Association* 71: 17–35.

Fellegi, I.P. and Sunter, A.B. (1969). A theory for record linkage. *Journal of the American Statistical Association* 64: 1183–1210.

Gill, L. (2001). *Methods of Automatic Record Matching and Linking and Their Use in National Statistics*, National Statistics Methodological Series, No. 25. Office for National Statistics. ISBN: ISBN 1 85774 420 9.

Goldstein, H., Carpenter, J., Kenward, M.G., and Levin, K.A. (2009). Multilevel models with multivariate mixed response types. *Statistical Modelling* 9 (3): 173–197.

Goldstein, H., Harron, K., and Wade, A. (2012). The analysis of record-linked data using multiple imputation with data prior values. *Statistics in Medicine* https://doi.org/10.1002/sim.5508.

Gutman, R., Afendulis, C.C., and Zaslavsky, A.M. (2013). A Bayesian procedure for file linking to analyze end-of-life medical costs. *Journal of the American Statistical Association* 108 (501): 34–47.

Herzog, T.N., Scheuren, F., and Winkler, W.E. (2010). Record linkage. In: *Wiley Interdisciplinary Reviews: Computational Statistics* (eds. D.W. Scott, Y. Said and E. Wegman), 535–543. New York, NY: Wiley.

Hof, M.H.P. and Zwinderman, A.H. (2012). Methods for analyzing data from probabilistic linkage strategies based on partially identifying variables. *Statistics in Medicine* 31: 4231–4232.

Hof, M.H.P. and Zwinderman, A.H. (2015). A mixture model for the analysis of data derived from record linkage. *Statistics in Medicine* 34: 74–92.

Hogan, H.H. and Wolter, K. (1984). Research plan on adjustment. https://www.census.gov/srd/papers/pdf/rr84-12.pdf (accessed 27 July 2020).

Karapiperis, D. and Verykios, V. (2014). Load-balancing the distance calculations in record linkage. *SIGKDD Explorations* 17 (1): 1–7.

Kawai, H., Garcia-Molina, H., Benjelloun, O., et al. (2006). P-Swoosh: Parallel Algorithm for Generic Entity Resolution. Stanford University CS technical report.

Kim, G. and Chambers, R. (2012a). Regression analysis under incomplete linkage. *Computational Statistics and Data Analysis* 56: 2756–2770.

Kim, G. and Chambers, R. (2012b). Regression analysis under probabilistic multi-linkage. *Statistica Neerlandica* 66 (1): 64–79.

Kolb, L. and Rahm, E. (2013). Parallel entity resolution with Dedoop. *Datenbank-Spektrum* 13 (1): 23–32. http://dbs.uni-leipzig.de/file/parallel_er_with_dedoop.pdf.

Kovar, J.G. and Winkler, W.E. (1996). Editing economic data. In: *JSM Proceedings, Survey Research Methods Section*, 81–87. Alexandria, VA: American Statistical Association, American Statistical Association (also http://www.census.gov/srd/papers/pdf/rr2000-04.pdf).

Lahiri, P. and Larsen, M.D. (2000). Model-based analysis of records linked using mixture models. In: *JSM Proceedings, Survey Research Methods Section*, 11–19. American Statistical Association: Alexandria, VA.

Lahiri, P.A. and Larsen, M.D. (2005). Regression analysis with linked data. *Journal of the American Statistical Association* 100: 222–230.

Larsen, K. (2005). Generalized Naïve Bayes classifiers. *SIGKDD Explorations* 7 (1): 76–81. https://doi.org/10.1145/1089815.1089826.

Larsen, M.D. and Rubin, D.B. (2001). Iterative automated record linkage using mixture models. *Journal of the American Statistical Association* 79: 32–41.

Little, R.A. and Rubin, D.B. (2002). *Statistical Analysis with Missing Data*, 2e. New York, NY: Wiley.

Meng, X. and Rubin, D.B. (1993). Maximum likelihood via the ECM algorithm: a general framework. *Biometrika* 80: 267–278.

Neter, J., Maynes, E.S., and Ramanathan, R. (1965). The effect of mismatching on the measurement of response errors. *Journal of the American Statistical Association* 60: 1005–1027.

Newcombe, H.B. and Kennedy, J.M. (1962). Record linkage: making maximum use of the discriminating power of identifying information. *Communications of the ACM* 5: 563–567.

Newcombe, H.B. and Smith, M.E. (1975). Methods for computer linkage of hospital admission – separation records into cumulative health histories. *Methods of Information in Medicine* 14 (3): 118–125.

Newcombe, H.B., Kennedy, J.M., Axford, S.J., and James, A.P. (1959). Automatic linkage of vital records. *Science* 130: 954–959.

Nigam, K., McCallum, A.K., Thrun, S., and Mitchell, T. (2000). Text classification from labeled and unlabeled documents using EM. *Machine Learning* 39: 103–134.

Rao, J.N.K. (1997). Developments in sample survey theory: an appraisal. *The Canadian Journal of Statistics/La Revue Canadienne de Statistique* 25 (1): 1–21.

Scheuren, F. and Winkler, W.E. (1993). Regression analysis of data files that are computer matched. *Survey Methodology* 19: 39–58.

Tancredi, A. and Liseo, B. (2011). A hierarchical Bayesian approach to matching and size population problems. *The Annals of Applied Statistics* 5 (2B): 1553–1585.

Tancredi, A. and Liseo, B. (2015). Regression analysis with linked data, problems and possible solutions. *Statistica* 75 (1): 19–35. https://doi.org/10.6092/issn.1973-2201/5821. https://rivista-statistica.unibo.it/article/view/5821/0.

Winkler, W.E. (1988). Using the EM algorithm for weight computation in the Fellegi–Sunter model of record linkage. In: *JSM Proceedings of the Section on Survey Research Methods*, 667–671. Alexandria, VA: American Statistical Association, also at http://www.census.gov/srd/papers/pdf/rr2000-05.pdf.

Winkler, W.E. (1989). Near automatic weight computation in the Fellegi–Sunter model of record linkage. In: *Proceedings of the Fifth Census Bureau Annual Research Conference*, 145–155. Washington, D.C.: U.S. Census Bureau.

Winkler, W.E. (1990a). String comparator metrics and enhanced decision rules in the Fellegi–Sunter model of record linkage. In: *JSM Proceedings of the Section on Survey Research Methods*, 354–359. American Statistical Association. Available at: www .amstat.org/sections/srms/Proceedings/papers/1990_056.pdf.

Winkler, W.E. (1990b). On Dykstra's iterative fitting procedure. *Annals of Probability* 18: 1410–1415.

Winkler, W.E. (1993). Improved decision rules in the Fellegi–Sunter model of record linkage. In: *JSM Proceedings of the Section on Survey Research Methods*, 274–279. American Statistical Association, also http://www.census.gov/srd/papers/pdf/ rr93-12.pdf.

Winkler, W.E. (1995a). Matching and record linkage. In: *Business Survey Methods* (eds. B.G. Cox, D.A. Binder, B.N. Chinnappa, et al.), 355–384. New York: Wiley.

Winkler, W.E. (1995b). SPEER Economic Editing Software. Unpublished technical report.

Winkler, W.E. (2000). Machine learning, information retrieval, and record linkage. In: *JSM Proceedings of the Section on Survey Research Methods*, 20–29. American Statistical Association. http://www.amstat.org/sections/SRMS/Proceedings/ papers/2000_003.pdf, also available at http://www.niss.org/sites/default/files/ winkler.pdf.

Winkler, W.E. (2002). Record linkage and Bayesian networks. In: *Proceedings of the Section on Survey Research Methods*. American Statistical Association.

Winkler, W.E. (2003). A contingency table model for imputing data satisfying analytic constraints. In: *Proceedings of the Section on Survey Research Methods*. American Statistical Association. CD-ROM, also http://www.census.gov/srd/papers/pdf/ rrs2003-07.pdf.

Winkler, W.E. (2004). Approximate string comparator search strategies for very large administrative lists. In: *Proceedings of the Section on Survey Research Methods*. American Statistical Association.

Winkler, W.E. (2006a). Automatic estimation of record linkage false match rates. In: *Proceedings of the Section on Survey Research Methods*. American Statistical Association. CD-ROM, also at http://www.census.gov/srd/papers/pdf/rrs2007-05 .pdf.

Winkler, W.E. (2006b). Statistical matching software for discrete data. Computer software and documentation.

Winkler, W.E. (2008). General methods and algorithms for imputing discrete data under a variety of constraints. http://www.census.gov/srd/papers/pdf/rrs2008-08 .pdf (accessed 30 July 2020).

Winkler, W.E. (2010). General discrete-data modeling methods for creating synthetic data with reduced re-identification risk that preserve analytic properties. http:// www.census.gov/srd/papers/pdf/rrs2010-02.pdf (accessed 30 July 2020).

Winkler, W.E. (2013). Cleanup and analysis of sets of national files. In: *Proceedings of the Bi-Annual Research Conference*. Federal Committee on Statistical Methodology.

Winkler, W.E. (2014). Matching and record linkage. In: *Wiley Interdisciplinary Reviews: Computational Statistics*. https://doi.org/10.1002/wics.1317; https://wires .wiley.com/WileyCDA/WiresArticle/wisId-WICS1317.html.

Winkler, W.E. and Hidiroglou, M. (1998). Developing analytic programming ability to empower the survey organization. http://www.census.gov/srd/papers/pdf/rr9804 .pdf (accessed 30 July 2020).

Winkler, W.E. and Scheuren, F. (1991). How Computer Matching Error Effects Regression Analysis: Exploratory and Confirmatory Analysis. U.S. Bureau of the Census, Statistical Research Division technical report.

Winkler, W.E. and Thibaudeau, Y. (1991). An Application of the Fellegi–Sunter Model of Record Linkage to the 1990 U.S. Census. U.S. Bureau of the Census, Statistical Research Division technical report 91-9. http://www.census.gov/srd/ papers/pdf/rr91-9.pdf.

Winkler, W.E., Yancey, W.E., and Porter, E.H. (2010). Fast record linkage of very large files in support of decennial and administrative records projects. In: *Proceedings of the Section on Survey Research Methods*. American Statistical Association, CD-ROM.

Wright, J. (2010). Linking Census Records to Death Registrations. Australia Bureau of Statistics report 131.0.55.030.

Yan, W., Xue, Y, and Malin, B. (2013). Scalable load balancing for map-reduced based record linkage. *2013 IEEE 32nd IEEE International Performance Computing and Communications Conference (IPCCC)*, San Diego, CA (6–8 December 2013). https://doi.org/10.1109/PCCC.2013.6742785.

Yancey, W.E. and Winkler, W.E. (2004). BigMatch Record Linkage Software. Statistical Research Division research report.

Yancey, W.E. and Winkler, W.E. (2009). BigMatch Record Linkage Software. Statistical Research Division research report.

6

Assessing Uncertainty When Using Linked Administrative Records

Jerome P. Reiter

Department of Statistical Science, Duke University, Durham, NC 27708-0251, USA

6.1   Introduction

Linking individual records in one database to those in another is a cornerstone of policy-making and research, particularly in public health and the social sciences. For example, linking subjects in a planned medical study to records in administrative databases, such as electronic health records (EHRs) and Medicare claims data, can enable researchers to evaluate long-term outcomes, as well as outcomes not measured in the planned study, without expensive de novo primary data collections. It also can allow researchers to incorporate important covariates not collected in the planned study, thereby reducing effects of unmeasured confounding and facilitating more nuanced estimation of treatment effects. Additionally, many government agencies use record linkage extensively, both for official purposes and in response to secondary researchers' requests to link government data to other databases.

Often researchers link records by matching on combinations of variables present in both databases, such as birth dates, names, addresses, diagnosis codes, or demographics, because unique identifiers like social security numbers or Medicare IDs are unavailable in at least one of the datasets, e.g. because of privacy restrictions. There is an extensive literature on record linkage techniques (e.g. Belin and Rubin 1995; Fellegi and Sunter 1969; Herzog, Scheuren, and Winkler 2007; Jaro 1989; Larsen and Rubin 2001), but linkage processes are inevitably imperfect, creating a source of error that introduces additional uncertainties into statistical analyses. Linkage error is rarely, if ever, taken into account in inference or data dissemination.

In this chapter, we review methods that attempt to account for uncertainty arising from imperfect record linkage. We begin in Section 6.2 with a high-level

Administrative Records for Survey Methodology, First Edition.
Edited by Asaph Young Chun, Michael D. Larsen, Gabriele Durrant, and Jerome P. Reiter.

description of key sources of uncertainty. In Section 6.3, we discuss approaches to handle uncertainty via statistical modeling. Here, we focus on problems involving record linkage of two files, although related ideas arise with more than two files and with de-duplication in one file. Finally, we conclude in Section 6.4 with a description of some open problems and practical implications.

6.2 General Sources of Uncertainty

The additional uncertainty from imperfect record linkage arises from two primary sources. First, the linkage process may result in incorrect matches, e.g. two records are claimed to belong to the same individual but actually belong to distinct individuals. Second, the linkage process may result in incomplete matching, e.g. the matched dataset under-represents some subgroups of individuals. We now describe these types of errors.

6.2.1 Imperfect Matching

Many record linkage techniques result in a single dataset with a set of best matches according to some criterion. When perfect unique identifiers are unavailable, this set can include false matches. For example, suppose we want to link a survey to an administrative database. The survey includes an individual named Jerome Reiter who is 48 years old. The administrative database contains two records, one with name Jerry Reiter who is 48 years old (the right match) and another with name Jerome Reiter who is 46 years old (the wrong match). Assuming for simplicity that the files contain no duplicates, the linking algorithm is likely to choose one of these two as a match. It easily could choose the wrong match, e.g. by giving high weight to exactly matching the name. A related issue arises when individuals are in one file but not in the other. For example, Jerome Reiter aged 48 may not be in the administrative database, so that both potential links are incorrect.

In some contexts, multiple records from one file could be equally plausible matches for a record in the other file. For example, if the only available linking variables include age, sex, and some coarse medical diagnosis code – so that there is no string variable like name available for matching – dozens of records could have identical values on the linking variables. In such cases, selecting a single best match at random creates opportunities for matching errors (in Section 6.3.2 we discuss an alternative that uses the non-linking variables to inform the selection of matches in this case). It is also possible that some of these categorical variables are measured with error, in which case linking errors can arise even for two records that appear to match perfectly on all linking fields.

Measurement errors could affect the blocking variables as well. Blocking in record linkage refers to subsetting units into groups, or blocks, according to a set of key variables, such as state of residence. Linking variables are compared only for pairs of records in the same block. Measurement errors on blocking variables could make matching errors more likely. For example, if the record that belongs to an individual in the first file is in a different block than the record for that individual in the second file, the blocking makes it impossible to get the correct match for that individual.

False matches can cause problems for inferences based on the linked data (Neter, Maynes, and Ramanathan 1965; Scheuren and Winkler 1993). Suppose the goal of linkage is to connect individuals in a completed randomized experiment involving a medical treatment with long term health outcome data from EHRs, for example, incidence of at least one cardiac arrest in the 10 years after the end of the experimental study period. Suppose that Jerome Reiter age 48 received the treatment. The EHR indicates that Jerry Reiter age 48 (the true match) did not have any incidences, whereas Jerome Reiter aged 46 (the wrong match) had at least one incidence. If the record linkage algorithm selects the wrong match, the treatment will appear, inaccurately, to have been ineffective for this individual. Aggregated over many individuals, such errors could completely distort the estimate of the treatment effect, as well as its standard error. Similar issues can plague any analysis, including regression modeling.

On the other hand, false matches need not always cause problems for inferences. For example, if instead both potential matches had no incidences of cardiac arrest, we get the same point estimate of the average treatment effect regardless of which match is selected. As this example suggests, the impact of matching errors is specific to the analyst's inference.

6.2.2 Incomplete Matching

When matching a planned study to administrative data, often it is sensible to assume that all, or nearly all, individuals in the planned study are part of the administrative database. However, in some contexts this is not the case. For example, suppose one has a data file comprising public school students' scores on standardized tests (and other demographics), and a second data file comprising information on all babies born in a particular state, such as the babies' birth weight and gestational age, characteristics about the babies' mother and father, etc. Suppose one seeks to link the two data files to investigate whether test scores vary with birth weight. Even with perfect unique identifiers, there will be individuals in both files who cannot be matched. They may have been born in another state and moved into the school system, and hence are not on the birth record file. They may have been born in the state but moved out of the state, or have attended private schools in the state, and hence are not in the educational records.

Another type of incomplete matching arises when some individuals lack accurate linking variables. For example, some hospitals may allow personally identifiable information to be available to researchers for matching, but others may choose not to do so for privacy reasons or because patients have not authorized that use of their data; or, variation in administrative practices across states may result in individuals' names in some demographic groups, say non-native English speakers, to be collected with high measurement errors. In both these cases, the linkage algorithm may not have enough information for the linkage to go forward for individuals in these groups, resulting in incomplete matching.

Incomplete matching can act as a selection mechanism resulting in biased inferences. For example, suppose for some reason the low birth weight babies who score highly on tests have low match rates compared to low birth weight babies who do not score highly on tests. Then, the association between low birth weight and low test scores could appear stronger than warranted. Even when such selection mechanisms are not present, so that the linkage process is ignorable (Rubin 1976) with respect to the analysis, i.e. the linked data can be treated as a representative sample for purposes of the particular analysis, using only the linked cases sacrifices partially complete information. This results in inflated variances in inferences.

Selection bias can result even with perfect and complete matching of the files in hand. For example, suppose we link administrative records of children who have been screened for lead poisoning to the education database, and suppose we find matches for every child who has been screened. This still can be a biased sample. The students who get tested for lead poisoning may not be representative of the population of students as a whole; for example, their families may be more likely to live in older housing (lead screening is often required for kids who live in houses built in the 1960s or earlier due to the use of lead paint and water pipes) or to participate in other public assistance programs. If this selection bias correlates with the outcome of interest and is not somehow controlled for, resulting inferences on the linked data can be biased.

6.3 Approaches to Accounting for Uncertainty

Researchers have developed a variety of approaches to account for additional uncertainties in record linkage. Here, we review some of these methods, focusing on three main strategies, namely (i) treating the matrix that indicates who matches whom as a parameter in a statistical model, (ii) embedding the linkage in a particular modeling task, and (iii) imputation of incomplete links to address biases from incomplete matching. This review does not cover every method used to account for uncertainty – notably, it does not cover machine learning approaches to record linkage – and new research on the subject appears frequently.

6.3.1 Modeling Matching Matrix as Parameter

Throughout the remainder of the chapter, we assume that an analyst seeks to link two files, File A comprising n_A records and File B comprising n_B records, using linking variables present in both files. We assume each individual to appear at most once in each file. In this setting, record linkage involves estimating the unknown matching matrix. To describe this matrix, for any record i in File A and j in File B, let $c_{ij} = 1$ when record i and record j belong to the same individual, and let $c_{ij} = 0$ otherwise. Let $C = \{c_{ij} : i = 1, \dots, n_A; j = 1, \dots, n_B\}$ be the matching matrix.

Most record linkage techniques, e.g. those based on the Fellegi and Sunter (1969) framework, create a single point estimate of C, typically by satisfying some loss criterion. Naturally, this point estimate has uncertainty. Unfortunately, it is usually not straightforward to characterize that uncertainty analytically, as the sampling distribution of C is hard to write down. One way to get around this dilemma is to use one of a number of Bayesian approaches (Fortini et al. 2001; Larsen 1999; Tancredi and Liseo 2011; Sadinle 2017). These rely on Markov chain Monte Carlo techniques to come up with numerical approximations to the (posterior) distribution of C.

To illustrate Bayesian approaches, we consider the relatively uncomplicated setting described by Fortini et al. (2001). Suppose every record k in File A or File B has L linking variables, $(d_{k1}, \dots, d_{kL})$ for $l = 1, \dots, L$ and all pairs of records (i, j) where i comes from File A and j from File B. Define the binary comparison variable e_{ijl} such that $e_{ijl} = 1$ when $d_{i1} = d_{j1}$ and $e_{ijl} = 0$ otherwise. Here, the analyst can compute $e_{ij} = \{e_{ij1}, \dots, e_{ijL}\}$ for all (i, j) pairs.

The Bayesian modeling approach specifies a data model for the (vector-valued) random variables E_{ij} for all (i, j) given the unknown C, coupled with a prior distribution on C. In general, this model allows for the possibility that one or more $e_{ijl} = 0$ when $c_{ij} = 1$, for example due to recording or measurement errors in the different files, and the possibility that $e_{ijl} = 1$ for many or even all l when $c_{ij} = 0$. Thus, the usual approach in Bayesian record linkage is to specify (at least) one model for cases where $c_{ij} = 1$, i.e. the matches, and another model for cases where $c_{ij} = 0$, i.e. the non-matches. Mathematically, we can write this as

$$E_{ij} \mid c_{ij} = 1 \sim f\left(\theta_m\right) \tag{6.1}$$

$$E_{ij} \mid c_{ij} = 0 \sim f\left(\theta_u\right) \tag{6.2}$$

where θ_m are parameters for the model corresponding to matches and θ_u are parameters for the model corresponding to non-matches. For example, a typical model (Fortini et al. 2001; Sadinle 2017) is to assume independent Bernoulli distributions; for $l = 1, \ldots, L$, assume $\Pr(E_{ijl} = 1 \mid c_{ij} = 1) = \theta_{ml}$ and $\Pr(E_{ijl} = 1 \mid c_{ij} = 0) = \theta_{ul}$. Hypothetically, one can specify any data model, although computational concerns often lead to the types of conditional independence assumptions used in mixture model implementations (Herzog, Scheuren, and Winkler 2007) of record linkage.

The second step is to specify a prior distribution on C, which can be done in many ways. One of the best performing prior distributions is described in Sadinle (2017). First, we specify a model for the number of matches, i.e. for $n_m = \sum_{ij} c_{ij}$; this can be expressed through prior beliefs on the proportion of matches, which can be modeled using a simple β distribution with parameters tuned to reflect prior expectations. Second, conditional on knowing n_m records match, we assume all possible bipartite matchings are equally likely. Sadinle (2017) shows via simulation that this sensible prior distribution has good performance. A key advantage of using such prior distributions is that they easily can enforce one-to-one matching.

The resulting model can be estimated using relatively straightforward Gibbs sampling techniques. The result is a posterior distribution on C that summarizes the uncertainty in the linkage (under the posited model). For example, with enough posterior draws, for any record i in File A, one can estimate the posterior probabilities that any record j in File B is its match; simply, one computes the percentage of times record j is matched to record i across the runs. We note that the Bayesian approach extends beyond binary comparison vectors. It can be used to handle levels of agreement, as well as missing values in the linking variables (Sadinle 2017).

Treating C as a parameter in a Bayesian model opens possibilities for other record linkage strategies. Tancredi and Liseo (2011) incorporate a measurement error modeling framework into record linkage. Using notation from our previous example to illustrate, they assume each d_{kl}, where $l = 1, \ldots, L$ and k is any record in File A or B, is a potentially noisy version of a true value μ_{kl}. The errors in d_{kl} arise from some posited distribution. For example, when d_{kl} has $D_l > 2$ levels, we might use

$$\Pr\left(d_{kl} = \mu_{kl} \mid \mu_{kl}\right) = \pi_l \tag{6.3}$$

$$\Pr\mid(d_{kl} = d)\mid\mu_{kl}, d \neq \mu_{kl}) = \left(1 - \pi_l\right)\left(1/\left(D_l - 1\right)\right)$$

and assume that the measurement errors are made independently across linking variables. This model says that the dataset records the true value with probability π_l and that, when an incorrect value is recorded, its value is completely random. Such measurement error models are somewhat simplistic, but they can result in good performance (Tancredi and Liseo 2011; Manrique-Vallier and Reiter 2018). The model is completed by relating the distribution of μ_{kl} to C. The details are somewhat complicated and so not reviewed here.

The Bayesian measurement error model approach to record linkage is also used by Steorts, Hall, and Fienberg (2016). Their innovation is to allow for multiple records within the same file to belong to the same individual, a scenario generally ruled out in other approaches. The key is to assume that each record is a potentially noisy representation of some individual, who may be represented multiple times. For example, at one extreme all $N = n_A + n_B$ records in File A and File B could correspond to unique individuals (so that the individuals in the files are disjoint), and at the other extreme all N records could correspond to a single individual. Of course, the number of unique individuals is not known and must be estimated from the data. This is done by defining latent variables λ_k for $k = 1, \ldots, N$, where λ_k is an integer from 1 to N. All records with the same value of λ_k are deemed to come from a common unique individual.

For each latent individual t, one can define the vector of unobserved true linking variables, $\mu_t = (\mu_{t1}, \ldots, \mu_{tL})$. Steorts, Hall, and Fienberg (2016) assume that, given λ_k and μ_{λ_k}, the reported values follow models like those in (6.3), and each $\mu_{\lambda_{kl}}$ follows an independent multinomial distribution. The model is completed by prior distributions on all parameters, including all λ_k. The model can be estimated using Markov chain Monte Carlo techniques. The resulting posterior distribution for $\Lambda = \{\lambda_k : k = 1, \ldots, N\}$ can be used to summarize posterior uncertainty over which records link to the same individual, both across and within files.

All of these Bayesian approaches generate a posterior distribution of the linkage structure, whether through C or Λ. In traditional applications of record linkage, however, analysts work with a single linked file. Indeed, the authors of the papers described in this section all present ways to find some optimal set of linkages. However, this seemingly sacrifices a major selling point of the Bayesian approach, as using a single linked dataset makes it difficult to characterize uncertainty in any subsequent analysis.

One possibility is to create multiple linked datasets using an ample number of approximately independent draws from the posterior distribution of C (or Λ) to construct each dataset. Then, analysts can run complete-data analyses on each plausibly linked dataset. At minimum, analysts can see how sensitive results are to different plausibly linked datasets. When they are overly sensitive to the point of changing substantive conclusions, this may indicate that results from the linkage process have too much variability to be trusted.

Alternatively, it is conceptually sensible to think that results from the analyses of the multiple plausibly linked files could be combined using some version of multiple imputation inference (Rubin 1987; Reiter and Raghunathan 2007; Larsen 1999; McGlincy 2004; Goldstein, Harron, and Wade 2012). To date, the theoretical validity of multiple imputation inferences for record linkage has been largely unexplored, although simulation studies suggest promising results. One would expect issues of uncongeniality (Meng 1994) to feature prominently, which complicates the practicality of when to expect inferences to have acceptable properties. Nonetheless, as with multiple imputation in missing data contexts, as a general-purpose tool it may result in inferences with better properties than simply using a single linked dataset.

6.3.2 Direct Modeling

The record linkage approaches in Section 6.3.1 do not use the variables that are measured in only one of the files. However, such variables can help inform the linkage process. For example, suppose two variables X and Y are known to be highly correlated, yet only available in File A and File B, respectively. Given two potential matches in File B, say j and j_0, to some record i in File A, it makes sense to choose the record with the reported value of Y_j or Y_{j_0} that respects the correlation with X_i most faithfully.

Of greater relevance to our purposes, one also can embed the record linkage process directly in an analysis model, generally some regression involving (potentially multivariate) X and Y, to incorporate uncertainty in the inferences. Methods for doing so take two main forms. The first is to use estimates of the matching probabilities to adjust point estimates of regression coefficients; we call this the adjustment approach. The second is to layer a regression model on top of some model for the linkage structure, such as those in Section 6.3.1; this can be done most naturally in hierarchical Bayesian modeling frameworks. We now review these two general strategies, starting with the adjustment approach.

Suppose that one seeks to estimate the regression, $y_i = x_i'\beta + \varepsilon_i$, where $\varepsilon_i \sim N(0, \sigma^2)$ for all i. Scheuren and Winkler (1993, 1997) recognize that regression with incorrectly linked data is actually based on the outcome z_i rather than y_i, where

$$\Pr\left(z_i = y_i\right) = q_{ii}; \Pr\left(z_i = y_j\right) = q_{ij} \quad \text{for } i \neq j \tag{6.4}$$

Here, q_{ij} is the probability that record j is the correct link for record i, where j ranges over all records in File B that could be matched to record i in File A. Using ordinary least squares to regress Z on X results in a biased estimate of β, since $E\left(z_i\right) = w_i'\beta$, where $w_i = \sum_j q_{ij} x_j'$. Scheuren and Winkler (1993) suggest an approach to approximate the bias in a naive estimate of β, based on estimating various q_{ij} from the outputs of the linkage procedure. For computational convenience, they restrict the set of plausible matches for case i to the two records with the largest and the second largest values of q_{ij}. Removing the estimated bias can result in reasonably accurate (although not unbiased) estimates of β, particularly when the top two values of q_{ij} dominate the remaining ones for most i and, of course, when one has reliable estimates of the linking probabilities.

The adjustment approach was extended by Lahiri and Larsen (2005), who used the formulation in (6.4) to derive an unbiased estimator of β. They compute $\hat{\beta}_{LL} = (W'W)^{-1}W'Z$, where Z is the vector of outcomes in the linked file and W is the matrix $(w'_1, \ldots, w'_n)$. They also present routines for estimating linkage-adjusted standard errors. This approach assumes that File A is fully contained in File B (or vice versa). It also requires estimates of all q_{ij}, which can be cumbersome, but not necessarily prohibitive, for large datasets.

These adjustment approaches require estimates of the linkage probabilities. One approach is to use estimated probabilities from the record linkage model, e.g. the posterior probabilities that each $c_{ij} = 1$ from one of the Bayesian models. It is important to note, however, that these probabilities depend on the model assumptions implied in the linkage model, which often are chosen for convenience (e.g. conditional independence). Hence, in practice, adjustment approaches with estimated q_{ij}s sourced from some linkage model can be unbiased with respect to the linkage model, and not necessarily with respect to an underlying population process. In some cases, it may be possible to estimate actual linkage probabilities, for example by using an evaluation sample in which true match status is known. Here, one typically has to make assumptions about how these validation cases relate to the rest of the data, e.g. they are representative. For examples of adjustment approaches that use validation cases, see Chambers (2008), Kim and Chambers (2009, 2011), and Chipperfield, Bishop, and Campbell (2011), among others.

The second approach, using hierarchical models, is exemplified in the work of Gutman, Afendulis, and Zaslavsky (2013). Each record in File A and File B is placed in one of T blocks defined by a cross-classification of categorical variables available on both files. For blocks that include only one case from each of File A and File B, the two records are called a match by default. For any block t that contains more than one record from each file, we define a latent matching matrix C_t (as in Section 6.3.1) within the block. Once we know $c_{tij} = 1$ for some pair (i, j) within block t, we have (y_i, x'_i) for that individual. For blocks with unequal numbers of records from each file, records without a match miss either Y or X; this values are treated as missing data. With this framework, Gutman, Afendulis, and Zaslavsky (2013) assume that

$$y_i \mid x_i, C_t \sim N\left(x'_i \beta, \sigma^2\right) \tag{6.5}$$

and

$$C_t \propto 1 \tag{6.6}$$

The prior distribution of each C_t is uniform over the number of possible permutations of matches within the block t. The model fitting algorithm, implemented by Markov chain Monte Carlo techniques, iterates the steps (i) sample a value of (β, σ^2) given C and imputations of the missing Y and X for blocks with unequal numbers of records from each file, (ii) propose new matches given parameter estimates and the imputations, and (iii) impute new values of the missing data given the draw of C and the parameters. The resulting posterior distribution provides inferences about (β, σ^2) that incorporate the uncertainty due to imperfect linkage.

A distinctive feature of the model in (6.5, 6.6) is that it simultaneously does both the record linkage and parameter estimation, whereas adjustment methods obtain estimates of linkage probabilities from separate procedures. This feature allows the relationship between Y and X to inform the linkage process. Of course, a downside of this approach is that a poorly specified model can impact both the linkage and the parameter estimation. Hence, it is crucial to have a large number of representative blocks with only one possible match, also known as seeds, so that one can specify a regression model with reasonable confidence in its ability to describe the relationship between Y and X. These seeds are also important for pinning down the relationship between Y and X to a sensible space. Such seeds can be available, for example, in applications where social security numbers are available on both files. Matching on social security numbers combined with other information like dates of birth and gender can create a unique matching key. Assuming this key is recorded accurately, one might be able to generate many seeds before trying to match records without social security numbers.

The model of Gutman, Afendulis, and Zaslavsky (2013) assumes that the variables used to construct blocks are measured without errors. This may not be the case, which could result in biased estimates (as could be the case for any approach that uses blocking). Dalzell and Reiter (2018) propose a model that allows the blocking variables to be measured with error, using a measurement error model along the lines of (6.3), which incorporates this additional source of uncertainty.

The model of Gutman, Afendulis, and Zaslavsky (2013) also generates a posterior distribution of C. As such, the model could be used as the engine for generating multiple imputations of linked datasets. The properties of this use, e.g. what happens when the model is mis-specified, are largely unstudied.

6.3.3 Imputation of Entire Concatenated File

When File A and File B are of different sizes, necessarily some records do not have complete data after the matching process. The most obvious approach is to use multiple imputation to fill in missing items for non-matched cases (as well as any item missing for all cases). Here, a general strategy is to estimate a predictive model for the missing values using the linked dataset (assuming it is a single linked dataset) and create completed versions of the non-matched cases. This is implicitly what is done in the methodology of Gutman, Afendulis, and Zaslavsky (2013), albeit only for variables used in the model fitting. The validity of such an approach relies on a critical assumption: the relationships in the linked data should apply for the non-linked data. Unfortunately, this assumption is largely untestable absent other data. This suggests that users analyze the sensitivity of inferences to different specifications of the model used to impute missing data for non-matched cases, for instance by using pattern mixture models (Little 1993). Such analyses are rarely done in practice.

When multiple linked datasets are available, for example via multiple draws of C from a Bayesian model, it is less clear how to go about dealing with incomplete linkage. Conceptually and practically, it makes sense to treat each plausibly linked dataset as a separate imputation problem, as the parameter distributions could differ across the plausibly linked datasets. Here, nested imputation approaches (Shen 2000; Reiter 2004, 2008) could provide a path forward. In a nested imputation approach, the analyst would create $m > 1$ possible linked datasets using draws of C, estimate an imputation model for each plausibly linked dataset, and use each model to create $r > 1$ completed datasets (i.e. with non-matched values filled in) within each nest. It is not immediately clear whether existing multiple imputation inferential methods for nested data are appropriate for this two-stage approach.

6.4 Concluding Remarks

6.4.1 Problems to Be Solved

The methods in Section 6.3 only scratch the surface on the challenges inherent in accounting for uncertainty in record linkage. The most pressing research area, particularly for Bayesian models, is computation. Algorithms based on Markov chain Monte Carlo (MCMC) sampling can be unacceptably slow for high dimensional problems. Hence, much research is needed to develop efficient computational strategies, such as ones based on approximations to full MCMC sampling. Related, practitioners would benefit enormously from general-purpose software routines for implementing various techniques. The second key research area is dealing with practicalities of linking complex surveys with weights and unit/item nonresponse. For example, when the concatenated file has incomplete linkages, how should agencies construct survey weights? Approaches akin to those in Rubin (1986) could provide solutions. A third intriguing research area is to fuse the best features of different record linkage strategies. For example, the approach of Gutman, Afendulis, and Zaslavsky (2013) uses only categorical blocking variables. Can we identify ways to include string variables in the model, or possibly use the latent variable modeling approach of Steorts, Hall, and Fienberg (2016) to allow duplicates in the algorithm? Finally, as noted multiple times in the chapter, the theory underpinning multiple imputation with record linkage is underdeveloped.

6.4.2 Practical Implications

Accounting for uncertainty in record linkage is far from a solved problem. However, we can glean some practical implications from research done thus far, particularly for organizations seeking to share linked datasets with secondary users. First, it seems prudent for analysts to consider using Bayesian methods, given their versatility and relative simplicity to develop. Bayesian methods result in draws of plausible values of the matching matrix, which can allow assessment of uncertainty in inferences due to the linkage process. For organizations sharing linked data, Bayesian approaches provide a means to make multiple imputation of plausibly linked datasets available to secondary users. This carries some additional storage costs, bookkeeping, and user training, but none of these are especially prohibitive. All have been overcome when using multiple imputation for missing data. Second, in the absence of multiple imputation approaches, it seems prudent for organizations sharing data to release estimated matching probabilities and information about linkage errors from validation samples. Without such information, it is difficult to apply many of the regression adjustment techniques, making it difficult for secondary users to propagate uncertainty through inferences. Third, the many open questions emphasize the importance of continued research. Such research arguably is most effective when it involves collaborative teams across sectors, including academia, government, and industry.

6.5 Exercises

1 Refer to the method of Scheuren and Winkler (1993) described in Section 6.3.2. Derive an expression for the bias of the naive estimate of β using the ordinary least squares regression of Z on X, where Z is defined in (6.4).

2 Suppose that we want to link two files, one containing a binary indicator of treatment T and another containing an outcome variable Y. When we match correctly, the expected value of Y for cases in the treated group equals $\mu + \tau/2$, and the expected value of Y for cases in the control group equals $\mu - \tau/2$. Thus, the true effect of the treatment equals τ. However, when we match incorrectly, the expected value of Y equals μ. Suppose that each of n individuals in the file that includes T has probability p of being linked correctly to a much larger database containing Y. Imperfect matching is done on variables other than T and Y. You then estimate the treatment effect

as $\bar{y}_t - \bar{y}_c$, the difference in sample averages for the treated and control group in the linked data. Assume equal numbers of treated and control records.

a) Suppose that $p = 1$. What is the expected value of $\bar{y}_t - \bar{y}_c$? How about when $p = 0$?

b) Derive a mathematical expression for the expected value of $\bar{y}_t - \bar{y}_c$ for arbitrary p.

Acknowledgment

This research was supported by a grant from the National Science Foundation (SES-11-31897).

References

Belin, T.R. and Rubin, D.B. (1995). A method for calibrating false-match rates in record linkage. *Journal of the American Statistical Association* 90: 694–707.

Chambers, R. (2008). Regression analysis of probability-linked data. Official Statistics Research Series, 4. Available from http://www.statisphere.govt.nz/official-statistics-research/series/default.htm.

Chipperffeld, J.O., Bishop, G.R., and Campbell, P. (2011). Maximum likelihood estimation for contingency tables and logistic regression with incorrectly linked data. *Survey Methodology* 37: 13–24.

Dalzell, N. and Reiter, J.P. (2018). Regression modeling and file matching using possibly erroneous matching variables. *Journal of Computational and Graphical Statistics* 27 (4): 728–738.

Fellegi, I.P. and Sunter, A.B. (1969). A theory for record linkage. *Journal of the American Statistical Association* 64: 1183–1210.

Fortini, M., Liseo, B., Nuccitelli, A., and Scanu, M. (2001). On Bayesian record linkage. *Research in Official Statistics* 1: 179–198.

Goldstein, H., Harron, K., and Wade, A. (2012). The analysis of record-linked data using multiple imputation with data value priors. *Statistics in Medicine* 31: 3481–3493.

Gutman, R., Afendulis, C., and Zaslavsky, A.M. (2013). A Bayesian procedure for file linking to analyze end-of-life medical costs. *Journal of the American Statistical Association* 108: 34–47.

Herzog, T.N., Scheuren, F.J., and Winkler, W.E. (2007). *Data Quality and Record Linkage*. Springer.

Jaro, M.A. (1989). Advances in record linkage methodology as applied to matching the 1985 census of Tampa, Florida. *Journal of the American Statistical Association* 84: 414–420.

Kim, G. and Chambers, R. (2009). Regression analysis under incomplete linkage. Technical report. Centre for Statistical and Survey Methodology, University of Wollongong, Working Paper 17-09.

Kim, G. and Chambers, R. (2011). Regression analysis under probabilistic multi-linkage. Technical report. Centre for Statistical and Survey Methodology, University of Wollongong, Working Paper 10-11.

Lahiri, P. and Larsen, M.D. (2005). Regression analysis with linked data. *Journal of the American Statistical Association* 100: 222–230.

Larsen, M. (1999). Multiple imputation analysis of records linked using mixture models. In: *Proceedings of the Survey Methods Section*, 65–71. Ottawa: Statistical Society of Canada.

Larsen, M.D. and Rubin, D.B. (2001). Iterative automated record linkage using mixture models. *Journal of the American Statistical Association* 96: 32–41.

Little, R.J.A. (1993). Pattern-mixture models for multivariate incomplete data. *Journal of the American Statistical Association* 88: 125–134.

Manrique-Vallier, D. and Reiter, J.P. (2018). Bayesian simultaneous edit and imputation for multivariate categorical data. *Journal of the American Statistical Association* 112: 1708–1719.

McGlincy, M.H. (2004). A Bayesian record linkage methodology for multiple imputation of missing links. In: *JSM Proceedings of the Section of Survey Research Methods*, 4001–4008. Alexandria, VA: American Statistical Association.

Meng, X.L. (1994). Multiple-imputation inferences with uncongenial sources of input (disc: P558-573). *Statistical Science* 9: 538–558.

Neter, J., Maynes, E.S., and Ramanathan, R. (1965). The effect of mismatching on the measurement of response error. *Journal of the American Statistical Association* 60: 1005–1027.

Reiter, J.P. (2004). Simultaneous use of multiple imputation for missing data and disclosure limitation. *Survey Methodology* 30: 235–242.

Reiter, J.P. (2008). Multiple imputation when records used for imputation are not used or disseminated for analysis. *Biometrika* 95: 933–946.

Reiter, J.P. and Raghunathan, T.E. (2007). The multiple adaptations of multiple imputation. *Journal of the American Statistical Association* 102: 1462–1471.

Rubin, D.B. (1976). Inference and missing data (with discussion). *Biometrika* 63: 581–592.

Rubin, D.B. (1986). Statistical matching using file concatenation with adjusted weights and multiple imputations. *Journal of Business & Economic Statistics* 4: 87–94.

Rubin, D.B. (1987). *Multiple Imputation for Nonresponse in Surveys*. New York: Wiley.

Sadinle, M. (2017). Bayesian estimation of bipartite matchings for record linkage. *Journal of the American Statistical Association* 112: 600–612.

Scheuren, F. and Winkler, W.E. (1993). Regression analysis of data files that are computer matched. *Survey Methodology* 19: 39–58.

Scheuren, F. and Winkler, W.E. (1997). Regression analysis of data files that are computer matched – Part II. *Survey Methodology* 23: 157–165.

Shen, Z. (2000). Nested multiple imputation. PhD thesis. Harvard University, Department of Statistics.

Steorts, R.C., Hall, R., and Fienberg, S.E. (2016). A Bayesian approach to graphical record linkage and de-duplication. *Journal of the American Statistical Association* 111: 1660–1672.

Tancredi, A. and Liseo, B. (2011). A hierarchical Bayesian approach to record linkage and population size problems. *The Annals of Applied Statistics* 5: 1553–1585.

7

Measuring and Controlling for Non-Consent Bias in Linked Survey and Administrative Data

Joseph W. Sakshaug[1,2]

[1] *Institute for Employment Research, Department of Statistical Methods Research, Nuremberg, Germany*
[2] *Ludwig Maximilian University of Munich, Department of Statistics, Munich, Germany*

7.1 Introduction

7.1.1 What Is Linkage Consent? Why Is Linkage Consent Needed?

Linking surveys to administrative databases is a common practice in the health and social sciences. Numerous large-scale surveys conducted around the world supplement their primary data collections with linkages to a variety of administrative sources, such as social security records, healthcare utilization, billing records, education records, among others. Even though these record sources are primarily intended to capture processes related to the provision of services and administrative functions and not scientific research, they often contain valuable information of interest to substantive researchers – information which may not be easily and accurately collected from surveys. For example, researchers seeking answers to questions related to the impact of educational attainment on income and labor market outcomes may require detailed information about one's lifetime education, earnings, and employment history. Other substantive research questions related to lifetime healthcare spending and end-of-life care costs may necessitate an extensive breakdown of one's healthcare episodes and itemized expenditures. Collecting such detailed longitudinal information from surveys to address these and other relevant research topics can be a burdensome and error-prone process for respondents, which makes the use of administrative measures of these constructs a potentially more attractive alternative.

Due to the highly sensitive and confidential nature of administrative records, however, accessing and linking such records to surveys requires agreement from multiple parties, including the administrative data owners and key stakeholders,

Administrative Records for Survey Methodology, First Edition.
Edited by Asaph Young Chun, Michael D. Larsen, Gabriele Durrant, and Jerome P. Reiter.

and in many cases, the survey respondents themselves. The process by which permission is sought from respondents is commonly referred to as the *informed consent* process. This process is generally intended to ensure that respondents are knowledgeable about the planned uses of their data and any potential risks and benefits that may result should they agree to the linkage, along with assurances on how the data will be protected and safeguarded against malicious use. Altogether this process is meant to provide a reasonable basis from which respondents can make an informed decision.

The specific content of a consent request can vary from study to study, but some legal guidelines exist. For example, the U.S. Privacy Act of 1974 lists several required elements of informed consent, including, among others, an explanation of the study procedures and their purpose, any foreseeable risks and discomforts, contact details to request further information about the study, and a declaration that expressing consent is voluntary and can be revoked at any time (U.S. Privacy Act of 1974). Independent research ethics boards may require that further elements be included in the informed consent request in addition to those mandated by law.

Obtaining informed consent from respondents prior to linking their survey and administrative records is often mandated by research ethics boards and/or jurisdictional legislation. For instance, in the United States, health data regulations stated within the Health Insurance Portability and Accountability Act of 1996 require that individual authorization be obtained prior to linking survey and administrative data (The HIPAA Privacy Act, 45 C.F.R. 164.501). Similar legislation exists in other countries, which requires explicit consent from survey respondents to link to administrative records (e.g. EU General Data Protection Regulation). Many large-scale, nationally representative surveys, including the US Health and Retirement Study, the English Longitudinal Study of Ageing, The UK Household Longitudinal Study, and the German study "Labour Market and Social Security" ask for explicit consent from respondents to link their survey responses to various types of administrative records, including health records, social security records, employment records, education records, and welfare benefit records, among others. Depending on the legal arrangement, some panel studies approach respondents for linkage consent only if they did not provide consent in a previous wave, while other panel studies approach all respondents at every wave of data collection regardless of previous consent.

7.1.2 Linkage Consent Rates in Large-Scale Surveys

Despite careful efforts by surveys to reassure respondents about the intended uses of their linked data and data protection protocols, a portion of respondents still deny the record linkage request. Reviews of the linkage consent literature show that linkage consent rates vary substantially from study to study. In a meta-analysis of health linkage studies, da Silva et al. (2012) observed significant variation in consent rates, ranging from 39% to 97%. In an analysis of linkage studies drawn from a variety of different discipline areas, Sakshaug and Kreuter (2012) found that linkage consent rates varied between 24% and 89%. In an examination of linkage consent rates, Fulton (2012) noted a declining trend in linkage consent in several studies, including the U.S. Current Population Survey (90% to 76% between years 1994 and 2003), the U.S. National Health Interview Survey (85% to 50% between years 1993 and 2005), and the U.S. Survey of Income and Program Participation (88% to 65% between years 1996 and 2004).

There are different reasons why respondents may agree or disagree to the linkage consent request. Sala, Knies, and Burton (2014) asked respondents in the Innovation Panel of The UK Household Longitudinal Study about factors that went into their decision when previously prompted to consent to link their survey data with federal administrative data on National Insurance contributions, benefits and taxes, savings, and pensions. Among those who agreed to the linkage, the most prominent factor cited was "Being helpful with the research" whereas those who disagreed to the linkage predominantly cited "Concerns about sharing of confidential data." There are also suggestions in the literature that the linkage consent decision is related to the survey participation decision. For example, it is common for individuals who previously expressed their initial reluctance to take part in the survey, or those who refused to answer financial or other sensitive questions, to refuse the linkage consent request (Larsen, Roozeboom, and Schneider 2012). Interviewer observations of uncooperative behavior exhibited prior to the linkage request have also been shown to be associated with linkage non-consent (Sakshaug et al. 2012).

Other factors that may play a role in the linkage consent decision include survey sponsorship, topic of the administrative records, and mode of data collection. For example, Fulton (2012, p. 85) found in an analysis of 162 linkage consent requests that government-sponsored surveys achieved consent rates that were about 22% points higher, on average, compared to surveys with a nongovernment sponsor. Fulton (2012, p. 88) also analyzed the impact of record topic on linkage consent rates using the 162 linkage requests, but found no statistically significant difference in linkage consent rates between the most common record topics: health records and income/employment records. Regarding mode of data collection, there is evidence suggesting that interviewer-administered survey modes yield higher linkage consent rates than self-administered modes (Fulton, 2012; Sakshaug et al. 2017), although observed interviewer characteristics do not appear to have a strong influence on consent rates (e.g. Sakshaug et al. 2012; Sala et al. 2012). The effects of survey sponsorship and interviewer presence on linkage consent rates appear to suggest that similar mechanisms operate for the linkage consent decision as they do for the survey participation decision.

7.1.3 The Impact of Linkage Non-Consent Bias on Survey Inference

The apparent decline in linkage consent rates is an issue for researchers interested in performing analysis on linked data. Linkage non-consent not only reduces the analytic sample size of the linked dataset and reduces statistical power, but it can also induce bias in estimates obtained from the linked data if individuals who consent are systematically different from those who do not consent based on the study variables of interest. Evidence drawn from the linkage consent literature suggests that linkage non-consent is not a strictly random process and systematic differences often exist between consenters and non-consenters based on the collected survey items. Indeed some of the most common survey items used in substantive analyses are correlated with linkage consent, including demographics (e.g. age, gender, foreign citizenship), health measures, and economic characteristics (e.g. income, welfare benefit receipt) (Mostafa, 2016; Sakshaug et al. 2012; Sala et al. 2012; Knies and Burton 2014; Jenkins et al. 2006; Young et al. 2001; Haider and Solon 2000). However, it should be noted that these oft-cited correlates of consent show inconsistencies across studies. For example, age has been shown to have positive and negative associations with consent (Dahlhamer and Cox 2007; Young et al. 2001; Banks et al. 2005; Jenkins et al. 2006). Health measures have also exhibited a conflicting relationship with consent (Haider and Solon 2000; Young et al. 2001; Dahlhamer and Cox 2007). These inconsistencies could be a result of a host of different survey design factors, including the target population under study, mode of data collection, the type of administrative data being requested for linkage, and the consent rate itself. Nevertheless, these studies suggest that linkage non-consent bias occurs in practice and can be an issue when analyzing these types of variables.

7.1.4 The Challenge of Measuring and Controlling for Linkage Non-Consent Bias

Compared to other sources of bias in the Total Survey Error framework (Groves et al. 2009; Biemer 2010), such as nonresponse bias and measurement error bias, linkage non-consent is a relatively understudied source. Moreover, it can be argued that survey organizations invest significantly more resources into controlling non-response and measurement error biases relative to linkage non-consent bias. This is evidenced in large-scale survey documentation which often contains comprehensive details about response rates, use of different measurement modes to elicit more accurate responses, and efforts to convert nonrespondents and adjust for nonresponse via weighting or other statistical procedures. In contrast, documentation related to linkage consent is typically limited to details regarding the consent procedure and overall consent rate. As such, deliberate efforts to study and minimize linkage consent bias at the design stage are rare in survey research.

Part of the challenge in dealing with linkage non-consent bias is that there is little empirical research that sheds light on the possible magnitude of linkage consent bias, and a lack of experimentation into ways in which this source of bias can be controlled. As a result, standard "best practice" guidelines for dealing with this source of bias are lacking. Furthermore, although survey organizations typically provide a set of weights that adjust for survey nonresponse, rarely are weights provided to data users to adjust for non-consent bias in linked data analyses. Thus, the decision to measure and/or adjust for consent bias is left to the discretion of the researcher. A further complication is that administrative records for the non-consenting respondents are typically not made available for research purposes, making the assessment of non-consent bias in administrative variables difficult to carry out.

Despite these challenges, there are strategies researchers can implement to better understand the magnitude of linkage non-consent bias and whether this source of bias is likely to impact one's analytic objectives. Furthermore, some progress has been made in implementing strategies to reduce the risk of non-consent bias in surveys, either through careful design of the linkage consent request or statistical procedures applied post-data collection by the survey organization or the data analyst. In this chapter, these strategies for measuring linkage non-consent bias and controlling for its effects are reviewed. Specifically, the empirical literature is summarized, highlighting examples where attempts have been made to measure and/or control for linkage non-consent bias, noting the methods used and their associated strengths and limitations. Practical guidance is provided to ensure readers are equipped to deal with linkage non-consent bias in their own research.

7.2 Strategies for Measuring Linkage Non-Consent Bias

7.2.1 Formulation of Linkage Non-Consent Bias

In simplest terms, an estimate of linkage non-consent bias for a sample mean derived from a set of survey respondents can be calculated using the following expression:

$$\text{Linkage Non} - \text{Consent Bias} = \overline{Y}_c - \overline{Y}_r = \left(\frac{m}{r}\right)\left[\overline{Y}_c - \overline{Y}_m\right]$$

where $\overline{Y}_c$ is the estimated mean based on the set of c respondents who provided linkage consent, $\overline{Y}_r$ is the estimated mean based on all r respondents irrespective of linkage consent, and $\overline{Y}_m$ is the estimated mean based on the set of m respondents who did not provide linkage consent.

Similar to other errors of non-observation in surveys (e.g. nonresponse, noncoverage), linkage non-consent bias contains two components: a rate and difference in means. The first component is the rate of linkage non-consent, defined as the ratio of non-consenting respondents and all respondents in the survey. The second component expresses the difference in means between respondents who consent to record linkage and respondents who do not consent. Based on this formulation, there would be no bias if the linkage non-consent rate or the difference between the consenting and non-consenting respondent means is zero.

It is important to note that this bias formulation assumes that survey respondents are the only subjects who receive the consent request. This is the most common situation in practice; survey non-respondents, for example, are rarely asked for consent to use administrative records. However, there are some exceptions. One exception is when the survey invitation is coupled with a request for consent to access administrative records as is very common in health studies. In some studies, survey participation may be contingent on linkage consent. In this situation, a nonresponse may reflect unwillingness to allow access to administrative records, unwillingness to take part in the survey, or both. But disentangling these two causes is not always possible. Another exception is when survey nonrespondents are reapproached in a nonresponse follow-up study and asked for consent to access administrative records in lieu of taking part in the full survey interview (see e.g. Sakshaug and Eckman, 2017).

7.2.2 Modeling Non-Consent Using Survey Information

A major advantage of studying linkage non-consent bias as opposed to other sources of survey bias (e.g. nonresponse) is that all variables collected from the interview may be used to investigate differences between consenting and non-consenting respondents. These variables provide a potentially rich source of auxiliary information available for studying mechanisms of linkage consent and adjusting for its biasing effects.

The majority of linkage consent research studies take advantage of this situation by utilizing the collected survey data to investigate possible discrepancies between consenting and non-consenting respondents. For example, in a bivariate context one might examine whether consent rates differ between respondents who identify themselves as being unemployed or not in the labor force and respondents who identify themselves as being employed. This simple approach is commonly used to assess whether respondents' demographic, health, economic, and other characteristics play a role in their decision to give consent. More complex analyses examine associations and relationships between the consent outcome and the collected survey variables in a multiple regression framework (Sala, Burton, and Knies 2012; Jenkins et al. 2006; Al Baghal 2015; Larsen, Roozeboom, and Schneider 2012). In this setting a typical setup involves the use of a logistic (or probit) regression model to explicitly model the binary consent outcome as a function of the collected survey variables. The primary objective here is to identify whether associations exist between the consent outcome and specific survey variables after controlling for other survey variables and minimizing confounding effects.

Identifying whether specific survey variables are associated with the linkage consent outcome (either in a bivariate or multivariate context) is useful, not only for determining whether such variables likely play a role in people's decision to consent, but also for identifying which variables are likely to be affected by non-consent bias in linked data analyses. This approach is also useful in identifying the direction of the bias. For example, if older people are more likely to consent than younger people, then there would be a positive age bias. However, there are some limitations associated with this approach. In particular, these association-based analyses do not provide a precise quantification of the magnitude of the non-consent bias for a particular variable or estimate, such as average healthcare costs. A regression model itself can be used to quantify a person's likelihood of consent in terms of odds or probabilities, but for data users who are interested in knowing the extent to which their estimates are likely to be shifted by non-consent bias, this information is not directly transferable from the model. Rarely are direct estimates of non-consent bias published in linkage applications.

Another limitation of utilizing survey information to assess non-consent bias is that it does not provide a direct indication of whether bias is present in the linked-administrative variables or estimates derived from these variables. In contrast to survey variables which are available for both consenting and non-consenting respondents, rarely are administrative records made available for the non-consenting respondents to assess bias. Thus, analyzing non-consent bias in administrative variables can be a challenge. To the extent that the survey variables are correlated with the administrative variables it may be possible to make an indirect inference about possible bias in the administrative variables. For example, if persons with higher reported incomes consent to linkage at a higher rate than persons with lower incomes, then administrative variables related to one's income may also be susceptible to non-consent bias. There have been some efforts to estimate non-consent biases in administrative variables for the full set of respondents, which we review in Section 7.2.3.

7.2.3 Analyzing Non-Consent Bias for Administrative Variables

As previously mentioned it can be difficult to analyze non-consent bias in administrative variables because administrative records are rarely made available for the non-consenting portion of the respondent pool. Nonetheless, there are some published examples of studies that have analyzed linkage non-consent bias in administrative variables for all responding cases (e.g. Sakshaug and Kreuter 2012; Sakshaug and Huber 2016; Sakshaug and Vicari 2018; Sakshaug et al. 2019). These studies all share a common feature in that the survey samples were drawn from an administrative database connected to the records for which linkage consent was sought. For example, Sakshaug and Kreuter (2012) used the German panel study "Labour Market and Social Security (PASS)" to estimate non-consent bias in a selection of administrative variables. The PASS study is comprised of two independent samples: a general population sample and welfare benefit recipient sample. The benefit recipient sample is drawn directly from a federal employment database which identifies persons who are currently receiving welfare benefit, whereas the general population sample is drawn by a commercial vendor with access to municipality registration lists. The authors examined the benefit recipient refreshment sample from wave 5 of the PASS, which asked all respondents for consent to link their survey responses with administrative employment records directly connectable to the sampling frame used to draw the benefit recipient sample. About 80% of the sample agreed to the linkage consent request. The consent agreement prohibited linkage of the survey responses to administrative records for the remaining 20%, but the researchers received legal permission to link the binary consent outcome indicator, which was deemed to be a process variable, to the administrative records for all respondents. They then performed an aggregate estimation procedure to obtain estimates of non-consent bias for the administrative variables. The results revealed a few statistically significant non-consent biases, but in general the magnitudes of the biases were small (i.e. less than 10% relative bias) and smaller than other survey biases estimated in the study, including nonresponse and measurement error bias.

Sakshaug and Huber (2016) performed a similar non-consent bias analysis in a panel survey of employees in Germany. Respondents in the survey were sampled directly from employer records maintained by the German Federal Employment Agency, the agency which also sponsored the survey. Respondents were asked for consent to link their survey responses to administrative employment records for which about 90% of respondents agreed. The agency's legal team approved the merging of the binary consent indicator to administrative employment records of the individuals who responded in the survey and the same aggregate estimation procedure was used to examine the magnitude of non-consent biases in cross-sectional and longitudinal administrative variables. Again, the results showed that non-consent bias was generally smaller than nonresponse bias – with the caveat that the consent rate was higher than the nonresponse rate (as was also the case in the above PASS wave 5 study). Another notable result from this study was that the strategy of asking previous non-consenters for consent again in subsequent waves not only improved the linkage consent rate in the study, but also reduced non-consent bias for most administrative variables. This finding is discussed further in Section 7.3.4.

7.3 Methods for Minimizing Non-Consent Bias at the Survey Design Stage

Existing methods for minimizing the risk of linkage non-consent bias tend to be divided into two groups: (i) those that aim to minimize the non-consent rate (i.e. the first component of the aforementioned linkage non-consent bias expression) through careful design and administration of the linkage consent request; and (ii) those that correct for imbalances between consenting and non-consenting respondents (i.e. the second component of the bias expression) using statistical adjustment procedures and covariate information. The two groups of methods can also be respectively divided into those that are typically implemented at the survey design stage and are under the control of the survey planner, and those that can be implemented post-data collection and prior to data analysis. Methods that aim to maximize the linkage consent rate are covered in this section; methods of statistically adjusting for non-consent bias are covered in Section 7.4.

7.3.1 Optimizing Linkage Consent Rates

Efforts to maximize the linkage consent rate have largely focused on identifying specific design features of the consent request that influence the consent decision. Such features include the location in the questionnaire at which the linkage consent question is placed or administered to respondents, the wording or framing of the linkage consent question, whether active or passive consent procedures are used, and design features related to the collection of linkage consent in panel studies where consent is sought at multiple time points. There has been some experimental work done on these features which we review in the following subsection.

7.3.2 Placement of the Consent Request

The placement of the linkage consent question in the survey questionnaire has only recently been identified as a potential factor that influences the consent decision. Traditionally, the linkage consent question has been administered as a survey item that is secondary to the majority of the questionnaire content and thus administered at (or near) the very end of the interview. There are two different perspectives on why placing the linkage consent question at the end of the questionnaire has historically been the preferred approach. The first perspective is based on the presumption that respondents are likely to have greater rapport and trust with the interviewer and survey organization over the course of the interview. Thus, the conventional wisdom is that respondents are more likely to be amenable to a relatively sensitive data sharing request toward the end of the interview, when levels of rapport and trust are expected to be at their peak. However, evidence of a relationship between measures of rapport and linkage consent has been mixed (Jenkins et al. 2006; Sala, Burton, and Knies 2012). The second perspective on waiting until the end of the interview to ask for linkage consent is borne out of concern that respondents might be put off by the sensitive data request, which could negatively impact their willingness to complete the remainder of the interview and/or participate in follow-up interviews. Studies investigating the validity of this argument are lacking; though, there is some evidence suggesting that linkage consent requests that utilize a passive (or "opt-out") procedure do not reduce participation rates or increase attrition (Das and Couper 2014).

In contrast to both perspectives, there are multiple arguments that suggest that placing the linkage consent question at the end of the questionnaire is suboptimal with respect to maximizing the consent rate. One argument is that, by the end of the interview, respondents have already provided information on an array of topics and thus may feel that a further request for personal data is unnecessary. An alternative argument is that consent should be requested at (or near) the beginning of the questionnaire soon after respondents have agreed to take part in the survey. Social psychological theories regarding "foot-in-the-door" effects suggest that once individuals agree to a previous small request they are more likely to agree to a larger subsequent request (Burger 1999). Immediately following up the survey participation request with a data linkage request may result in a similar effect as respondents are likely to begin the interview in a more positive mood compared to at the end of the interview when respondents are more likely to be fatigued or of the belief that enough information has been shared with the survey organization. Introducing the linkage request upfront may also help convey to respondents the importance of the linked data in meeting the goals of the research.

Only a few studies have experimentally explored the relationship between the placement of the linkage consent question and the resulting consent rate, and each one suggests that asking for linkage consent at the end of the interview is not an optimal strategy. Sakshaug, Tutz, and Kreuter (2013) found that asking for linkage consent at the beginning of the interview yields a higher consent rate than asking for consent at the end of the interview. In a replication study conducted by (Sakshaug et al. 2019), a similar placement effect was found leading to a higher rate of consent when requested at the beginning of the interview. The same result was found in a survey of establishments (Sakshaug and Vicari 2018). Sala, Knies, and Burton (2014) examined the impact of asking for linkage consent in the context of topic-related survey items (e.g. placing the consent-to-link employment records question adjacent to other employment-related questions). The authors found that this placement produced a higher consent rate than placing the consent question at the end of the questionnaire. Whether asking for linkage consent affects the quality of responses collected after the request is largely unknown, although Eckman and Haas (2017) find some evidence that persons who grant consent at the beginning of the interview may be more prone to misreporting later in the interview.

7.3.3 Wording of the Consent Request

Another feature of the linkage consent question that the survey designer has some control over is the wording (or framing) of the question. Certain elements of the linkage consent request must be stated explicitly in order to meet legal requirements and/or meet requirements set forth by the institutional ethics committee responsible for overseeing the study (e.g. assurance that the request is voluntary, list of risks and benefits of linkage, etc.). Nonetheless, the survey organization may have some flexibility in how these elements are worded and any additional elements about the linkage that may be useful for respondents to consider. Some survey designers exploit these liberties to frame the consent request in a way that makes the request more appealing to respondents by emphasizing certain elements of the linkage that are likely to be salient to respondents as they consider the request. One such approach is to emphasize the benefits of linkage to the respondent. For example, the survey organization might stress the importance of linkage in meeting the scientific goals of the study, or how linkage ultimately leads to a more parsimonious questionnaire and reduces data collection costs, which may be particularly salient in the context of a government-sponsored survey funded by taxpayers.

Emphasizing the benefits of linkage is thought to be a useful strategy for increasing respondent cooperation as opposed to a neutral appeal. However, the experimental evidence (while limited in scope) does not strongly support this claim. For example, Pascale (2011) did not observe an effect when different linkage benefits were mentioned to respondents at random in a telephone survey, including reduced costs, reduced respondent burden, and improved data accuracy. A null finding was also reported in the telephone study by Sakshaug, Tutz, and Kreuter (2013) in which a random subset of respondents was exposed to a time-saving argument ("In order to keep the interview as short as possible, we would like to…"). A possible explanation for these null findings is that interviewers did not read the benefit arguments as scripted in the questionnaire, or read them too quickly for respondents to notice. In a replication of the wording experiment reported in Sakshaug, Tutz, and Kreuter (2013), Sakshaug and Kreuter (2014) observe a positive effect of the time-saving argument in a web survey implementation, suggesting that wording effects may be perceived differently when administered in a visual, self-administered mode.

Instead of emphasizing the advantages of linking one's data, an alternative strategy that has been explored in the linkage consent literature is to emphasize the disadvantages of not linking one's data. Framing the decision in terms of losses instead of gains has roots in Prospect Theory (Kahneman and Tversky, 1979, 1984). These classic works demonstrated through several experiments that a person's decision-making is affected by whether the available alternatives are framed in terms of gains or losses. More specifically, they showed that people are risk-averse decision makers when they are presented with alternatives that are framed in terms of sure gains, and risk-seeking when the alternatives are framed in terms of sure losses. Operationalizing this theory in a linkage consent context can take various forms. For example, framing the linkage consent decision in terms of losses might involve mentioning to respondents that, without their consent, surveys would be longer and costlier to administer, place a greater burden on its respondents, or would negatively impact the study's ability to achieve its research objectives.

As is the case with most studies of linkage consent, the experimental evidence on "loss framing" is limited with just a few studies to report; though, the available evidence indicates that loss framing is indeed an effective means of obtaining linkage consent. Kreuter, Sakshaug, and Tourangeau (2016) reported in a telephone survey that framing the linkage consent question in terms of losses ("The information you have provided so far would be *much less valuable* to us if we can't link it to...") yielded a 10-percentage point increase in the consent rate compared to gain-framing the request ("The information you have provided so far would be a lot more valuable to us if we could link it to..."). A similar effect was found by Sakshaug et al. (2019) in a web survey, but not in a telephone survey, and only when the consent question was administered at the end of the questionnaire. This study also reported that the effect of placement of the linkage consent question in the questionnaire was larger than the effect of framing in both surveys. Mixed results of framing were also reported in a telephone survey by Sakshaug, Wolter, and Kreuter (2015).

7.3.4 Active and Passive Consent Procedures

A fundamental decision in the design of any linkage consent study is whether active (opt-in) or passive (opt-out) consent procedures will be used. Active consent procedures require that respondents take some form of action in order to express their willingness to consent to the linkage. The taken action may involve signing a consent form, ticking a box in a web or paper survey, or verbally agreeing to the request over the telephone. In some applications, a unique identifier (e.g. social security number) is requested in order to identify and link an individual's administrative record, and provision of this identifier is a means of expressing one's consent to the linkage under an active consent procedure. Passive consent procedures, on the other hand, only require action from the respondent if they wish to object to the linkage; otherwise, consent is implied. Common methods of "opting out" of the linkage include contacting the survey organization or notifying the interviewer during the interview. In some cases, a signature may be needed to document one's wish of opting out.

Both types of consent procedures (active and passive) can influence the consent rate in different ways. In the context of obtaining consent to participate in research studies, passive consent procedures generally yield higher consent rates than passive consent procedures (Ellickson and Hawes 1989; Range, Embry, and MacLeod 2001; Junghans et al. 2005). However, there is significant controversy over whether passive consent procedures are ethical in all research contexts and whether they indeed convey the true wishes of the individual (Hollmann and McNamara 1999). In the linkage consent context, the majority of studies implement an active consent procedure with consent expressed either verbally or via a signed consent form. Studies that implement a passive consent procedure typically provide respondents with instructions on how to opt-out – though, the exact instructions for opting out tend to vary from study to study. Evidence of the effects of active and passive record linkage consent is sparse. Das and Couper (2014) observed a passive consent rate of about 95% in an online survey of the Dutch population – a very high consent rate compared to the studies reviewed in Section 7.1.2. Sala, Knies, and Burton (2014) document the negative effects of asking for a signature in an active consent procedure implemented in a British survey, noting that almost 4% of respondents agreed to record linkage but refused to sign the consent form.

7.3.5 Linkage Consent in Panel Studies

Panel studies that reinterview the same set of respondents over time pose interesting challenges for obtaining record linkage consent. Many panel studies seek consent for access to past, present, and future record information, but the frequency at which consent is requested over the course of the panel can vary. Although variations exist, there are two general models of obtaining linkage consent that panel studies may ascribe to for linkage to a particular data source, which I refer to as the *single-consent* model and the *repeated-consent* model. In the single-consent model, respondent consent is needed only once over the course of the panel study for past, present, and future record access over the course of respondents' participation in the panel study. Consenting respondents may be reminded over the course of the study that they can withdraw their consent at any time to prevent future linkages, but they will (usually) not be explicitly asked again for consent in future waves. In the repeated-consent model, the linkage consent request is repeated periodically on the same respondents regardless of whether they provided consent in an earlier wave. This is done to ensure whether respondents who previously consented are still happy for the linkages to occur. The timing of the repeat consent request may occur every consecutive wave or between non-adjacent waves depending on the administrative data type, legislative requirements, and/or other requirements set forth by the administrative data custodian.

Different strategies have been proposed for optimizing linkage consent rates under each consent model. For instance, in the single-consent model, consent rates can be maximized by continuously following up with non-consenters in subsequent waves and repeating the consent request. Some evidence suggests that, not only does this help to convert previous non-consenters and increase the consent rate over time, it also helps to reduce consent bias (Sakshaug and Huber, 2016). Under the repeated-consent model, Sala, Knies, and Burton (2014) compared an independent versus dependent strategy for optimizing the linkage consent rate when repeat consent requests are made to respondents over time. The dependent strategy consisted of reminding respondents of their consent decision from an earlier wave, whereas the independent strategy did not offer this reminder. The experimental study found that respondents are generally consistent with their earlier decision if they are reminded of it. Thus, the authors recommend the strategy of implementing the dependent consent question only to those respondents who consented in the earlier wave and administering the independent question to those who earlier refused the linkage request; although, the authors note that not all ethics committees may sign off on this tailored strategy.

7.4 Methods for Minimizing Non-Consent Bias at the Survey Analysis Stage

Linkage consent rate optimization strategies, such as those described above, offer survey designers the possibility to administer the consent request in a way that may be more attractive to survey respondents. However, the empirical evidence suggests that such methods are only modestly effective in increasing consent rates, and generally fail to achieve 100% consent from all respondents. Survey organizations and the researchers who intend to analyze the linked data are thus left to decide whether further efforts are needed to minimize non-consent bias. Strategies for minimizing linkage non-consent bias post-survey data collection may involve the use of post-survey statistical adjustment procedures (e.g. weighting, imputation) that are commonly used to adjust for unit or item nonresponse bias. Survey organizations routinely apply statistical adjustment methods that account for differential nonresponse, but rarely do they apply methods that adjust for linkage non-consent bias. Typically, the decision of whether to adjust for non-linkage consent bias is left to the discretion of the data analyst. In this section, some adjustment methods that survey organizations and analysts might consider for the purpose of minimizing linkage non-consent bias are reviewed.

7.4.1 Controlling for Linkage Non-Consent Bias via Statistical Adjustment

A significant advantage of controlling for linkage non-consent bias after the survey data has been collected is the availability of substantial covariate information that can be utilized for adjustment purposes. Because the linkage consent question is typically administered only to respondents during the survey interview, answers to all other survey questions collected during the interview may be used to evaluate whether differences between consenting and non-consenting respondents exist on a range of survey variables, and it is possible to utilize these variables to inform statistical adjustment procedures. The availability of potentially hundreds of survey variables that may be used for adjustment of linkage non-consent bias is a more appealing situation than that of unit nonresponse in large-scale surveys where the number of available auxiliary covariates tends to be small.

7.4.2 Weighting Adjustments

One method that can exploit the large amounts of covariate information collected in survey interviews to statistically control for linkage non-consent bias is a weighting adjustment. Weighting adjustments can be used to compensate for discrepancies between consenters and non-consenters in much the same way that they are used to compensate for discrepancies between survey respondents and nonrespondents. One procedure for creating a weighting adjustment for linkage non-consent involves the estimation of propensity scores from a regression model (e.g. logit, probit), where the dependent variable is the binary linkage consent outcome. The model is then regressed on a selection of independent variables measured from the survey interview. This procedure follows the approach outlined in Section 7.2.2 where regression models are fitted using the survey data to identify variables that are affected by linkage non-consent bias. Extending this approach further, estimated probabilities of consent can then be produced from the model and used to generate weights by taking their inverse. The resulting linkage consent weight produced from this inverse propensity score weighting procedure can then be combined with the survey weight, and adjusted to reproduce the weighted distribution for the target sample. This is the procedure used by the US Health and Retirement Study which produces cross-wave linkage weights for analysis of linked social security records (Health and Retirement Study [HRS] 2017). Large variability in the linkage weights can occur, as was observed in an application of linking National Health Interview Survey to Medicare data (Larsen, Roozeboom, and Schneider 2012). General strategies for reducing the variability of the weights may be considered, including weight trimming (see e.g. Potter 1988; Elliott 2008), which is used to produce the HRS linkage weights. The final combined weight can then be used in subsequent analyses involving the linked survey and administrative data.

Detailed investigations of the effectiveness of weighting adjustments for reducing linkage non-consent are rare. One example is that of Yang, Fricker, and Eltinge (2019) who included the creation of an inverse propensity score weighting adjustment in the U.S. Consumer Expenditure Survey. No administrative data were actually linked to the survey, rather the evaluation assessed the impact of the weights on survey estimates derived from consenting respondents by comparing them to the unweighted estimates and the estimates derived from the full sample of respondents (both consenters and non-consenters). The results were mixed as some of the weighted estimates shifted closer to the (unbiased) full-sample estimates, yielding a reduction in linkage non-consent bias; however, for other estimates, the weights shifted the values further away from the full-sample estimates, yielding an increase in the non-consent bias.

Larsen, Roozeboom, and Schneider (2012) compared propensity score weighting with post-stratification to adjust for linkage bias in an application of linking National Health Interview Survey to Medicare data. They showed that the propensity score procedure produced weights that were substantially more variable than those produced under the post-stratification procedure. Both sets of weights produced slightly different estimates of Body Mass Index and Height, although bias was not assessed for these items. Nonetheless, the authors caution of a potential bias–variance tradeoff when using weights to adjust for record non-linkage.

7.4.3 Imputation

The failure to obtain linkage consent from all survey respondents can be viewed as a missing data problem. Even if large amounts of interview data are collected from survey respondents, only a subset of the data of these respondents (i.e. those who consent) can ultimately proceed to the next stage of being linked to one or more administrative databases. In essence, the missing data component consists of administrative records corresponding to the non-consenting cases (and any other administrative records that could not be linked for other reasons). Put another way, the non-linked administrative variables represent a subset of incompletely observed variables from the combined survey and administrative data set. One way to "fill in" these incompletely observed administrative variables is through imputation. Imputation procedures, which have been extensively studied in the missing data literature (see e.g. Rubin 1987; Little and Rubin 2002), offer a way of adjusting for linkage non-consent bias by replacing the missing administrative values with plausible values generated from a predictive model. The predictive model can be fit using survey variables available for the linkage consenters and non-consenters and linked administrative variables available for the linkage consenters. Thus, the imputation model can account for both data sources in the generation of predictive values. Multiple imputation is recommended to measure the variance due to imputation uncertainty.

There are several advantages of using imputation to adjust for linkage non-consent bias. First, as the imputation procedure is performed on the linked data, it preserves associations between the survey and administrative variables, unlike weighting procedures that do not take into account administrative information. Second, separate imputation models can be fit for each administrative variable in order to account for different variable types and to maximize model fit for a particular variable (see e.g. Raghunathan et al. 2001). This is in contrast to weighting adjustments that apply the same global adjustment to all variables. Another advantage is that the full set of respondents and not just those who consent to linkage can be retained in any subsequent analyses.

A published example of using imputation to adjust for non-linkage bias is reported in Zhang, Parker, and Schenker (2016). In their case study, survey data collected from respondents to the U.S. National Health Interview Survey were linked to administrative Medicare claims data using a deterministic linkage procedure. Due to record linkage refusal not all survey respondents were eligible for linkage, and some could not be linked to the administrative data. The article considers the problem of dealing with missingness in the administrative variables for the non-linked survey respondents. Specifically, the administrative variable of interest in the study is mammography screening status as well as the specific Medicare plan that beneficiaries are enrolled in; both of which are missing for the non-linked cases. The authors apply a multiple imputation procedure to impute the missing values for these variables. The procedure was repeated 10 times resulting in 10 imputed data sets which were analyzed separately and their inferences combined in a way that accounts for the variation between the different sets of imputed values (Rubin 1987). Three different imputation procedures were compared, and all three produced lower rates of mammography screening among women with a particular beneficiary plan compared to the rates derived from the linked data without imputation. Gessendorfer et al. (2018) investigated a related-type of imputation, known as statistical matching, where they identified units in the administrative database that were statistically similar to the non-consenting survey respondents and merged those administrative records to the survey data. They showed that the method was effective in reducing non-consent bias in univariate estimates but worsened the bias in multivariate estimates.

7.5 Summary

Compared with other sources of error in surveys, linkage non-consent bias is understudied and guidelines for measuring and controlling for this source of bias are not widespread. This chapter reviewed a series of methods and strategies for measuring and controlling for linkage non-consent biases in linked survey and administrative data. The techniques covered are relevant for (i) survey organizations which are interested in designing the linkage consent request to optimize the consent rate and reduce the risk of linkage non-consent bias; and (ii) survey organizations and users of the linked data who wish to measure the extent to which linkage non-consent bias exists, and possibly adjust for its occurrence, after the survey data has been collected. Key features of the chapter are summarized below.

7.5.1 Key Points for Measuring Linkage Non-Consent Bias

- Most studies measuring linkage non-consent bias rely on large amounts of information collected during the survey interviews to identify systematic differences between consenters and non-consenters. Utilizing the survey data to identify differences between these two groups is relatively straightforward to perform as it does not require access to the linked administrative data. However, linkage non-consent biases identified in the survey variables may not be indicative of non-consent biases in the linked administrative variables.
- Estimating linkage non-consent bias in the administrative variables is less straightforward as it requires access to the administrative records and identification of which records belong to consenting and non-consenting respondents. Few studies are able to acquire administrative information for non-consenting cases with the exception of survey studies that draw their samples from population registers that can be linked to relevant administrative databases.

7.5.2 Key Points for Controlling for Linkage Non-Consent Bias

- Procedures for controlling for linkage non-consent bias fall under two categories: (i) those that attempt to minimize the risk of consent bias by altering the design and administration of the linkage consent question in ways that potentially make the request more attractive to survey respondents; and (ii) those that attempt to reduce systematic differences between consenters and non-consenters by employing statistical adjustment strategies after the survey data have been collected.

- There are various strategies for optimizing the linkage consent rate in surveys, including (i) placing the linkage consent question at (or near) the beginning of the questionnaire or adjacent to topic-related questions; (ii) framing the linkage consent question in a way that emphasizes the negative consequences of not linking one's data as opposed to the positive benefits of linkage; (iii) employing passive consent procedures rather than active consent procedures (if permitted by legal regulations and ethics review committees); and (iv) in panel surveys, repeating the consent request to respondents who did not consent in an earlier wave. Furthermore, it is advisable to remind respondents of their previous consent decision if they provided consent in an earlier wave, but not if they previously refused.

- Statistical adjustment procedures, including weighting and imputation, have been used in linkage consent studies to adjust for systematic differences between consenters and non-consenters. Both have advantages and disadvantages. Weighting adjustments do not necessarily require access to administrative records and thus can be performed prior to linkage, but their reliance on survey data alone may make them less effective for reducing non-consent bias in the linked administrative variables. Imputation, on the other hand, takes into account data from both survey and administrative sources and preserves relationships between them when generating the predicted values. Imputation also retains all eligible respondents, increasing the case base for any subsequent analyses performed on the linked data. However, unlike weighting adjustments, the linked data are needed in order to perform the imputations.

7.6   Practical Implications for Implementation with Surveys and Censuses

In practice, measuring and controlling for linkage non-consent bias should always begin by reviewing the local legal regulations with respect to data sharing, data linkage, and data protection and consulting with ethics review committees responsible for overseeing the conduct of the study to determine the extent to which strategies for measuring and controlling for linkage non-consent bias are possible for a given study. Meeting with the relevant administrative data custodians to discuss the planned strategies is also recommended.

A major question of practical importance is whether survey organizations should play a more prominent role in addressing the linkage non-consent issue than what is currently being done. Even though survey organizations routinely produce weights that adjust for unit nonresponse, are known to release imputed data products for item nonresponse (see e.g. Schenker et al. 2006), and offer guidelines for data users on how to incorporate these features into their analysis, none of these approaches is routinely used for addressing linkage non-consent.

One advantage of addressing the non-consent issue at the level of the survey organization is that data users would then be able to use the recommended adjustments and/or imputed data products, instead of relying on their own ad-hoc devices of accounting for linkage non-consent or, worse yet, not applying any devices and implicitly assuming the absence of such bias. This would ease the burden on data users who may not be skilled in the application of weighting and/or imputation procedures. Moreover, addressing linkage non-consent at the survey organization level would increase the consistency and reproducibility of inferences derived from the linked data.

Of course, any additional efforts implemented by the survey organization come with added costs. The decision of whether the potential benefits justify the added costs is an important discussion to be had among stakeholders, including survey organizations, survey sponsors, and data users.

7.7 Exercises

1 Describe at least two ways in which administration of the linkage consent question could be designed in order to maximize the consent rate?

2 Does a linkage consent rate of 70% guarantee a smaller linkage non-consent bias than a linkage consent rate of 50%?

3 If the linkage consent rate is 65% and the mean of some estimate is 6.5 for consenting respondents and 8.0 for non-consenting respondents, what is the estimate of linkage non-consent bias?

4 If the linkage consent rate is 80% and the mean of some estimate is 5.0 for consenting respondents and 15.0 for non-consenting respondents, what is the estimate of linkage non-consent bias?

5 Describe at least one method of adjusting for linkage non-consent bias that can be applied after the survey data have been collected?

References

Al Baghal, T. (2015). Obtaining data linkage consent for children: factors influencing outcomes and potential biases. *International Journal of Social Research Methodology* 19 (6): 623–643.

Banks, J., Lessof, C., Taylor, R., Cox, K., and Philo, D. (2005). Linking survey and administrative data in the English longitudinal study of ageing. *Presented at the Meeting on Linking Survey and Administrative Data and Statistical Disclosure Control*, Royal Statistical Society, London, May.

Biemer, P.P. (2010). Total survey error: design, implementation, and evaluation. *Public Opinion Quarterly* 74 (5): 817–848.

Burger, J.M. (1999). The foot-in-the-door compliance procedure: a multiple-process analysis and review. *Personality and Social Psychology Review* 3 (4): 303–325.

Dahlhamer, J. M., and Cox, C. S. (2007). Respondent consent to link survey data with administrative records: results from a Split-Ballot Field Test with the 2007 National Health Interview Survey. *Proceedings of the Federal Committee on Statistical Methodology Research Conference.* https://nces.ed.gov/FCSM/pdf/2007FCSM_ Dahlhamer-IV-B.pdf

Das, M. and Couper, M.P. (2014). Optimizing opt-out consent for record linkage. *Journal of Official Statistics* 30 (3): 479–497.

Eckman, S. and Haas, G.-C. (2017). Does granting linkage consent in the beginning of the questionnaire affect data quality? *Journal of Survey Statistics and Methodology* 5 (4): 535–551.

Ellickson, P.L. and Hawes, J.A. (1989). An assessment of active versus passive methods for obtaining parental consent. *Evaluation Review* 13 (1): 45–55.

Elliott, M.R. (2008). Model averaging methods for weight trimming. *Journal of Official Statistics* 24 (4): 517–540.

Fulton, J. A. (2012). Respondent consent to use administrative data. University of Maryland, Joint Program in Survey Methodology, PhD dissertation; College Park, MD: 2012. https://drum.lib.umd.edu/bitstream/handle/1903/13601/Fulton_umd_ 0117E_13741.pdf;sequence=1

Gessendorfer, J., Beste, J., Drechsler, J., and Sakshaug, J.W. (2018). Statistical Matching as a Supplement to Record Linkage: A Valuable Method to Tackle Nonconsent Bias?. *Journal of Official Statistics*, 34(4): 909–933.

Groves, R.M., Fowler, F.J., Couper, M.P. et al. (2009). *Survey Methodology*, 2e. New York: Wiley.

Haider, S. and Solon, G. (2000). *Non Random Selection in the HRS Social Security Earnings Sample.* Santa Monica, CA: RAND Corporation. https://www.rand.org/ pubs/drafts/DRU2254.html.

Health and Retirement Study (HRS); (2017). Restricted data social security weights: data description and usage. Technical Report, Institute for Social Research, University of Michigan. https://hrs.isr.umich.edu/sites/default/files/restricted_ data_docs/SSWgtsDDv31.pdf

Hollmann, C.M. and McNamara, J.R. (1999). Considerations in the use of active and passive parental consent procedures. *The Journal of Psychology: Interdisciplinary and Applied* 133 (2): 141–156.

Jenkins, S.P., Cappellari, L., Lynn, P. et al. (2006). Patterns of consent: evidence from a general household survey. *Journal of the Royal Statistical Society: Series A (Statistics in Society)* 169 (4): 701–722.

Junghans, C., Feder, G., Hemingway, H. et al. (2005). Recruiting patients to medical research: double blind randomised trial of 'opt-in' versus 'opt-out' strategies. *BMJ* 331 (7522): 940.

Kahneman, D. and Tversky, A. (1979). Prospect theory: an analysis of decision under risk. *Econometrica* 47 (2): 263–291.

Kahneman, D. and Tversky, A. (1984). Choices, values, and frames. *American Psychologist* 39 (4): 341–350.

Knies, G. and Burton, J. (2014). Analysis of four studies in a comparative framework reveals: health linkage consent rates on British cohort studies higher than on UK Household Panel Surveys. *BMC Medical Research Methodology* 14: 125.

Kreuter, F., Sakshaug, J.W., and Tourangeau, R. (2016). The framing of the record linkage consent question. *International Journal of Public Opinion Research* 28 (1): 142–152.

Larsen, M.D., Roozeboom, M., and Schneider, K. (2012). Nonresponse adjustment methodology for NHIS-medicare linked data. In: *Proceedings of the Joint Statistical Meetings, Section on Survey Research Methods*, 4078–4087. American Statistical Association.

Little, R.J.A. and Rubin, D.B. (2002). *Statistical Analysis with Missing Data*. Wiley.

Mostafa, T. (2016). Variation within households in consent to Link Survey Data to administrative records: evidence from the UK Millennium Cohort Study. *International Journal of Social Research Methodology* 19 (3): 355–375.

Pascale, J., (2011). Requesting Consent to Link Survey Data to Administrative Records: Results from a Split-Ballot Experiment in the Survey of Health Insurance and Program Participation (SHIPP). *Center for Survey Measurement, Research and Methodology Directorate (Survey Methodology #2011-03)*. U.S. Census Bureau. Available online at https://www.census.gov/srd/papers/pdf/ssm2011-03.pdf

Potter, F. (1988). Survey of procedures to control extreme sampling weights. In: *Proceedings of the Survey Research Methods Section of the American Statistical Association*, 453–458.

Raghunathan, T.E., Lepkowski, J.M., Van Hoewyk, J., and Solenberger, P. (2001). A multivariate technique for multiply imputing missing values using a sequence of regression models. *Survey Methodology* 27 (1): 85–95.

Range, L.M., Embry, T., and MacLeod, T. (2001). Active and passive consent: a comparison of actual research with children. *Ethical Human Sciences and Services* 3 (1): 23–31.

Rubin, D.B. (1987). *Multiple Imputation for Nonresponse in Surveys*. Wiley.

Sakshaug, J.W. and Eckman, S. (2017). Are survey nonrespondents willing to provide consent to use administrative records? Evidence from a nonresponse follow-up survey in Germany. *Public Opinion Quarterly* 81 (2): 495–522.

Sakshaug, J.W. and Huber, M. (2016). An evaluation of panel nonresponse and linkage consent bias in a survey of employees in Germany. *Journal of Survey Statistics and Methodology* 4 (1): 71–93.

Sakshaug, J.W. and Kreuter, F. (2012). Assessing the magnitude of non-consent biases in linked survey and administrative data. *Survey Research Methods* 6 (2): 113–122.

Sakshaug, J.W. and Kreuter, F. (2014). The effect of benefit wording on consent to link survey and administrative records in a web survey. *Public Opinion Quarterly* 78 (1): 166–176.

Sakshaug, J.W., Couper, M.P., Ofstedal, M.B., and Weir, D.R. (2012). Linking survey and administrative records: mechanisms of consent. *Sociological Methods & Research* 41 (4): 535–569.

Sakshaug, J.W., Tutz, V., and Kreuter, F. (2013). Placement, wording, and interviewers: identifying correlates of consent to link survey and administrative data. *Survey Research Methods* 7 (2): 133–144.

Sakshaug, J.W., Hülle, S., Schmucker, A., and Liebig, S. (2017). Exploring the effects of interviewer- and self-administered survey modes on record linkage consent rates and bias. *Survey Research Methods* 11 (2): 171–188.

Sakshaug, J.W., Schmucker, A., Kreuter, F., Couper, M.P., and Singer, E. (2019). The effect of framing and placement on linkage consent. *Public opinion quarterly*, 83(S1): 289–308.

Sakshaug, J.W., and Vicari, B.J. (2018). Obtaining record linkage consent from establishments: the impact of question placement on consent rates and bias. *Journal of Survey Statistics and Methodology*, 6(1): 46–71.

Sakshaug J.W., Wolter S. & Kreuter F. (2015), Obtaining Record Linkage Consent: Results from a Wording Experiment in Germany. Survey Insights: Methods from the Field. Retrieved from https://doi.org/10.13094/SMIF-2015-00012

Sala, E., Burton, J., and Knies, G. (2012). Correlates of obtaining informed consent to data linkage: respondent, interview, and interviewer characteristics. *Sociological Methods & Research* 41 (3): 414–439.

Sala, E., Knies, G., and Burton, J. (2014). Propensity to consent to data linkage: experimental evidence on the role of three survey design features in a UK Longitudinal Panel. *International Journal of Social Research Methodology* 17 (5): 455–473.

Schenker, N., Raghunathan, T.E., Chiu, P.-L. et al. (2006). Multiple imputation of missing income data in the National Health Interview Survey. *Journal of the American Statistical Association* 101 (475): 924–933.

da Silva, M.E.M., Coeli, C.M., Ventura, M. et al. (2012). Informed consent for record linkage: a systematic review. *Journal of Medical Ethics* 38 (10): 639–642.

Tenth Book of the German Social Code (2018). Social Administrative Procedures and Social Privacy. Section 75: Erhebung, Verarbeitung oder Nutzung von Sozialdaten im Auftrag. http://www.gesetze-im-internet.de/sgb_10/BJNR114690980.html

The HIPAA Privacy Act (n.d.), 45 C.F.R. 164.501

U.S. Privacy Act of 1974, PL 93-579, December 31, 1974; 5 U.S.C.-552a.

Yang, D., Fricker, S., and Eltinge, J. (2019). Methodsfor exploratory assessment of consent-to-link in a household survey. *Journal of Survey Statistics and Methodology*. 7: 118–155.

Young, A.F., Dobson, A.J., and Byles, J.E. (2001). Health services research using linked records: who consents and what is the gain? *Australian and New Zealand Journal of Public Health* 25 (5): 417–420.

Zhang, G., Parker, J.D., and Schenker, N. (2016). Multiple imputation for missingness due to nonlinkage and program characteristics: a case study of the National Health Interview Survey linked to medicare claims. *Journal of Survey Statistics and Methodology* 4 (3): 319–338, https://doi.org/10.1093/jssam/smw002.

Part III

Use of Administrative Records in Surveys

8

A Register-Based Census: The Swedish Experience

Martin Axelson, Anders Holmberg, Ingegerd Jansson and Sara Westling

Statistics Sweden, Box 24300, Stockholm SE-104 51, Sweden

8.1 Introduction

During 2011, national censuses were carried out within all countries of the European Union (EU). Extensive measures were taken to ensure comparability between member states, and tables as well as metadata and quality descriptions were eventually made available at EU level.

In Sweden, the 2011 census was, for the first time, fully based on registers. In order to facilitate the production of census statistics, the existing system of registers for production of household and housing statistics was further developed and evaluated, and is now an important part of the production of official statistics.

In this chapter, we set out to describe important circumstances and methodological issues that had to be considered during the development. The chapter will give the reader a picture of methodological and practical concerns with the use of administrative data for censuses. In our discussion we address two largely coinciding goals that were pursued; the first and overarching goal was to ensure the census output and its quality; the second was to seize the opportunity to create a sustainable production environment and a register-based system for household and housing statistics of sufficient quality and at a relatively low cost. Although the two goals overlap there are important differences that did affect underway prioritizations. The pursuit of the census goal was guided by a fixed deadline and a predefined statistical output, and the need to find solutions for first-time issues that occurred while developing the system and simultaneously using the registers. The goal of an improved register-based system, on the other hand, was inexplicit and had a long-term focus. Such a system should provide a capability that satisfies flexible sets of output statistics and over time it can benefit from continuous improvements in a reusable register infrastructure.

Administrative Records for Survey Methodology, First Edition.
Edited by Asaph Young Chun, Michael D. Larsen, Gabriele Durrant, and Jerome P. Reiter.

Already at an early stage in the development, budget, and organizational priorities made it clear that the census goal was to be the driving factor. Other benefits were regarded as desirable, but not certain, spin-off effects. In a sense the census output was a potential but not deliberate stepping stone towards a coherent statistical system for household and housing statistics. This meant that several options for building the system had to be discarded (or postponed) because of constraints in time and budget. On the other hand, the progress of the household and housing statistics system gained from the census activities would have been slower without them. Today, Statistics Sweden has the data, the methodology, and the infrastructure in place for the register-based parts of such a system. In Sections 8.2–8.6 we will describe the experiences and the methodological efforts that we believe were the most important for this development, in order to achieve the census goal as well as putting the pieces of today's system in place.

We set the scene by describing how administrative data have been used for statistical purposes and how this has bearing on the development and solution for the Swedish 2011 census. In this context we explain the basis of the Swedish register-based census with its requirements on output and the implications of the processes underlying the construction of the necessary new register sources. After this introductory background, we describe and motivate the methodological steps we took to ensure, to measure, and to understand the quality of the census delivery. The main step in this undertaking was an evaluation of important household information. Since this is the main source of information about quality of the census as well as the register part of a household and housing statistics system, we describe this in detail. We conclude with summarizing remarks of our experiences.

8.2 Background

Sweden has a long tradition of keeping track of the population demographics. It dates back to 1688, when the church was first required to keep records of births, deaths, and households in all parishes. In 1749, Tabellverket was published, the first continuing census in the world. Twenty years later, Tabellkommissionen was founded, the first statistical agency in the world and the predecessor of Statistics Sweden (Jorner 2008).

As illustrated in the report Register-based statistics in the Nordic countries (UNECE 2007), administrative registers have long been utilized for the production of official statistics. Registers of the population, businesses, buildings, as well as subject matter registers, have for many years formed the basis of the official statistics production. However, while Denmark (since 1977), Finland (since 1980), and Norway (since 2001) have had a register of dwellings in place for many years, it did not exist in Sweden until 2011.

The Swedish system of statistical registers relies on three base registers kept by Statistics Sweden: the Total Population Register (TPR), the Real Property Register (RPR), and the Business Register (BR). The base registers are linked to various subject-matter registers and rely on the unique identification of physical persons, businesses, and dwellings. The comprehensive identification of physical persons, businesses, buildings, and dwellings is an essential part of a register-based system.

Unique identification of persons and businesses has been in use in Sweden for many years for various administrative purposes. The personal identification number was introduced in 1947 and is assigned to every person registered in Sweden. The number is unique and carries information about date of birth, gender, and (for persons born before 1990) region of birth. The identification number remains unchanged throughout the lifetime of a person. It is extensively used for administrative purposes at the state and local level (for example by the tax authorities, for different kinds of social services, and in all personal contact with health care institutions), including private economy (i.e. bank accounts, insurance, etc.) and dealings with private businesses.

Between 1960 and 1990, traditional censuses were conducted in Sweden every fifth year, with data being collected by a mandatory self-administered mail-out mail-back questionnaire. In 1995, the Swedish parliament decided that the next census should be completely register-based. For several reasons, among them political concerns of privacy, the necessary legal regulation was not in place until more than ten years later. At that time, a fully register-based census was not feasible due to the lack of a unique link at the dwelling level between the TPR and the RPR. In 2007, the Swedish parliament passed a bill on the creation of a dwelling register (DR) and population registration on dwellings. This included both multi-dwelling buildings and single houses. Together with population registration on dwellings, a dwelling identification key was the missing link in a complete Swedish system of registers and thus a fully register-based census was not possible before 2011.

8.3 Census 2011

European census programs were developed already in 1980, 1990, and 2001, but for the 2011 census, the work to harmonize European census outputs was done on a larger scale. The objective was to make comparable, detailed data for all European countries publicly available in a user-friendly way.

The EU legislation on population and housing censuses respects that the EU member states have developed different methods which they consider to be best suited to the administrative practices and traditions of their country. The Regulation of the European Parliament and of the Council on population and housing censuses (Regulation (EC) No 763/2008), which regulated the 2011 census, is concerned with output rather than input harmonization.

To achieve comparability between member states and to assess quality, certain conditions had to be met. This was specified in three additional regulations.

- Regulation (EC) No 1201/2009 contains definitions and technical specifications for the census topics and their breakdowns. The census topics include geographic, demographic, economic, and educational characteristics of persons, as well as household, family, and housing characteristics.
- Regulation (EU) No 519/2010 specifies the data output, including the combinations of variables in detailed multidimensional tables ("hypercubes"), specification of missing data, definitions, and metadata on the data sources and on methodological issues.
- Regulation (EU) No 1151/2010 concerns transmission of a quality report containing a description of the data sources and of the quality of the results. In this context, member states described the extent to which the data sources and the method chosen were relevant to the major features of the census. The regulation also specifies the technical format for transmission of data.

Results from all (at that time) 27 member states were made publicly available in March 2014 through a common interface using a hub solution,[1] developed, and maintained by Eurostat, the statistical office of the EU.[2] Documentation of metadata and the quality of data is also available. Statistics Sweden did not publish any other statistics based on the Swedish census data, but microdata are available for research, upon approval of Statistics Sweden.

A total of 60 tables for each country are published at the Eurostat web site, displaying the total population of persons, households, families, and dwellings. The geographical breakdowns are defined according to the nomenclature of territorial units for statistics (NUTS) as defined by Eurostat, and there is a strong focus on the NUTS 2 level (areas with between 800 000 and 3 million inhabitants). Even though only selected breakdowns of variables are combined, some of the tables are very large and detailed.

At the time, Eurostat decided that each country should make its own decisions on feasible measures for risk assessment and methods to protect persons and households from disclosure. Data that were delivered to Eurostat were assumed to be properly protected and no measures for statistical disclosure control were taken by Eurostat before publication. This freedom of choice probably affects the comparability between countries, but it was deemed necessary due to country specific regulations on confidentiality.

---

1 https://ec.europa.eu/CensusHub2/query.do?step=selectHyperCube&qhc=false
2 http://ec.europa.eu/eurostat/about/overview

8.4   A Register-Based Census

Although well experienced in register-based statistics and the use of administrative data for statistical purposes, Statistics Sweden faced new challenges with a completely register-based census. Register-based statistics differ from traditional surveys in several respects. A register-based survey utilizes a register created and maintained for administrative purposes. In a register-based system, a statistical register must first be created where the administrative data are edited and transformed to best meet multiple statistical purposes and the aims of multiple surveys.

The population register is of crucial importance for a register-based census. It implicitly defines the census population, i.e. the census population is defined as all persons included in the register. Households and families are formed using only information from registers. Of critical importance is the dwelling identification key and that all persons are registered with such a key, linking persons to dwellings and thus making it possible to form register-based households and families.

In this context, missing dwelling identification keys was considered as one of the most important issues for the census quality. Without the key, no linking between persons and dwellings is possible. If the key is missing for any person in the household, this will cause error in the household composition. If the key is missing for all persons in a household, this will result in missing households in the census population. The process to assign dwelling identification keys is described in Section 8.4.3 and some possible causes for errors are discussed in Section 8.7.

8.4.1   Registers at Statistics Sweden

Statistics Sweden maintains a number of statistical registers based on administrative information, either created using administrative data from external sources, or with Statistics Sweden as the responsible holder of the register. As previously indicated, the TPR, the BR, and the RPR are referred to as base registers since they make up the core of the register system. The base registers serve multiple purposes; they are not only used as sources for statistics, but also as sampling frames and as auxiliary information, for example for imputation and calibration.

The TPR is based on administrative data from the Swedish Tax Agency and contains a large amount of information, for example identifying information of persons (personal identification number, name, and address), location of registered address (region, municipality, and parish), and demographic information (age, sex, marital status, citizenship, deaths, births, etc.), as well as information on kinship and immigration, etc.

The BR is kept and maintained by Statistics Sweden, following a government decision in 1963. The register contains all businesses, government offices, and organizations in Sweden, as well as their workplaces. The main sources are the

Swedish Tax Agency, the Swedish Companies Registration Office, and the postal service for change of address. It contains information about legal units, such as their identifying information and geographic location, number of employees, and NACE[3]-code.

The RPR is a longitudinal register that contains information at the microdata-level about all properties, buildings, and addresses in Sweden. The register is based on administrative data from the Swedish land surveying authority, which by law is responsible for maintaining a register containing information on addresses, buildings, and property tax assessment.

8.4.2 Facilitating a System of Registers

The system of statistical registers relies on the three base registers. The base registers are linked to various subject matter registers, such as registers of employment, occupation, and education.

The system is dependent on the possibility to uniquely link information to objects. Figure 8.1 is a schematic description of the registers and the links between them. The personal identification number links a person to data on for example occupation and education. The personal identification number and the identification number of businesses provide the prerequisites for creating

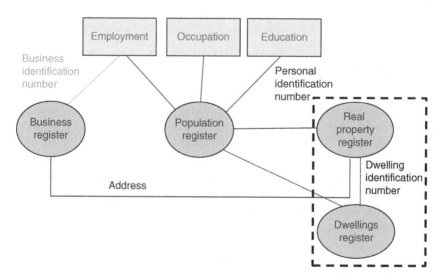

Figure 8.1 Registers and links in the system of registers.

---

3 NACE is an acronym for "Nomenclature statistique des activités économiques dans la Communauté européenne," which translates to "Statistical classification of economic activities in the European Community."

statistics on employment. Businesses and persons are linked to the RPR by the address of the house or building where they are located or reside. We refer to the dwelling information as a separate register, but the dwelling information is in practice incorporated with the RPR so that complete linking between persons and properties is facilitated.

For those living in detached or semi-detached houses, the address is unique within the municipality. Before 2011, for multi-dwelling buildings, the address would give information about the entrance and possibly which floor, but not about the specific apartment. Thus, since cohabiting without being married is common in Sweden, the available register information on marital status and parent–child relations was not sufficient for forming households based on administrative data only. However, with the introduction of the additional register of dwellings and the population registration on dwellings, a unique identification key was given to all dwellings (apartments and houses), opening for new opportunities to create register-based statistics on households and housing conditions.

8.4.3   Introducing a Dwelling Identification Key

Constructing a register of dwellings is a cumbersome procedure. The process used in Sweden differed between two types of dwellings; those in one/two dwelling and semi-detached houses, and those in multi-dwelling buildings. For the first type of dwellings, updated addresses in the RPR were used to construct a unique dwelling identification key, automatically linked to all persons registered at the same dwelling address in the TPR. For multi-dwelling buildings a more cumbersome and costly procedure was used. The process is briefly outlined in the following steps:

1) The local administration checked and updated all addresses within a municipality and thus issued formal addresses, for example to buildings that had only traditional, informal names, or to buildings sharing the same address.
2) The land surveying authority collected information from land property owners about dwellings on the property.
3) The property owner labeled apartments in a block of apartments and submitted the dwelling numbers to the land surveying authority where the administrative register of addresses and dwellings is kept. The labeling was done according to a predefined protocol in order to ensure that the dwelling numbers are unique within a multi-dwelling building and that the apartments are numbered following the same principle in all buildings. The dwelling numbers are four-digit numbers and carry information about the floor and the ordering of apartments on a specific floor. As an example, the dwelling number 1201 means that the dwelling is the first apartment to the left of the stairs on the second floor. The dwelling number is part of the unique dwelling identification key that is used in the RPR.

4) The property owner informed the residents of their dwelling numbers.
5) For each municipality, a formal decision that a dwelling register was established for the municipality had to be taken by the Swedish government.
6) Using a mailed questionnaire to every adult in the country, the Tax Agency asked about their address of residence, including the dwelling number. Thus, persons living in apartments were expected to know their dwelling number and provide this information to the Tax Agency, upon which the administrative population register was updated.

The population registration on addresses and dwelling identification started in September 2010 and the last forms where sent out by the Tax Agency in March 2011. All steps were subject to error, and particularly the dwelling number was expected as prone to being missing or incorrect. For further reference, see Axelson et al. (2010), Hedlin et al. (2011), Hedlin, Holmberg, and Jansson (2011), or Andersson et al. (2013).

The data collection resulted in 96.6% of the Swedish population being registered with a dwelling identification key. The goal was to reach 95% in each of Sweden's 290 municipalities, but 17 municipalities did not reach the goal. The largest rate of missing registration on dwellings for a single municipality was 14%.

8.4.4 The Census Household and Dwelling Populations

In a register-based census, the population register defines the census population, but persons with missing dwelling information were still present in the TPR. On the reference date of the census (31 December 2011), about 320 000 persons were not registered with a dwelling identification key and could not be connected to a dwelling. These are counted as missing, even though it is possible that keys were missing because persons were nonresidents on Census day.

Without a dwelling identification key, there were no linking possibilities between the DR and the TPR, and whether a person belonged to a household or a family could not be derived from the registers. The creation of households and families in a register-based census is entirely based on information from registers. Persons registered at the same dwelling form a dwelling household (not necessarily the same as a household in terms of a housekeeping unit). From the information in the registers, household, and family variables are derived, such as size of family or household, and type of family or household. This requires some information in addition to where persons are registered (i.e. legal marital status or child and parent relations), in combination with some basic rules (i.e. there must be at least two people to make up a family, two married couples living together count as two families, children with divorced parents can only be counted as members of one household, etc.). There can be more than one family in a household, but never more than one household in a family.

After careful consideration, it was decided that no imputation of missing dwelling identification keys should be carried out. One important reason was that the fraction of missing keys was considered as low, less than 3.5%, and it was not anticipated that this would seriously affect the quality of the census results. In addition, there was no useful auxiliary information from other sources to aid imputation (recall that the previous Swedish census was carried out in 1990). The census regulation does not require imputation, but leaves the decision to the member states.

Matching persons without dwelling keys to apartments that appeared to be unoccupied would have been another option. However, 48% of the persons with missing dwelling keys had addresses in buildings with no apartments according to the RPR, implying that the numbering of apartments failed for the whole building. The remaining 52% were registered at addresses where the buildings had at least one registered apartment. The registered apartments might be used for matching, based on the assumption that the relationship between size of apartment and size of household is similar for persons with and without registered dwelling keys, but this would have required a large effort with doubtful quality of the results. Imputation at macro level was not considered, it would not have been worth the effort to construct an imputation system that would render consistent results for all 60 hypercubes.

Considering the above, and the fact that compensating for missing data was not required by Eurostat, it was decided that missing data would be documented and reported, but not adjusted for.

The total number of dwellings in the census was just above five million. Of these, about 900 000 did not have any registered inhabitants. Note that this includes holiday houses, and that one person (by law) can only be registered at one dwelling. However, there was also an unknown number of people living in dwellings where they were not registered, for example young people still registered with their parents but living elsewhere or people living in group quarters but not being registered there. About 40% of the group quarters had no registered inhabitants. Thus, the census overestimated the number of unoccupied dwellings, and underestimated the number of people living in group quarters.

This tendency was confirmed when data from a survey on unoccupied dwellings in multi-dwelling buildings were matched to the register. The survey targeted rental units on the open market and had two parts, a total survey of municipal housing companies and a sample survey of private companies. The survey and the register had slightly different reference dates, 1 September 2011 and 31 December 2011, respectively, but the comparison still showed that the register overestimated the number of unoccupied dwellings.

8.5 Evaluation of the Census

8.5.1 Introduction

Assessing the quality of a register-based survey poses problems that differ from those encountered when measuring quality of sample survey data. Typically, variance-based indicators of quality cannot be used, and thus other types of indicators have to be defined and measured.

A main difference between the two types of surveys is that with a register-based survey, data collection is primarily devised for administrative purposes which may differ from the statistical purposes. Hence, effects that the statistical agency should be aware of might include inadequate definitions of variables and reference periods, or lack of relevance and validity. Data transformed from administrative to statistical purposes should meet the aims of multiple surveys and be fit for many different purposes, such as being the primary source of data collection and ad hoc aggregates, serve as a sample frame, or be used as auxiliary information in estimation.

As yet, there is no unifying theory for administrative data as a source of data collection, but there is a clear notion of the necessity of such theory. Some references are Wallgren and Wallgren (2007), Zhang (2011, 2012), Reid, Zabala, and Holmberg (2017), and Hand (2018).

For the Swedish census, the quality of the DR was of utmost importance since that was the only part of the system of registers that had not been in use previously. Further, the DR is a prerequisite for future household and housing statistics. In order to assess the quality, a study comparing the register data with household data was carried out. Such validating studies were done for all Swedish censuses between 1960 and 1990, and all of the studies followed the same methodology (Andersson and Holmberg 2011). The idea is to obtain additional observations from a sample survey and measure to what extent answers are consistent with the register information at the reference time of the census. Based on the analysis of data and subsequent processing (reconciliation), some quality measures such as the net and gross errors of categorical register variables can be estimated.

The evaluation of the 2011 census differed from the previous evaluations to some extent. The evaluation of the 1990 census was done by telephone interviews and included the variables household type and size, but also additional variables on occupation, education, and some dwelling characteristics.

An evaluation survey may not only aim for quality measures of the data in the registers used for the census (see for example Office for National Statistics 2012). To a certain extent, the design of the evaluation survey can be tailored to pick up information which can be used to improve census statistics. Groups of people with characteristics known to be partially missing in registers, or where a lower quality in the register is suspected, can be targeted in the sampling design. If the evaluation survey is successful, data about these groups can be used as auxiliary information in census calculations. The sampling design in the Swedish census evaluation was targeted for this, but as it turned out, this information was not used in the census calculations. However, important knowledge about these groups was gained.

The evaluation survey was designed for several possible studies but some of them were not carried out due to practical reasons. Limited time was one reason. The evaluation survey had to be launched close to the completion dates of the register sources, and the results had to be finalized in time for census publication.

Due to cost restrictions and concerns about response burden, the length of the survey questionnaire and the data collection had to be limited. Of the census variables, only some carefully chosen vital register variables were included in the survey. The evaluation focused mainly on the accuracy of the registration on dwelling identification number. In the end, the analyses were limited to two variables, household size and type.

However, the questionnaire was designed to facilitate a number of studies which were considered but not carried out at the time. All of them are more or less related to the introduction of the DR and population registration on dwellings. Some examples:

- In the derivation of relations within households, rules for forming cohabiting couples were applied. The rules were determined using information from the Labor Force Survey, but their effect and accuracy in the population register were largely unknown.
- People without household affiliation according to the register were of special interest. There may be a number of reasons for missing household affiliation, and analyzing this group further would have been useful to understand the consequences on the quality, at household level as well as aggregated levels. There were concerns including for example if people with missing household information were due to overcoverage or if they belonged to a household, and if there were any systematic patterns in type and size of these households.
- A register household includes all the persons registered at the same dwelling. In most cases, where people are registered as living (de jure living) coincides with where they have their de facto living. Registered living differs from the census definition (according to Eurostat) of actual living, so it would have been of interest to investigate if there were discrepancies between the two concepts. Systematic differences between groups were suspected, and perhaps also between geographical regions.

Although no studies of the above kind have yet been carried out, the data still exist and allows for studies to be performed, e.g. as part of the preparation for the next census.

Table 8.1 Variables used for stratification.

Variables	Categories
Dwelling id-key exists in TPR	Yes
	No
Municipality of residence (combined with dwelling id-key exists in TPR)	Stockholm region (Yes)
	Göteborg region (Yes or No)
	Malmö region (Yes or No)
	Municipality with more than 70 000 inhabitants excl. Stockholm, Göteborg, and Malmö (Yes or No)
	Other municipalities (Yes or No)
	Stockholm county area (No)
	Norrtälje (No)
Age class	18–34 years
	35–74 years
	75 years or older
Type of dwelling	(semi)-detached house
	Apartment block
	Communal establishment, not dwelling house, or missing value
Number of families in dwelling according to TPR	1–2
	3 or more
	No dwelling id-key

8.5.2 Evaluating Household Size and Type

The study focused on how complete and how accurate the population registration on dwelling identification key was, and how this might influence household type and household size (the design of the study is also described in Axelson et al. 2012). It is important to keep in mind that the TPR shows where a person has his or her registered address, which is not necessarily the same as where the person actually lives. The survey asked for actual living, which was considered as the only reasonable thing to ask for. In this context, accuracy is thus the extent to which the registered address is equivalent to the actual place of living.

8.5.2.1 Sampling Design

A stratified random sample of 15 000 persons was drawn from the TPR. A total of 108 strata were defined by five variables, with categories as defined in Table 8.1.

The variable Municipality of residence (combined with dwelling id-key exists in TPR) is essentially the size of the municipalities (the three largest cities in Sweden, other fairly large municipalities, and small municipalities). The combination with the existence of a dwelling identification key is necessary because the Stockholm area had a high proportion of missing dwelling keys, and in particular the municipality of Norrtälje was problematic in this respect.

The number of families in dwelling according to TPR defines categories where there is likely to be false values in the register.

The sample was allocated to strata by a two-step procedure. The first step was to allocate approximately 1/3 of the sample to the group with no dwelling key in the TPR. The TPR still contained valid address information and telephone numbers, even if the dwelling identification key was missing, so that these persons could be contacted. The motive to oversample the persons with missing dwelling keys was to collect more information about the characteristics of this group. The group was of particular interest since it was suspected that its members had missing values, both in the register-based statistics and if they were selected for sample surveys. In the second step, a sequential scheme with proportional allocation using both minimum and maximum constraints in each stratum was used in each of the two groups of step one.

8.5.2.2 Data Collection

Data collection started in January and ended in May 2012. A sequential mixed-mode approach with the order web-paper-CATI was used for the data collection. Respondents were given the opportunity to respond online or by a mailed questionnaire. Interviews, for follow-up as well as for reconciliation, were conducted by telephone.

The respondents were asked to confirm whether they were living at the address they were registered at on 31 December 2011 (the reference day of the census). If the address given was incorrect, they were asked for a correct address. Other questions concerned if the dwelling was owned or rented, how many dwellings there were at the same address, and how many other persons were living at the same address on the reference day. For all others living in the same dwelling, name, sex, year of birth, and whether the person in question was living together with parent or spouse/partner was asked for.

The total weighted response rate was 65%. In all strata, the response rate was at least 40%. As anticipated, nonresponse was higher among those who were not registered with a dwelling identification key in the TPR. Nonresponse was also higher among younger people (18–34 years). As can be expected, nonresponse was higher when there was more than one family registered at the same dwelling. This was probably an indication of a higher amount of incorrect information in the register for those records.

8.5.2.3 Reconciliation

When quality is to be assessed by means of an evaluation study, an important question (almost of a philosophical nature) is what to view as correct information, if anything? Another issue that has to be handled is missing data in the evaluation study. The evaluation of the register has itself to be evaluated. Which data are correct; the sample data, the register data, or neither data source? Discrepancies can differ in magnitude; how large a discrepancy can be accepted before it should be viewed as a serious flaw to data? When discrepancies are serious enough, there must be methods to investigate and, if possible, correct them.

In the Census 2011 evaluation study, if household size or household type differed between survey and register, there was a recontact by telephone in order to establish the true value. Almost 3000 respondents were contacted and 85% of them agreed to confirm the answers they gave in the questionnaire in the initial data collection of the evaluation study. The re-interview focused on household size and type of household. In 65% of the cases, the initial data collection gave the true value, in 25% the register gave the true value, and in the remaining 10% of the cases neither the initial data collection nor the register was correct.

8.5.2.4 Results

The following are examples of questions that were of basic interest for evaluation:

- The proportion of incorrect or partially incorrect addresses. In particular, the proportion of false dwelling identification keys was of interest.
- The proportions of persons or households renting or owning their dwelling
- Estimated deviations from the register in the number of household members in a dwelling
- Estimated deviations from the register in household composition

Household composition was of primary interest since household was a target unit for the census and for future production of official statistics where a clear picture of the number and types of households is vital. The survey data were compared with the register data for estimates of net and gross error.

For a categorical register variable Y, both overall gross error and gross errors of the respective categories can be computed. It is derived by comparing the sample data and the register data in a two-way $J*J$-table, where J is the number of categories of Y. With the observations from the survey in the rows, and those of the register in the columns, the off-diagonal proportion in the two-way table equals the proportion of misclassified objects in the register, that is, the overall gross error of Y. The proportion of gross error in a category is given by $GE_j = (N_j + N_{j.} - 2N_{jj})/N_j$ where N_{jj} is the number of objects of category j in both the survey and the register. N_j and $N_{j.}$ indicate summation over rows and columns, respectively. The notation is illustrated in Figure 8.2.

Category according to survey	Category according to register					
	1	...	j	...	J	
1						
...						
j			N_{jj}			$N_{j.}$
...						
J						
			$N_{.j}$			

Figure 8.2 Notation for a J^*J-table.

Gross errors reflect a particular aspect of reliability that does not concern the quality of estimates from the register data, but is of interest for reliability in general. The size of the gross error of the register variables is relevant when data are used in different kinds of analysis, for example for forecasts and projections, or to study changes with flow models. Large gross errors may destroy the possibility of using the data for analysis and it may also be a sign of poor measurement quality. Users of the registers should take care when aggregates are calculated for subpopulations or when the information is used to construct frames and stratification in sample surveys. The net error proportion is given by $NE_j = (N_{.j} - N_{j.})/N_{.j}$. It gives information about the error in a category when the register is used for aggregates.

Table 8.2 shows some results for household size. The number of correctly classified households is the number of households where the register and the evaluation study agree. Gross error is the number of households wrongly included in a category plus the number wrongly excluded from a category. Net error is the number of households according to the register minus the number estimated from the evaluation study. The proportion of correctly classified households and the relative net error are calculated relative to the number of households in the register.

The results indicate that the numbers for smaller households are underestimated and the numbers for larger households are overestimated by the register.

Table 8.3 is the corresponding table for household type. Results are similar to Table 8.2, i.e. the numbers for larger households (multi-person, two or more families) are overestimated and the numbers for smaller households are underestimated in the register. One explanation of the figures could be that younger people and students fail to register their new address as they move away from their parents. Note however that the larger types of households are less common and thus the number of such households in the sample is low.

8.5.3 Evaluating Ownership

The census variable on type of ownership differed in categories and definitions from what is commonly used in Sweden and supplied by the registers. Thus this

Table 8.2 Results from evaluation study: household size.

Size of household	1	2	3	4	5	6–10	10<	Total
Number correct	1 410 283 ± 15 366	1 212 330 ± 13 768	396 320 ± 16 524	426 208 ± 15 092	126 055 ± 10 684	35 237 ± 6386	143 ± 121	3 606 574 ± 32 662
Proportion correct	98.0 ± 1.1	95.5 ± 1.1	78.4 ± 3.3	83.9 ± 3.0	71.4 ± 6.1	45.2 ± 8.2	7.3 ± 6.2	90.7 ± 0.8
Gross error	256 571 ± 45 277	306 984 ± 35 894	199 400 ± 24 748	146 163 ± 20 854	67 824 ± 12 387	45 666 ± 6787	1814 ± 121	---
Net error	−199 233 ± 45 233	−192 778 ± 35 853	18 352 ± 24 801	17 755 ± 20 934	33 210 ± 12 431	39 760 ± 6782	1 814 ± 121	−281 119 ± 33 799
Relative net error	−13.8 ± 3.1	−15.2 ± 2.8	3.6 ± 4.9	3.5 ± 4.1	18.8 ± 7.0	51.0 ± 8.7	92.7 ± 6.2	−7.1 ± 0.8

Table 8.3 Results from evaluation study: household type.

Household composition	1 adult, no children	1 adult with children	2 adults no children	2 adults with children	Multi-person households	2 or more families	Total
Number correct	1 410 281 ± 15 365	263 768 ± 13 814	1 037 597 ± 9798	949 246 ± 18 511	38 687 ± 10 014	11 038 ± 3754	3 710 616 ± 31 328
Proportion correct	98.0 ± 1.1	88.1 ± 4.6	97.7 ± 0.9	93.6 ± 1.8	34.3 ± 8.9	21.3 ± 7.2	93.3 ± 0.8
Gross error	267 860 ± 46 619	64 927 ± 17 389	197 325 ± 30 473	118 991 ± 21 557	91 124 ± 13 304	44 465 ± 4450	---
Net error	−210 517 ± 46 625	6362 ± 17 335	−149 317 ± 30 465	9786 ± 21 561	56 890 ± 13 311	37 325 ± 4462	−249 471 ± 32 734
Relative net error	−14.6 ± 3.2	2.1 ± 5.8	−14.1 ± 2.9	1.0 ± 2.1	50.5 ± 11.8	71.9 ± 8.6	−6.3 ± 0.8

Table 8.4 Results from evaluation study: Type of ownership.

Type of ownership	Owner-occupied dwelling	Dwelling in cooperative ownership	Rented dwelling
Number correct	$1\,631\,015 \pm 24\,254$	$797\,256 \pm 20\,036$	$1\,421\,220 \pm 31\,590$
Proportion correct	96.5 ± 1.4	94.4 ± 2.4	95.3 ± 2.1
Gross error	$99\,994 \pm 27\,865$	$132\,585 \pm 28\,793$	$227\,599 \pm 41\,964$
Net error	$18\,794 \pm 27\,253$	$-37\,932 \pm 28\,883$	$-87\,318 \pm 42\,062$
Relative net error	1.1 ± 1.6	-4.5 ± 3.4	-5.9 ± 2.8

variable called for a rather complicated derivation, and it was necessary to evaluate its quality. A question about type of ownership was added to the evaluation study, and the data from the survey were compared to the derived census variable. Table 8.4 shows results from the comparison. The relative net error rate indicates that the quality of the data was sufficient, but the results should be interpreted with caution since some assumptions were made when translating the Swedish situation to fit the Eurostat definition.

8.5.4 Lessons Learned

Although the evaluation study in the end had a more limited scope than initially intended, several important experiences came out of it. One of them was the difficulty to construct a web/paper question designed to collect information about the household status. We used a template from a previous census evaluation, which was done by telephone, but it still took quite some time to test and decide on the final design. This was not only because of slightly different goals compared to previous evaluations, it was also due to technical difficulties in the web questionnaire environment. Collecting information in order to identify type of household and its members is most likely simpler to do by interview.

Secondly, and not unexpectedly, we observed that the nonresponse was higher in the groups where the Tax Agency also had difficulties collecting information, and where we suspect errors in the register. The stratification variable Number of families in dwelling according to TPR had a category with three or more families per dwelling. This is very unusual in Sweden and might be an indication of registration errors. High nonresponse could be an effect of not reaching people if they do not live where they are registered.

A similar explanation can be brought up for the low response rate (57%) among those lacking the dwelling identification key. However, it does not explain the lower response rate in the youngest age strata, following a nonresponse pattern similar to other Swedish surveys. One theory is that, since young adults are more mobile, the register might not be updated fast enough and they are therefore harder to reach. Another theory is that their propensity to respond is low, including low propensity to report to the Tax Agency when moving. This could be studied by analyzing variables registering events of change in the TPR together with the evaluation survey. Hence, there is a suspicion of lower register quality in the younger adult population.

Although anticipated, we must stress that the analysis and reconciliation of the household composition demanded plenty of resources. For the Type of household variable, the various combinations of discrepancies that might exist between the register information and the survey mean that the work required a lot of manual handling. We also had to make judgments about which source to trust, and when.

The evaluation survey of the DR and the household statistics gave Statistics Sweden important knowledge about some of the errors and the quality of the register-based statistics. It provided information to at least describe the strengths and the weaknesses of the data, but it was never used to adjust the census statistics.

8.6 Impact on Population and Housing Statistics

Large investments were made in the Swedish register system due to Census 2011, but the usefulness of the system of registers stretches far beyond a single census. The complete system of registers has now been in place for more several years. It is of great importance for the production of official register-based statistics on population, households, and housing, as well as major surveys such as the Labor Force Survey. In some subject matter areas, the register information on household affiliation has made possible annual register-based statistics on households and housing, sometimes replacing sample surveys. An up to date and coherent system of registers further makes it possible not only to improve the official statistics on households and housing, but also to compile statistics on demand with improved longitudinal quality at low cost, increase the possibilities of producing statistics for smaller domains and special populations, and provide standardized register variables and populations.

One problem that has to be dealt with subsequently is the treatment of persons that cannot be linked to a dwelling, i.e. with missing household information. For those persons, a model-based household is imputed using registered relations in the TPR (e.g. marriage, parent–child relations). To date, slightly different approaches have been used in different subject matter areas, mainly because of differences in household definitions. As a consequence, the total number of households and the distribution of type of household may differ slightly. See for

Table 8.5 Percentage of coverage and number of persons with missing dwelling identification key.

Year	2011	2012	2013	2014	2015	2016	2017
Percentage of coverage	96.6	96.8	96.8	97.2	97.3	97.9	97.8
Number of persons with missing id-key	320 000	300 000	305 000	270 000	260 000	210 000	220 000

example statistics on households' housing,[4] Income and tax statistics,[5] and Population statistics.[6]

Even after imputation, some non-linked persons remain which the model cannot account for. The consequence of this is either undercoverage of households, if the person lives alone or in a household with only non-linked persons, or errors in household variables such as size and type, if the person lives in a household with linked persons.

When administrative sources are used for production of official statistics, close cooperation with the responsible actors is crucial to ensure high quality statistics. Statistics Sweden is continuously cooperating with the Tax Agency, the land surveying authority, county councils, and municipalities in order to continue to improve the quality of the administrative registers that are used as input to the TPR and the DR. Following a government decision, each Swedish municipality has responsibility for their part of the DR, hence they are obliged to update and maintain the registers when there are demolitions, new constructions, and reconstructions of real properties.

As residents move to a different apartment or single house, they are required to inform the Tax Agency. Overall, this is a well-functioning procedure since almost all public services utilize the administrative registers, and there is an incentive for people to keep the information updated in order to get information and public services. The coverage in the statistical registers has improved accordingly since 2011 (see Table 8.5).

The overcoverage of persons in the TPR was investigated in 2014, using available register information. Persons living in Sweden leave imprints in different administrative registers as they work, pay tax, study, move to a new home, etc. Thus, a

---

4 http://www.scb.se/en/finding-statistics/statistics-by-subject-area/household-finances/income-and-income-distribution/households-housing/
5 http://www.scb.se/en/finding-statistics/statistics-by-subject-area/household-finances/income-and-income-distribution/income-and-tax-statistics/
6 http://www.scb.se/en/finding-statistics/statistics-by-subject-area/population/population-composition/population-statistics/

person registered as living in Sweden who no longer leaves such traces is likely to have emigrated without reporting this to the tax authority. A model was developed to calculate the overcoverage. Over time, the share that remained as overcoverage decreased from year to year, but the decrease slowed down gradually, indicating that a small group of people remains as overcoverage over a long period (Statistics Sweden 2015).

8.7   Summary and Final Remarks

We have tried to describe not only the work that was carried out for the 2011 Census and the knowledge that was gained, but also methodological and practical issues that we believe are important when using administrative data for statistical purposes.

Census 2011, and the decision that this was to be the first fully registered-based census in Sweden, became a driving force for the final development of the complete register-based system which is now the basis for a large part of the production of population and housing statistics. Of crucial importance was the development of a register of dwellings and registration of the Swedish population on dwelling identification number. This process of completion was complicated and took a long time, but made it possible to form households based on register information only.

The quality of the household information was of crucial importance for the quality of the census data. The evaluation survey gave important knowledge and helped in assessing the quality of the census.

We end by highlighting a few points that are of general importance when administrative data are used for census purposes. They concern quality, maintenance, error sources, and the need for unique identification of objects.

With the goal to maximize the quality of register-based statistics given available resources, there are many methodological issues that need to be considered. Certain characteristics are always present in a register-based census. There will be relevance errors when the register sources fail to perfectly capture the census' target concepts. Other potential problems arise from the underlying administrative procedures that feed the registers with data. Deficiencies in those procedures will lead to missing values, erroneous values, coverage problems, and identification and unit errors as defined by Zhang (2012). Furthermore, some of the computational rules that are used when data from different sources are matched and processed are based on assumed circumstances rather than deterministic true relations. This adds to the uncertainty in the results which a register-based census should develop approaches to handle. All of the above matters have a methodological component and preferably the work to minimize their impact and improve the statistical quality should be done continuously, and not only in projects leading up to another census round.

Although censuses in Sweden are far in between, a scheme for continuous quality improvement and quality assurance is possible to run. Once the register and production system for the census is in place, methodological issues can be addressed in parts with relatively small means as a part of the maintenance of the system. In this chapter, we have not addressed how Sweden maintains and improves its ability for future register-based census statistics, our focus has been on the practical and methodological questions that were specific when making the first register-based census.

Since the Swedish register-based census required the construction of a new register, practically all of the traditional error sources in surveys except the sampling errors were present. Examples are errors due to missing values, processing mistakes, and measurement. There were a number of matters that made the methodological work challenging:

- Many different authorities were involved. There was a risk of errors due to lack of coordination and misunderstanding between the involved bodies.
- In addition, 290 different municipalities had to be informed that they had to be willing to provide information.
- There were different input processes for one- and two-dwelling buildings compared to multi-dwelling buildings.
- The registers differ in updating frequencies and reference periods.
- The real property owners and residents were respondents in the build-up of the system of registers. Their incentives to respond differed.
- The relevance of some of the register variables compared to the Eurostat requirements was not known from the start.
- The quality of the other necessary (subject matter) registers was not always known, and the effect on the quality of the statistics when joining them was largely unknown.

The large amount of data was an additional circumstance that added to the challenges.

The Swedish system of registers relies completely on unique identification of persons, households, and businesses. It is a prerequisite for forming register households. The evaluation survey was used to assess the quality of the household formation, but more work remains to be done. The coverage of the registration on dwelling identification number is constantly improving, but persons without household affiliation and persons not living where they are registered are examples of issues that need further investigation.

Although the evaluation survey gave important information, it is not without flaws. The survey suffered from nonresponse, notably higher in groups where the

Tax Agency had difficulties collecting information and where errors in the registers could be suspected. Moreover, it is not obvious which source is to be viewed as the correct information, the survey or the register. However, the evaluation survey also asked persons to clarify the factual circumstances when there were discrepancies between register and initial survey response. This provided valuable measures and insights for comparing register data and self-administered survey data.

References

Andersson C. and Holmberg A., (2011) Methods and design to evaluate the 2011 Census and future register based household and housing statistics. Discussion paper to the Advisory Scientific Board of Statistics Sweden. Meeting Sigtuna, Sweden (14–15 April 2011).

Andersson C., Holmberg A., Jansson I., Lindgren K., and Werner P. (2013). Methodological Experiences from a Register-Based Census. *Proceedings of the 2013 Joint Statistical Meetings*, Montreal (3–8 August 2013).

Axelson M., Hedlin D., Holmberg A., and Jansson I. (2010). Methodology in the Swedish register-based census. Paper presented at the 2010 International Methodology Symposium, Statistics Canada, Ottawa (26–29 October).

Axelson M., Holmberg A., Jansson I., Werner P., and Westling S (2012) Doing a register-based census for the first time: the Swedish experiences. Proceedings of the 2012 Joint Statistical Meetings, San Diego, CA (28 July–2 August 2012).

COMMISSION REGULATION (EC) No 1201/2009 of 30 November 2009 implementing Regulation (EC) No 763/2008 of the European Parliament and of the Council on population and housing censuses as regards the technical specifications of the topics and of their breakdowns. L 329/29, 15 December 2009. https://eur-lex .europa.eu/LexUriServ/LexUriServ.do?uri=OJ:L:2009:329:0029:0068:EN:PDF (accessed 17 August 2020)

COMMISSION REGULATION (EU) No 519/2010 of 16 June 2010 adopting the programme of the statistical data and of the metadata for population and housing censuses provided for by Regulation (EC) No 763/2008 of the European Parliament and of the Council. L 151/1, 17 June 2010. https://eur-lex.europa.eu/LexUriServ/ LexUriServ.do?uri=OJ:L:2010:151:0001:0013:EN:PDF (accessed 17 August 2020)

COMMISSION REGULATION (EU) No 1151/2010 of 8 December 2010 implementing Regulation (EC) No 763/2008 of the European Parliament and of the Council on population and housing censuses, as regards the modalities and structure of the quality reports and the technical format for data transmission. L 324/1, 9 December 2010. https://eur-lex.europa.eu/legal-content/EN/TXT/PDF/? uri=CELEX:32010R1151&from=EN (accessed 17 August 2020)

Hand, D.J. (2018). Statistical challenges of administrative and transaction data. *Journal of the Royal Statistical Society Series A* 181: 555–605.

Hedlin D., Holmberg A., Jansson I., and Lorenc B. (2011). The first fully register-based census in Sweden. Paper presented at the 2011 Joint Statistical Meetings, Miami Beach, FL (30 July–4 August 2011).

Hedlin D., Holmberg A. and Jansson I. (2011). Combining registers into a fully register-based census – some methodological issues. Paper presented at the NORC U.S. Census Bureau conference: Utilizing Administrative Data: Technical, Statistical and Research Issues, Washington, DC (27–28 October 2011).

Jorner, U. (2008) Summa summarum. SCB:s första 150 år. Statistiska centralbyrån.

Office for National Statistics (2012). Making a National Adjustment to the 2011 Census. 2011 Census: Methods and Quality Report.

REGULATION (EC) No 763/2008 OF THE EUROPEAN PARLIAMENT AND OF THE COUNCIL of 9 July 2008 on population and housing censuses. L 218/14, 13 August 2008. https://eur-lex.europa.eu/LexUriServ/LexUriServ.do?uri=OJ:L:2008:218:0014:0020:EN:PDF (accessed 17 August 2020)

Reid, G., Zabala, F., and Holmberg, A. (2017). Extending TSE to administrative data: a quality framework and case studies from stats NZ. *Journal of Official Statistics* 33: 477–511.

Statistics Sweden (2015). Overcoverage in the total population register. *Background facts* 2015: 1.

UNECE (2007). *Register-Based Statistics in the Nordic Countries – Review of Best Practices with Focus on Population and Social Statistics*. Geneva: United Nations Publications.

Wallgren, A. and Wallgren, B. (2007). *Register-Based Statistics – Administrative Data for Statistical Purposes*. New York: John Wiley.

Zhang, L.-C. (2011). A unit-error theory for register-based household statistics. *Journal of Official Statistics* 27: 415–432.

Zhang, L.-C. (2012). Topics of statistical theory for register based statistics and data integration. *Statistica Neerlandica* 66: 41–63.

9

Administrative Records Applications for the 2020 Census

Vincent T. Mule Jr, and Andrew Keller

U.S. Census Bureau, Decennial Statistical Studies Division, 4600 Silver Hill Road, Washington, DC 20233, USA

9.1 Introduction

Countries conduct censuses of their populations and housing to provide a snapshot of the inhabitants in relation to their totals and demographic characteristics. Censuses provide in-depth information at the national level but also provide much desired characteristics about smaller governmental units and smaller population groups. This chapter focuses on a proposed usage of administrative records being developed for the 2020 Census in the U.S. It also provides information on how countries from around the world utilize administrative records in different ways when attempting to conduct an enumeration of their population.

Most countries conduct a traditional census based on a full field enumeration by different modes of data collection. For some countries, administrative record coverage allows for traditional enumerations to not need to be conducted. The last survey of census methods conducted in 2009 by the United Nations indicated that, about 30 years after the first fully register-based census in Denmark in 1981 (Lange 2013), 12 countries were able to utilize administrative records to the extent that they could implement register-based censuses for the 2010 cycle.

The U.S. has used administrative records for some aspects of its decennial censuses of population and housing for many years. Section 9.2 highlights various ways the U.S. has utilized administrative record information in a traditional census. The U.S. is considering substitution of administrative records and third-party data in the possible place of field enumeration in the 2020 Census in the U.S. This application is described in Section 9.3. An important aspect of the

---

☆ The views expressed in this chapter are those of the authors and not necessarily those of the U.S. Census Bureau.

Administrative Records for Survey Methodology, First Edition.
Edited by Asaph Young Chun, Michael D. Larsen, Gabriele Durrant, and Jerome P. Reiter.

introduction of administrative record usage is assessing the potential changes in quality. Section 9.4 describes methods of evaluating quality of data from administrative records for census purposes. Of course, the practices and choices being considered in the U.S. are not the only options. Section 9.5 provides an overview of a few of the other applications of administrative records for either conducting or evaluating a county's census.

9.2 Administrative Record Usage in the U.S. Census

Since 1790, the United States has conducted a census every 10 years using traditional methods of conducting the full field enumeration of the population. As the decades have passed, the U.S. Census Bureau has used different modes of data collection to obtain responses from the public. In the 1970 Census, the U.S. Census Bureau introduced the usage of the postal delivery of questionnaires at the start of the enumeration by asking respondents to mail back the completed form. This allowed a reduction in the field visits by enumerators. While every address did not require a field enumeration, the 2010 Census still required the hiring of about 500 thousand enumerators to enumerate the 50 million addresses in the Nonresponse Followup (NRFU) operation that did not respond to the initial mailing.

In the past 40 years, the U.S. has considered how administrative records could be utilized for the U.S. census. Jabine and Scheuren (1985) summarized that the first goal listed was to "explore fully and develop the uses of major administrative record systems in the conduct and evaluation of the decennial population census and for current population estimates" (p. 381). Alvey and Scheuren (1982) proposed the idea of conducting a register-based census with documenting possible operational and technical problems. The Committee on the use of Administrative Records in the 1990 Census evaluated the proposal but determined that there were potential issues related to coverage, accuracy, and demographic information available for all people (U.S. Census Bureau 1995). However, for the 1990 Census, administrative records were used to enumerate overseas personnel and their dependents.

In the early 1990s, the Census Bureau provided to the Office of Management and Budget 14 possible designs for the 2000 Census. Several of these designs used administrative records for either the full enumeration source or combined with statistical estimation. The Panel on Census Requirements in the year 2000 and beyond studied alternative methods for conducting the census that included the usage of administrative records. For both a national register and an administrative records census, the panel concluded that both were not feasible replacements for the decennial census (Edmonston and Schultze 1995). Bauder and Judson (2003) document the administrative records experiment where an administrative record census was conducted for two counties in the state of Maryland and three counties in Colorado. Their results showed that matched occupied addresses, that the administrative record household count, matched the census enumeration 51% of the time.

For the 2010 Census, the Census Bureau utilized administrative records in the coverage follow-up operation. Govern and Coombs (2013) show that the operation selected 8.2 million returns for various reasons for an additional interview to address potential under or overcoverage. Of the 8.2 million, 641 425 were selected based on differences when matching the census return of the household to administrative records. These cases where at least one person was matched between the administrative records and the census returns and at least one person was identified on the administrative records but not on the census return. The coverage follow-up assessment indicated that the resulting interview added 28 219 people in 16 086 addresses and deleted 19 766 people in 15 738 addresses.

9.3 Administrative Record Integration in 2020 Census Research

For the 2020 Census planning, the U.S. Census Bureau had the challenge to conduct the next census at a lower cost per household after adjusting for inflation than the 2010 Census while maintaining high quality results. Utilizing administrative records and third-party data was one of the four innovation areas to reduce the costs of the door-to-door enumeration of addresses in the nonresponse follow-up operation. In this section, we provide an overview of how administrative records is integrated into the actual enumeration. For the 2020 Census, the Census Bureau has come up with methods for the partial usage of administrative record information where they are the strongest and thus allowing full field contacts to be conducted for the remaining addresses.

9.3.1 Administrative Record Usage Determinations

For this application, this work researched using administrative records and third-party data to reduce the number of times an address must be contacted during NRFU. The research has developed an enumeration mode approach where the usage of administrative records allows us to make the necessary determinations for each address.

For each address, the NRFU operation must determine if an address is one of the following three outcomes:

- *Occupied*: The address is a valid housing unit and is occupied on Census Day based on the residence rules. For occupied addresses, the roster of all of the people living at the address needs to be obtained along with the demographic characteristics of age, sex, race, Hispanic origin, relationship to householder, and whether the house is owned or rented.

- *Vacant*: The address is a valid housing unit but was unoccupied on Census Day. For vacant addresses, the reason for being vacant like for sale, for rent or seasonal vacancy needs to be obtained.
- *Nonexistent or delete*: The address does not meet the Census Bureau's definition of a housing unit and needs to be removed from the Census list.

9.3.2 NRFU Design Incorporating Administrative Records

This section covers the NRFU contact strategy related to administrative record occupied and vacant cases proposed for the 2020 Census. This strategy was laid out in the second release of the 2020 Operational Plan (U.S. Census Bureau 2015) and has been tested in census tests between 2013 and 2016.

Before the NRFU operation, an address will have received multiple mailings before and after Census Day. These include some combination of postcards, letters, and a paper questionnaire. The exact mailing strategy and ordering varies by geography. If the address does not respond to these mailings, then a decision needs to be made about how many times to contact the address during the NRFU operation.

Figure 9.1 shows the flowchart of the contact strategy related to administrative record cases for the NRFU operation discussed in the 2020 Operational Plan. This design was incorporated in the 2016 Census Test. Addresses determined to be vacant by our method received no contacts during the NRFU operation. While these units did not receive any NRFU visits, a postcard was mailed to them during the 2016 Census Test at the beginning of the NRFU operation. This allowed people at occupied addresses to self-respond by going online and filling out the internet questionnaire or dialing the questionnaire assistance phone number.

The remaining cases will receive an initial field visit. This visit allows each case to be resolved in several ways. It can be resolved by

- completing the interview with the household member
- determining the address to be vacant,
  or
- determining the address was not a housing unit.

If nobody in the household is home, the enumerator leaves a notice of visit. This notice of visit includes information that instructed persons in the household to respond:

- by going online and self-responding,
- dialing the questionnaire assistance number,

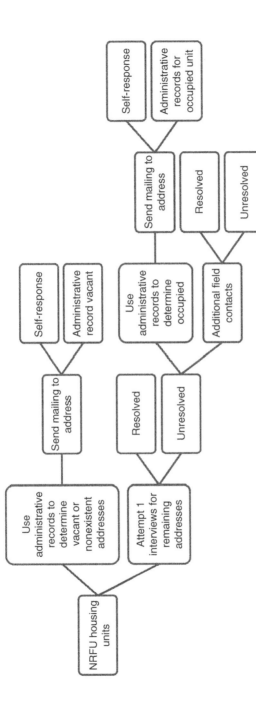

Figure 9.1 2016 NRFU contact strategy for administrative record cases.

or

- sending back the paper questionnaire that they received earlier.

For cases determined to be occupied by our method, they received only the initial visit in the 2016 Census Test. While they received only the initial visit, an additional postcard mailing was sent to the address. This postcard contained information detailing how the household could still go online or dial the questionnaire assistance number to self-respond. As shown, there are several ways before and during NRFU that the Census Bureau is attempting to obtain and use self-responses before having to use the administrative record determinations. This is not necessarily the final contact strategy for the 2020 Census. Other alternatives including conducting more fieldwork or utilizing the results of being able to deliver the additional mailings as part of the determination that addresses are vacant or do not exist can be utilized.

9.3.3 Administrative Records Sources and Data Preparation

This section summarizes the administrative record and third-party data sources used in this analysis. The U.S. has a decentralized statistical system. As a result, the Census Bureau has made the necessary agreements so that information about income tax payments, health care, and other government programs can be transferred to the Census Bureau. The Census Bureau also uses information obtained from third-party data vendors.

In order to allow linkage of the person from the different sources, the Census Bureau's Person Identification Validation System has been used to assign a Protected Identification Key (PIK) to each person's record on the administrative record and third-party files. The PIK is a one-time pad encryption of the person's Social Security or Tax Identification number. Most of the administrative record files received from other federal agencies were able to provide the Social Security or Tax Identification Number of the person. For these instances, the Person Validation System needs to verify the Social Security Number and assign the encrypted PIK in its place. For some files like the 2010 Census, the Social Security or Tax Identification Number was not collected. For this case, the Person Validation System uses record linkage techniques to compare the name, date of birth/age, sex, and address information to the name, date of birth/age, sex, and address information available for each PIK record. Based on the variable comparisons and the matching scores for the specific blocking pass, the PIK can be assigned to person records. For the 2010 Census, PIKs were able to be assigned to 90% of the census enumerations. Please see Layne et al. (2014) for more information about the Person Validation System processing. For administrative record and third-party files that have address information, the processing is able to assign our Master Address

Table 9.1   Administrative record and third-party sources.

Core data sources
IRS 1040 Individual Tax Returns
IRS 1099 Informational Returns
CMS Medicare Enrollment Database
IHS Patient Registration File
USPS Undeliverable-As-Addressed: NIXIE reasons From Census Mailings
Additional data sources used to corroborate information in core sources and/or used in models
IRS 1040 Individual Tax Returns
IRS 1099 Informational Returns
CMS Medicare Enrollment Database
IHS Patient Registration File
CMS Medicaid Statistical Information System
SSA Numerical Identification (NUMIDENT) file
HUD Public and Indian Housing Information Center
HUD Tenant Rental Assistance Certification System
HUD Computerized Homes Underwriting Management System
SSS Registration File
USPS National Change of Address
2010 Census
Census Bureau Master Address File
American Community Survey 5-year Block Group estimates
Census Bureau's Best Race and Best Hispanic Research

File Identification Number (MAFID). This allows us to identify any of the person's records from multiple sources which are associated with a given address.

Table 9.1 documents the administrative records from the Internal Revenue Service (IRS), Center for Medicare and Medicaid Services (CMS), Indian Health Service (IHS), U.S. Postal Service (USPS), Social Security Administration (SSA), Department of Housing and Urban Development (HUD), Selective Service System (SSS), and the 2010 Census. Our approach also used information from two third-party files: Veterans Service Group of Illinois (VSGI) and Targus.

Here is a summary of the processing steps of the administrative record and third-party data sources listed above. This is the approach used in the 2016 Census Test. The approach for the 2020 Census may be updated but will be similar.

- One of the first steps was to build rosters at each address using the person records found on the IRS 1040 Individual Tax Returns, IRS 1099 Informational returns, CMS Medicare Enrollment Database, IHS Patient Registration File.

Only persons rostered from these federal sources were used. The approach did not build rosters using third-party information.

- Only addresses where all of the records were assigned a PIK were used. PIKs are one-time encryption of Social Security or Tax Identification Numbers. Since the rosters were built from federal sources that provided us Social Security or Tax Identification Numbers, most could have the PIK assigned.
- The files in the Additional Data Sources column of Table 9.1 are used to indicate if the person rostered from the core data sources are found at that address on those other sources.
- For our processing, only household rosters where there were multiple sources that indicated that a family lived at an address were eligible to have their contact reduced. If only one source was providing the information, then the number of contacts was not reduced.
- For the roster people, the process obtained age and sex from the Numerical Identification (NUMIDENT) file. For Race and Hispanic origin, we utilize best race and best Hispanic origin assignments for the person. This research used administrative record and third-party information where available to assign seven values of race based on six race alone categories and two or more races. The research used similar administrative record and third-party information where available to assign people as Hispanic or Non-Hispanic. See Ennis et al. (2015) for a full explanation of how race and Hispanic origin are assigned to persons in the administrative record data.
- From the first and the second mailings, we obtained the detailed USPS Undeliverable-as-Addressed (UAA) Nixie reasons. These are the detailed reasons like "Vacant" or "No Such Number" that explain why the census mailing could not be delivered.
- For the methodology described in Sections 9.3.4–9.3.6, we used information from the master address file (MAF) and American Community Survey (ACS) five-year block group estimates as modeling covariates. For the ACS block groups, we used estimates including vacancy rate, mobility rate, Hispanic population rate, and Black population rate.

9.3.4 Approach to Determine Administrative Record Vacant Addresses

This section provides an overview of predictive modeling approaches to determine vacant addresses. Section 9.3.5 extends the approaches described in this section to determine nonexistent addresses. In researching approaches to identify addresses that were vacant, the approach needs to make a prediction about whether the address is vacant before fieldwork is conducted based on the available information. Other areas of statistical research have faced this challenge and have developed predictive model approaches. Some examples of predictive modeling approaches have been developed for health care to predict diabetes. Collins et al. (2011) summarizes the 39 studies that have developed 43 predictive models for type 2 diabetes. This research cultivated similar approaches of supervised learning to apply to our census application.

For this, we have a multinomial outcome that we want to make predictions about. This section utilizes the results of predicting two of those outcomes of whether the address is vacant or occupied. The Section 9.3.5 will show how this approach utilizes the third outcome to determine addresses that are not housing units.

For the predictive modeling, this approach utilizes the previous 2010 Census results. This provides a set of previous results where the multinomial outcome can be associated with information available for the housing unit. There are several possible predictive modeling approaches that can be utilized for multinomial outcome variables. These include multinomial logistic regression, decision tree classification, and random forest models. Multinomial logistic regression allows the nominal outcome variables to be modeled in relation to available covariates. This approach allows a parametric framework to be utilized to assess the relationships. Decision trees are classification algorithms that identify the key variables and produce a series of nodes like a tree that show those relationships. Based on the parameters specified, the trees can have various number of levels and final nodes. While a finite number of nodes may limit the number of unique predictions that can be made to cases, the decision trees have the advantage of producing a pictorial representation that is easier to explain which cases are being determined to be vacant. Random forests is a supervised modeling approach based on developing decision trees on multiple bootstrap samples of the training data. For this application, the processing uses multinomial logistic regression to allow capability of being able to include more covariates in the final model. This research showed similar results for this and the random forest approaches.

For the multinomial logistic regression, a model was fit on the 2010 training data. The model includes k covariates for the n observations. This determines the k β-coefficients based on the results of how the 2010 outcomes correspond to these.

For these regressions, the approach is able to utilize information from the U.S. Postal Service about whether the Census mailings were able to be delivered around Census Day. This approach also includes information about the presence of persons found at the address on the administrative record and third-party sources. The models also include information about the address from the Census Bureau's MAF and information about the neighborhood by using American Community Survey estimates of rates of vacancy, poverty, and other measures for the block group in which the address is located. The estimated β coefficients can then be utilized to make predictions about whether addresses that are eligible for the NRFU operation are occupied, vacant, or do not exist.

Now that predicted probabilities have been assigned to every case eligible for NRFU, the last part is to figure out how to use this information to decide which cases should receive fewer contacts. One way that this has been done for binary outcomes is to utilize Receiver Operator Curves that assess the specificity and sensitivity of the training model to help chose a probability cutoff. For this Census application, the approach attempts to identify vacant addresses, but it guards against potential undercoverage by unnecessarily classifying occupied units as vacant.

For this approach, the processing uses a distance function approach. The desire is for the administrative record vacant cases to have a probability of being vacant as close to 1 as possible and probability of being occupied as close to 0 as possible. This led to the approach to calculate the distance of those two predicted probabilities to the ideal points to make the administrative record vacant determination.

As a result, we define the Euclidian vacant distance function for vacant determination as:

$$d_h^{\text{vac}} = \sqrt{(1 - \widehat{p}_h^{\text{vac}})^2 + (\widehat{p}_h^{\text{occ}})^2}$$

where $\widehat{p}_h^{\text{vac}}$ is the vacant probability and $\widehat{p}_h^{\text{occ}}$ is the occupied probability for housing unit h. This continuous measure can be assigned to every address.

The cutoff to decide which cases should receive fewer contacts can be adjusted upward or downward. This is based on the tradeoff between reduction of the NRFU workload versus the allowable differences between the administrative record and census fieldwork determinations.

9.3.5 Extension of Vacant Methodology to Nonexistent Cases

The approach for determining administrative record vacant addresses was extended to be able to determine administrative record nonexistent addresses that can receive fewer contacts as well. The approach took advantage that of the modeling of the multinomial outcome space of addresses being classified as occupied, vacant, or delete.

This processing defines the nonexistent distance function based on the nonexistent probability and occupied probability discussed earlier. The approach substitutes the nonexistent probability for the vacant probability. In this instance, the ideal situation would be for an address to have a nonexistent probability of 1 and an occupied probability of 0.

This continuous measure is assigned to every address eligible for the NRFU operation. The cutoff can be adjusted upward or downward based on the trade-off between fieldwork reduction and allowable differences.

9.3.6 Approach to Determine Occupied Addresses

This section lays out the methodology to determine administrative record occupied addresses. A roster of people eligible for enumeration was developed based on using information from tax returns, tax information, and federal health care enrollment in Medicare or IHSs. Census applications for determining whether the roster developed for an address meets the quality expectation in reducing contacts is a challenge because there are multiple dimensions of quality that need to be assessed. Health care and other predictive modeling applications have advantages since they tend to assess binary outcomes about whether someone has a disease or if a new medicine was able to successfully treat a disease. For this application, there are several options. One is a comparison of population counts to see if they agree. A second is assessing whether the individual people found on the administrative record sources were counted at the address in the Census. A third is to compare the household composition of adults and children found on the administrative record sources to the composition of Census. Other comparisons can be done as well.

This application developed an approach that allows the results of multiple comparisons to be used to assess the address. The approach described in this section uses two models, but it can be extended to include additional dimensions of quality as well. The two approaches in this section are similar to those in Section 9.3.4 where predictive models are developed using 2010 census results.

Our first model is the person-place model. This model assesses whether the administrative record sources place the people at the same address as the Census enumeration would. This model predicts the probability of enumerating the administrative record person at the sample address if fieldwork were conducted.

This model compiles person-place pairs in administrative record files mentioned above and the 2010 Census person-place pairs to define the dependent variable of interest in the person-place model:

$$y_{ih} = \begin{cases} 1 & \text{if person } i \text{ is found in AR and 2010 Census at the same address } h \\ 0 & \text{otherwise} \end{cases}$$

The first step is interested in a predictive model for estimating the probability, $p_{ih} = P(y_{ij} = 1)$, that the 2010 Census and the administrative record roster data place the person at the same address. These probabilities are estimated via a logistic regression model. Morris (2014) documents initial work on this person-place model and investigates logistic regression, decision trees, and random forest methodologies for estimating the models. This research showed little difference in predictive power between the three methodologies.

The person-place model is fit at the person-level, but decisions are made at the housing unit-level. Therefore, the person-level predicted probabilities, $\hat{p}_{ih}$, are summarized for each address such that the housing unit-level predicted probability for address h is defined as:

$$\hat{p}_h^{\hat{p}p} = \min(\hat{p}_{1h}, \ldots, \hat{p}_{n_h h})$$

where n_h is the number of people at address h. This minimum criterion assigned to the housing unit the predicted probability for the person in the housing unit for which we have the lowest confidence which is a relatively conservative approach. The administrative record household count is defined as the sum of all individuals associated with the administrative record address, and each address has the associated predicted probability of having an administrative record and census address match. These are the predicted probabilities that are passed to the linear programming portion to decide which cases were determined to be occupied.

The second model is the household composition model. The household composition model predicts the probability that the sample address would have the same household composition determined by NRFU fieldwork as its pre-identified administrative record household composition. The results from the 2014 Census Test motivated the development of the household composition model. During that test, Keller et al. (2016) observed that units identified as occupied with administrative records were more likely to be occupied and have higher count agreement in NRFU if the household composition of the administrative records for the unit was a single adult, a two-person adult unit without kids, or a two-person adult unit with kids.

Each administrative record household roster is categorized in this manner:

- No administrative record persons
- 1 Adult, 0 Child
- 1 Adult, >0 Child
- 2 Adult, 0 Child
- 2 Adult, >0 Child
- 3 Adult, 0 Child
- 3 Adult, >0 Child
- Someone with undetermined age in household
- Other

This approach then creates a dependent variable from the household composition on the 2010 Census. The categorization is similar except that, in all units, all persons have an age. This is because this approach uses the Census Edited File as the basis for forming the census household composition. Since this data has age imputed, there are no missing values for age. This approach fits a multinomial logistic model with the 2010 Census household composition as the dependent variable over a sample of the data. The predicted probability for the housing unit is the multinomial probability associated with the administrative record household type. For every eligible NRFU address, this approach assigns a predicted household composition probability, $\hat{p}_h^{hhc}$.

For the occupied decision, this approach uses the two predicted probabilities as inputs. The first is the minimum person-place probability for the address. For the person-place probability, this probability should be as close to 1 as possible. The second is the household composition probability. This is the probability that, given there is observed a certain administrative record household composition, the census household would be enumerated with the same household composition. For this household composition probability, this would also like that to be as close to 1 as possible.

Based on this, the situation can be thought of as a two-dimensional plane where each probability is on one dimension with values between 0 and 1. Based on the two probabilities, each address would have a point in this two-dimensional space. The best administrative record cases would be those that have the shortest distance to the point where the person-place model equals one and the household composition equals one (i.e. the (1,1) point).

Based on this idea, we use the Euclidian occupied distance function for occupied determination:

$$d_h^{occ} = \sqrt{(1 - \widehat{p}_h^{pp})^2 + (1 - \widehat{p}_h^{hhc})^2}$$

This continuous measure is assigned to every address. The cutoff can be adjusted upward or downward. This is based on the trade-off between reduction of the NRFU workload versus the allowable differences between the administrative record and census fieldwork determinations. A possible extension if desired is to include additional measures in the distance calculation.

9.3.7 Other Aspects and Alternatives of Administrative Record Enumeration

For the 2020 Census, the administrative record information will only be used for enumeration if the address does not self-respond or is not resolved during the one visit of field work enumeration. For the person enumerated based on administrative records, the demographic characteristics of age, sex, race, Hispanic origin, and relationship need to be assigned. This information will need to be obtained from administrative record sources like past census responses or the Social Security NUMIDENT file. For other characteristics like race and Hispanic origin, the final determination of assignment will also need to account for potential changes to the question to possibly include the Middle Eastern or North African response option. For any of these characteristics that cannot be assigned directly based on administrative record information, imputation procedures will be implemented.

The administrative record procedures to reduce contacts can be executed throughout the NRFU data collection window. For this application, the Census Bureau received an additional delivery of IRS tax returns in the middle of the operation. The test processing used this new information so that updated predicted probabilities could be assigned to the cases still active in the NRFU operation. During this processing, we computed new distance function results. We then compared them to our previously determined cutoffs and were able to determine that additional cases could have their contacts reduced.

Another possible application that can be considered is to continue to assess whether you want to continue to try to obtain an interview for an address. Based on having the continuous distance function measures available for each address, this allows the organization to have different criteria during the data collection window. Based on cutoffs determined at the beginning of the fieldwork operation, the organization may decide to send cases to the field to try to obtain an interview. However, if the results of multiple attempts are that an interview still has not been completed, the organization may decide to lower the cutoff and stop the fieldwork on the case. For this instance, the case would shift data collection modes to receive a final postcard reminder. This would allow the fieldwork in the remaining part of the data collection window to be attempted on other cases that do not have that administrative record information available.

This remaining part of this section provides some possible alternative considerations of this application that could also be utilized. Table 9.1 listed the administrative record and third-party data sources that are currently being used in the Census Tests. The U.S. Census Bureau will make a final decision on the administrative record and third-party data files to be used in September 2018. The Census Bureau is pursuing agreements to obtain additional data sources. In the U.S., individual states implement programs that provide nutritional and other support for families that need assistance. Since these are implemented by individual states, the U.S. Census Bureau is working to obtain agreement and data about the addresses and people who are participating. These sources can provide coverage of harder-to-count populations that are not fully covered in other federal sources available.

The 2016 application uses the 2010 Census as the source of the training data. Instead of using the past Census as a training source, another possibility is to use more current data from an ongoing survey that is collected. For the U.S., another possibility would be to train the model using responses from the American Community Survey. The Census Bureau is researching if this can be used as a source of the training data for administrative occupied models. Nevertheless, a potential benefit from training on ACS data is the recency of the ACS data. For example, for the 2020 Census, administrative record determinations could be trained on 2018 ACS data as opposed to 2010 NRFU data. In addition, the ongoing nature of the ACS data could allow for the inclusion of administrative record datasets not necessarily available or could not provide 2010 vintage data for training.

Last, use of administrative record data may also reduce the status imputation workload. In the 2010 Census, count imputation was a one-time process performed following the completion of the NRFU operation. This ensured that each census address was provided a final status of occupied, vacant, or nonexistent. This approach could be run again using relaxed cutoffs to determine additional administrative record enumerations.

9.4 Quality Assessment

The development of possible administrative record models has been guided by doing internal validation of comparing models retrospectively against 2010 Census results. For example, running a simulation on 2010 Census data, we counted how many addresses identified as vacant by the model were actually vacant during the 2010 Census. Essentially, this type of analysis treats 2010 Census results as "truth." However, a difficulty underlying the evaluation of administrative record modeling is the inherent error in Census results. Although the analysis using the 2010 Census results as "truth" provides a solid basis for assessing model performance, it is not the only way model performance can be measured. It is possible that census quality could be improved using administrative record data that is not reflected by solely comparing model against 2010 Census "truth."

Mule et al. (2017) document a possible way of assessing accuracy of the introduction of administrative record usage into a census or a survey. This section provides some of the quality metrics utilized in our research using 2010 Census data. These metrics have fallen into two main types. The first is microlevel and summarizes the amount of agreement and differences for cases when comparing on an individual basis. The second is macro-level and shows the implications of our change on aggregate level results. Our research has shown both types to be beneficial in assessing quality implications. For both of these, we have found it useful to bring in the Census Coverage Measurement (CCM), other coverage evaluations, and assessments of the 2010 Census to help with the quality analysis.

In this section, we will utilize a retrospective analysis of applying these approaches on the 2010 Census. For this analysis, we determined about 10% of the addresses to be administrative record vacant and about 14% of the addresses to be administrative record occupied and about 0.1% determined to be administrative record deletes or not existent addresses.

9.4.1 Microlevel Evaluations of Quality

For the microlevel evaluations, we used several different methods to do the comparison. An initial approach was to compare the results of the administrative record cases to the 2010 Census NRFU outcomes. One method calculated percent agreement for our administrative record determinations. A second approach compared the population count between the administrative record and NRFU results. Our analysis determined whether the administrative record count was higher, same, or lower than the Census count. An extension of this approach that other researchers could use in their own applications would be to calculate Cohen's kappa or other statistics to measure reliability between two sources.

For the about 10% of the NRFU eligible cases determined to be administrative record vacant, Table 9.2 shows the comparison of occupancy status of the administrative record cases to the 2010 Census NRFU outcomes. Table 9.2 shows that 80% of the cases were both vacant in our determination and the Census NRFU outcome. There is 10% of the cases which the Census result was occupied. The first thought was that this could be a potential source of undercounts. In assessing this result, Cresce (2012) documents that the gross vacancy rates from the Housing Unit Vacancy Survey were higher than the 1990, 2000, and 2010 Census. This indicates that the Decennial census has had a trend of understating the number of vacant addresses. One consideration that agencies should take into account when reviewing the results is to assess how the results compare to past coverage results. A difference from the past census may not necessarily be an error. Also, to potentially minimize undercoverage, the NRFU contact strategy includes an additional mailing during the NRFU operation. This provides another opportunity for people that should be enumerated at these addresses to respond without being visited by an enumerator.

For the administrative record occupied cases, we also compared the population count based on the administrative record roster to the census result. Table 9.3 shows that the population count agreed 62% of the time. For 22% of the cases, the administrative record count was higher than the census result. For 16%, the administrative record count was lower than the census count. Census results that were vacant or nonexistent had a population count of zero for these comparisons. These results show that when the counts differ that the administrative record counts were higher more often.

A second approach was to quantify how often demographic characteristics could be determined for our administrative record enumerations. Our research

Table 9.2 2010 Census NRFU status outcomes for administrative record vacant cases.

2010 Census NRFU outcome	Percent
Occupied	10
Vacant	80
Nonexistent	10

Table 9.3 Count agreement for administrative record occupied cases.

Count agreement	Percent
Administrative record count higher	22
Same count	62
Administrative record count lower	16

has focused on the availability of age, sex, race, and Hispanic origin in administrative records or third-party sources. For our administrative record enumerations, we were able to utilize PIKs assigned to the person records on different administrative record and third-party sources. We were able to research how different sources could be utilized to assign characteristics.

Table 9.4 summarizes the ability to assign these characteristics. Based on the requirement that all roster people need to have age available, age was available 100% of the time. Sex was available almost 100% of the time as well. Both of these were obtained from the SSA NUMIDENT file. Combined race and Hispanic origin was able to be assigned 90% of the time at the person level. This assignment was based on being able to assign a person to having Hispanic. Race was based *either* on being able to assign someone to one of the six race categories alone or *on the person being associated two* or more races. For the persons who could not be assigned a race or Hispanic origin, it was most often the case that these persons were at an address where the information was available for someone else in the household. For example, we have the race and Hispanic origin for the adults in the household, but not the children.

9.4.2 Macrolevel Evaluations of Quality

The microlevel summations of quality tell only one part of the story of how administrative records and third-party data impact census quality. We also assessed how the administrative record usage impacts aggregate population results. Since we

Table 9.4 Characteristic availability for administrative record occupied cases.

Characteristic availability	Percent
Age	100
Sex	100
2010 Combined race and Hispanic origin	90

were researching changes to the NRFU operation for housing units, we focused our macro-level analysis on the population in housing units. The results in this section do not include the group quarters population.

In assessing the macro-level implications, we did a retrospective analysis of the 2010 Census. This was possible by implementing the administrative record processing to identify certain percentages of NRFU cases to determine to be administrative record occupied, vacant or nonexistent. Using the paradata from the 2010 Census enumeration, we can implement a simulated 2010 Census. Based on a proposed design, we could reduce the contacts for administrative record cases. In the simulation with using the 2010 contact history information, we could determine if the case would be resolved within those number of contacts or if the administrative record results would be utilized. We proceeded to quantify what the household population results would be with this administrative record usage for vacant and occupied addresses. This section shows results for our current design. For the administrative record occupied addresses, any case that was resolved on the first contact we used the census result, but for the remaining housing units we used the administrative record information. This assesses using administrative records and third-party information to enumerate about 5 million vacant and about 4.5 million occupied housing units.

When assessing simulated results, we examined both the numerical and distributional accuracy. One example of numerical accuracy was to compare the total population. Table 9.5 shows the total household population for the 2010 Census and our research scenario. To aid the assessment, the net coverage as compared to the 2010 CCM results are shown as well. For the 2010 Census, the household population was 300 758 000. Our research scenario had a total household population result of 300 294 000. This was about 464 000 below the 2010 Census result. To help assess, we compared these results to the 2010 CCM results. The 2010 Census has a national overcount that was not significantly different from zero. Our research scenario has 0.14% undercount that is also not significant. Both scenarios are not significantly different than the 2010 CCM results.

To analyze the effect of using administrative records on distributional accuracy, we looked at results by demographics. We chose characteristics where there was

Table 9.5 Quality evaluation of total household population.

	Household population count	Net coverage (Standard error)
2010 Census	300 758 000	0.01% Overcount (0.14%)
Research scenario	300 294 000	0.14% Undercount (0.14%)
Difference	464 000	

Table 9.6 Age distribution of research scenario and 2010 Census.

	Household population	0-4	5-9	10-17	18-29	30-49	50+
Scenario	300 294 000	6.7%	6.8%	11.2%	15.9%	27.3%	32.2%
2010 Census	300 758 000	6.7%	6.8%	11.1%	15.9%	27.3%	32.2%

Note: Due to rounding of, the percentages may not sum up to 100.

potential for undercoverage of some of the population in the administrative record usage like age. We proceeded to look at the distribution of the household population by age. Table 9.6 shows the distribution of the population by six age groupings. The results show that the distribution of the research scenario is very similar to the 2010 distribution. This result was showing that administrative record usage was not adversely impacting one group or another. This was good to see especially for the younger child populations where there have been undercounts in the census and undercoverage in administrative record and third-party sources. Similar distribution analysis was done for race and Hispanic origin.

We also performed this distributional analysis at the subnational level. In the U.S., we used estimates at small geographic areas from the American Community Survey about the percent of the population that were Hispanic, Non-Hispanic Black, in poverty, and other measures. We classified the block groups based on the range of concentration of the hard-to-count population in the block group (0–10% poverty, 10–20% poverty, etc.). For each grouping, we repeated the distributional analysis above so comparisons could be made to the 2010 Census. This analysis altered the results for only 9.5 million NRFU cases out of 131 million census total addresses so the assumption could have been that these changes would not make any differences in the distributions. However, by completing this analysis, this allowed us to observe the impact of using administrative records.

Keller and Konicki (2016) provides additional ways of using coverage survey results to analyze the administrative record cases. One example was that for addresses that were determined to be administrative record occupied, vacant, or nonexistent, they generated how often those person enumerations were correct, erroneous, or required all of their characteristics to be imputed.

While not used in our research, more extension simulation approaches could have been utilized to allow for stochastic results to include some inherent randomness. Our approach was deterministic in that it only replaced cases one time based on the 2010 contact strategy. While this can be done, there are challenges to being able to set the necessary parameter values, initial conditions, and correlations of subsequent outcomes to produce useful results.

9.5 Other Applications of Administrative Record Usage

Examples described in Sections 9.2 and 9.3 demonstrate how the U.S. has used administrative records in the past and is conducting research for expanded implementation in the 2020 Census. This section gives an overview of three alternative approaches for use of administrative records when conducting a census. They are presented in decreasing order of reliance on administrative records. A register-based census replaces traditional census data gathering with existing records. Adjustment of census results utilizes information in administrative records matched to census data for purposes of estimating and correcting for errors in enumeration. Evaluation compares census data and administrative records for the purposes of checking census results but stops short of making any official adjustments. To date, the U.S. has employed only the third option.

9.5.1 Register-Based Census

A register-based census uses information already provided to the government and removes the need for conducting the data collection component found in the traditional census. Several countries have been able to determine that the quality of their administrative records is high enough to replace new data collection for census purposes. This is a cost-effective solution that reduces the burden on the respondent.

Register-based censuses fall into two main categories. The first are those conducted by only using register information. Countries including Austria, Denmark, Finland, Norway, Slovenia, and Sweden recently conducted censuses this way. These countries developed registers of individuals and dwellings that could be used for their statistical census needs.

The second are those that combine register data with information obtained from surveys already conducted. Countries including Belgium, Iceland, and the Netherlands implemented this approach. Netherlands in 2011 used the central population register based on individual municipal population registers (Schute Nordholt 2014), van Zeijl (2014) for population counts. Since occupation and educational attainment that were not available from the register, the Netherlands used response data from their Labor Force Survey to generate estimates for these two topics. The Netherlands implemented repeated weighting of the survey data, through multiple applications of a survey regression estimator, as documented in Houbiers et al. (2003), to run their approach.

9.5.2 Supplement Traditional Enumeration with Adjustments for Estimated Error for Official Census Counts

Administrative records can be used to supplement traditional census methodology in the production of the official counts. Canada conducts a traditional census enumeration but also uses the results of a sample of people found from administrative record lists to produce their official census population counts. In 1961, Canada introduced their reverse record check methodology to measure the amount of undercoverage of the traditional census enumeration. This approach takes a sample of person found from one or more administrative record lists. For each sampled person, the reverse record check attempts to determine where the person should have been enumerated on Census Day. The methodology then checks the census enumeration for that address to attempt to classify the person as enumerated, missed, or out of scope. Based on the results of this sample, an estimate of the population undercoverage can be generated. In the 2011 Census, Canada estimated that the traditional census had a 4.07% undercoverage. This result was combined with the traditional count and the overcoverage estimate of 1.85% to produce the official results (Statistics Canada 2011). Wang (2017) provides more information about the plans and challenges for implementing this for the 2016 Canadian Census.

9.5.3 Coverage Evaluation

The U.S. and other counties have used administrative records to evaluate their censuses for many years. Since 1960, the U.S. has used demographic analysis to measure coverage. This approach estimates the population based on estimates of births, deaths, immigration, and emigration. Administrative record information collected on births, deaths, immigration, and emigration is used in the evaluation. A simple algebraic equation determines the population estimate.

$$\text{Population} = \text{Births} - \text{Deaths} + \text{Immigration} - \text{Emigration}$$

U.S. Census Bureau (2012) has more details about the production of the demographic analysis estimates. In December 2010, the Census Bureau published the initial results and then proceeded to produce a revised set in 2012 based on additional analysis. In the evaluation of the 2010 Census, the U.S. started with administrative record counts of births and deaths compiled by the National Center for Health Statistics for births prior to 2009. The initial estimates had to estimate the births in 2008 and 2009 since the data was not available for the December 2010 release. The number of births in 2010 before Census Day was estimated based on the number of births in 2010 and the distribution by month in 2009. U.S. Census Bureau has details about how race and Hispanic origin was assigned.

For the number of deaths, the estimates of deaths included the total number of deaths published by the National Center for Health Statistics. This included being able to use final data for the years 2008 and 2009. In addition, deaths for 2010 were estimated in similar manner as for births.

For the components of immigration and emigration, the approach utilized data from Census 2000, the American Community Survey, and mortality statistics from the National Center for Health Statistics. In the revised estimates, the approach was able to do separate estimates for Mexican and Non-Mexican migration flows based on information from the 2010 Census of Mexico.

For the evaluation of the coverage population aged 65 or older, the administrative record data for births was supplemented to address the concern about incompleteness of birth data before 1935. The approach used births plus estimates based on recent enrollment in the Medicare health care program.

Evaluating coverage for subpopulation groups is dependent on the availability of information for that desired subgroup. For age and sex, the births, and deaths for each year could be utilized to do this. To evaluate the coverage by race, the program produced estimates for two groups – Black and non-Black. The program revised the method to assign race to births for those between 1 April 1980 and 31 March 2010 to allow a multiple race distribution. For the population 29 and younger, this allowed estimates of the population for Black alone, Black alone or in combination, or not Black alone or in combination. Based on the capturing of ethnicity information on birth certificates since 1990, the Demographic Analysis program was able to produce population estimates for the Hispanic population for ages 19 and younger Overall, the 2010 program estimated an overcount of 0.1% for the total population. While the result was showing almost no coverage difference for the total population, the program found differing results for age. For children 0–4, the Demographic Analysis program estimated an undercount of 4.6%. For women who were aged 50 or older, the program estimated an overcount of 2.4%. In order to show uncertainty in the estimates, the U.S. generated five sets of estimates based on different assumptions of births, deaths, immigration, and emigration.

9.6   Summary

This chapter provided an overview about some of the possible usages of administrative records when conducting a census. While administrative records have a long history of being used in economic censuses, the usage of administrative records for population censuses has made great strides in the past 10 years. For traditional censuses, the use of administrative records and third-party data is allowing the possibility of contacts to be reduced during the data collection period while maintaining the statistical agencies' commitments to quality. These new usages allow the potential for reduced data collection costs and the possibility of allocating additional contacts to addresses where the information is not available. A challenge for these approaches is in the assignment of characteristics obtained during the enumeration. In the U.S. situation, characteristics like age or sex may be available from another source. For other characteristics like race, Hispanic origin, or housing tenure, the question remains about the ability to obtain those values without conducting an interview especially with the desired level of detail required for census reporting.

From a survey methodology perspective, this usage of administrative records in the 2020 Census is allowing the data collection to switch modes. In 2020, addresses determined to be occupied based on administrative records will switch in the NRFU operation after one visit. So while there was a reduction of fieldwork visits for these selected addresses, they were being sent an additional mailing that would allow them to participate. In addition, with people being allowed to respond online in 2020 during the NRFU period that was not available to the public in the 2010 Census, the addressee in this follow-up operation will be able to respond without having to provide the response to a field enumerator. The approaches used here to reduce contacts based on administrative record information could also be applied to surveys as well as censuses.

When considering the addition of administrative records for enumeration purposes, an important part of the process is to assess the quality implications of the proposed change. This chapter demonstrates some quality assessment tools that were integrated into our research. This work found it was beneficial to look at microlevel and macro-level implications and suggest others to do as well. In addition, this chapter included some alternatives to examine as well.

This chapter also highlights additional ways that administrative records can be used besides the new way being used by the U.S. for a traditional test. The increase in administrative record information available for the population and dwelling allows more countries to choose to conduct a register-based census. For those that conduct traditional censuses, we present two additional examples of how administrative records can be used for those. We showed how Canada has been able to use their Reverse Record Check methodology to account for undercoverage in their official census results. Another usage for traditional censuses is for evaluation purposes. We presented how the Demographic Analysis program in the U.S. is able to use administrative records about births and deaths to assess the coverage of the population.

9.7 Exercises

9.1 This work utilized counting people in the right place and obtaining the same household compositions as observed in administrative records. What other measures would you use as alternatives?

9.2 The U.S. has researched using past census responses and the American Community Survey as training data. What other alternatives could be used?

9.3 What are the other ways administrative record data can be utilized to evaluate censuses?

References

Alvey, W. and Scheuren, F. (1982). Background for an administrative record census. In: *JSM Proceedings, Social Statistics Section*, 137–152. Washington, DC: American Statistical Association.

Bauder, M. and Judson, D. (2003), Administrative records experiment in 2000 (AREX 2000) household level analysis, U.S. Census Bureau, https://www.census.gov/pred/www/rpts/AREX2000_Household%20Analysis.pdf (accessed 17 August 2020)

Collins, G., Mallett, S., Omar, O., and Yu, L. (2011). Developing risk prediction models for type 2 diabetes: a systematic review of methodology and reporting. *BMC Medicine* 9: 103.

Cresce. A (2012), Evaluation of gross vacancy rates from the 2010 Census versus current surveys: early findings from comparisons with the 2010 Census and the 2010 ACS 1-year estimates. Federal Committee on Statistical Methodology Research Conference.

Edmonston, B. and Schultze, C. (eds.) (1995). *Modernizing the U.S. Census.* Washington, DC: National Academy Press.

Ennis, S.R., Porter, S.R., Noon, J.M., and Zapata, E. (2015). When race and Hispanic origin reporting are discrepant across administrative records and third party sources: exploring methods to assign responses. Center for Administrative Records Research and Applications, working paper #2015-08. Washington, DC: U.S. Census Bureau. https://www.census.gov/content/dam/Census/library/workingpapers/2015/adrm/carra-wp-2015-08.pdf (accessed November 2017).

Govern, K. and Coombs, J. (2013), 2010 Census coverage followup assessment report, U.S. Census Bureau.

Houbiers, M., Knottnerus, P., Kroese, A. H., Renssen, R. H., Snijders, V. (2003) Estimating consistent table sets: position paper on repeated weighting. Discussion paper 03005, Statistics Netherlands.

Jabine, T. and Scheuren, F. (1985). Goals for statistical uses of administrative records: the next 10 years. *Journal of Business and Economic Statistics* 3: 380–391.

Keller, A. and Konicki, S. (2016), Using 2010 Census coverage measurement results to better understand possible administrative records incorporation in the decennial census," in *JSM Proceedings, Survey Research Methods Section.* Alexandria, VA: American Statistical Association.

Keller, A., Fox, T., and Mule, V.T. (2016), Analysis of administrative record usage for nonresponse followup in the 2014 Census Test. U.S. Census Bureau.

Lange, A. (2013). The population and housing census in a register based statistical system. In: *The Proceedings of the 59th World Statistics Congress of the International Statistical Institute*, 2315–2320. The Hague, Netherlands: International Statistical Institute.

Layne, M., Wagner, D. and Rothhaas, C. (2014), Estimating record linkage false match rate for the person identification validation system, CARRA working paper #2014-02.

Morris, D.S. (2014). A comparison of methodologies for classification of administrative records quality for census enumeration. In: *JSM Proceedings, Survey Research Methods Section*, 1729–1743. Alexandria, VA: American Statistical Association.

Mule, V.T., Keller, A., Konicki, S. and Kjeldgaard, I. (2017), The use of administrative records to reduce costs in the 2020 decennial nonresponse followup operation. Presented at the 2016 Federal Committee on Statistical Methodology Policy Conference.

Schute Nordholt, E. (2014), The Dutch Census 2011. Conference of European Statistics Stakeholders, Methodologists, Producers and Users of European Statistics.

Statistics Canada (2011) Census technical report: coverage. https://www12.statcan.gc .ca/census-recensement/2011/ref/guides/98-303-x/index-eng.cfm (accessed 17 August 2020)

U.S. Census Bureau (1995). *1990 Census of Population and Housing History Part B*. Washington, DC, https://www.census.gov/library/publications/1996/dec/cph-r-2 .html; https://www2.census.gov/prod2/cen1990/cph-r/cph-r-2.pdf.

U.S. Census Bureau (2012), Documentation for the revised 2010 demographic analysis middle series estimates. https://www2.census.gov/programs-surveys/ popest/technical-documentation/methodology/da_methodology.pdf (accessed 17 August 2020)

U.S. Census Bureau (2015). *2020 Census Operational Plan*. Washington, DC: Census Bureau. http://www2.census.gov/programs-surveys/decennial/2020/ programmanagement/planning-docs/2020-oper-plan.pdf. (accessed March 2016).

Wang, J. (2017), "The collection challenges of the 2016 reverse record check survey. 2017 Annual Meeting of the Statistical Society of Canada.

van Zeijl, J. (2014), From traditional to register-based censuses in the Netherlands. The National Academies of Science: International Conference on Census Methods.

10

Use of Administrative Records in Small Area Estimation

Andreea L. Erciulescu[1], Carolina Franco[2] and Partha Lahiri[3]

[1] *Senior Statistician, Westat, Rockville, Maryland, USA*
[2] *Principal Researcher, U.S. Census Bureau, Center for Statistical Research and Methodology (CSRM), Washington, USA*
[3] *Professor in the Joint Program in Survey Methodology and Department of Mathematics, at the University of Maryland, College Park, Maryland, USA, 20742*

10.1   Introduction

The steadily increasing demand for various socio-economic and health statistics for small geographical areas or geo-demographic groups has led to the implementation of small area estimation programs at some government agencies as well as to a flurry of research on related statistical methods.

Small area estimation typically involves the use of auxiliary information to improve upon traditional design-based methods for inference from survey data. These design-based methods rely only on the survey data from the domain of interest for inference about that domain, rather than seeking to model and exploit relationships between the survey data from all domains and the auxiliary information available. They are typically referred to in the small area literature as *direct* methods. The auxiliary information for small area estimation methods is often drawn from administrative records and censuses. In the literature, the term *small area* is described as a domain for which the survey data alone cannot provide reliable direct estimates (Rao and Molina 2015) either because the sample size is too small or because there is no sample at all for the domain. Small area estimation refers to techniques to "borrow strength" from auxiliary variables, typically through modeling. Though this definition of small area estimation links it directly to the use of data from surveys, it is not necessary to have survey data to develop techniques for estimation of population characteristics of small domains. Some techniques that derive estimates from administrative records without using survey data will be mentioned briefly. A broader definition of small

Administrative Records for Survey Methodology, First Edition.
Edited by Asaph Young Chun, Michael D. Larsen, Gabriele Durrant, and Jerome P. Reiter.
© 2021 John Wiley & Sons, Inc. Published 2021 by John Wiley & Sons, Inc.

area estimation also includes these techniques. In this chapter, we adopt this broader definition, though we focus primarily on techniques using survey data and administrative records.

Administrative records arise from the operation of government programs, not for the purpose of estimating population characteristics. Hence, their content, coverage, accuracy, reference period, definition of variables, etc., are determined by their use in program operation, not by their use for statistical purposes. Nonetheless, valuable information for statistical inference can often be extracted from administrative records. Due to advances in computing, government agencies can process administrative records and link them with sample survey and census records for statistical purposes in a fraction of the time and costs required for field data collection. Brackstone (1987) discussed potential uses of administrative records in the production of a wide range of official statistics and pointed out their merits and demerits.

There are various ways administrative records can be used to produce small area statistics. There are methods that are purely based on administrative registers, commonly referred to as register-based methods, and use no survey data at all. Such use of administrative records in small area estimation can be traced back to the eleventh century England and the seventeenth century Canada; see Brackstone (1987). Zhang and Fosen (2012) constructed register-based small area employment rates and evaluated the progressive measurement errors in the small area estimates, using historic data from the Norwegian Employer Employee Register (NEER). For a good review of register-based small area methods, the readers are referred to Zhang and Giusti (2016).

Demographers have been using administrative records such as birth, death, migration, housing records, etc., in conjunction with the population census for estimating population for small areas for a long time now. For a comprehensive review of demographic methods for small area estimation, see Ghosh and Rao (1994) and Rao (2003).

Zanutto and Zaslavsky (2002) developed a small area method using census data and administrative records in conjunction with the nonresponse follow-up survey to impute small area detail while constraining aggregate-level estimates to agree with unbiased survey estimates. They applied their method in the 1995 U.S. Decennial Test Census where small area estimation was necessary because nonresponse follow-up was conducted in only a sample of blocks, leaving the data incomplete in the remaining blocks.

The above methods do not directly model survey data. Area-level and unit-level models, which will be discussed in more length in Section 10.3, use modeling to capture the relationship between the auxiliary data and the survey data. Such models have been widely studied in the literature and have been applied by government agencies, as will be seen in Section 10.3.

The accessibility of different administrative data from different sources has brought new opportunities for statisticians to develop innovative SAE methods that can cut down costs and improve the quality of estimates. Hundreds of papers were written in the last three decades and conferences and workshops are now being organized every year to disseminate research in the small area estimation research community. A recent comprehensive review of various methods in small area estimation can be found in Rao and Molina (2015). In this chapter, we attempt to illustrate the benefit of model-based methods that extract information from administrative records using sample survey and census data. In practice, there may be a need to use complex hierarchical models to capture spatio-temporal variations. We, however, stay with more basic models for the sake of simplicity in exposition.

The chapter is organized as follows. Section 10.2 discusses the preparation of data, including the identification and processing of administrative records for use in small area estimation models. Section 10.3 discusses small area estimation models, including both area-level models in Section 10.3.1 and unit-level models in Section 10.3.2. Section 10.4 illustrates the concepts discussed in Sections 10.2 and 10.3 with an application involving 1993 county poverty rates for school-aged children, with covariates drawn from administrative records. Section 10.5 concludes, with exercises provided in Section 10.6.

10.2   Data Preparation

Small area estimation models are most effective when good auxiliary information is available – that is, additional information that can be used in conjunction with the survey data to improve inference. This information is often drawn from administrative records, though censuses, other surveys, or past estimates from the same survey can also be used in modeling. The challenges in the identification and preparation of covariates from administrative records are not always emphasized in the literature. This section highlights some of the practical considerations for implementing a small area estimation program using administrative records. We focus here on the issues surrounding preparation of data from administrative records for use in small area estimation models, and on the qualities that make for good covariates. Examples from small area estimation programs used to produce official statistics are used as illustrations.

Data from administrative records often have the advantages that they cover large parts of the population and are relatively inexpensive. The data are collected for other purposes, so no additional cost for data collection or additional respondent burden needs be incurred, although there are some costs for obtaining the proper agreements to use the data and for preparing them. However, administrative records may not accurately represent the population for which inference is desired or measure the quantity of interest directly. For example, about 88% of the U.S. population is included in tax records from the Internal Revenue Service (IRS), the agency that collects taxes in the U.S. Selected data items from tax records are provided to the Census Bureau for its use in statistical purposes. These data are kept in the strictest confidentiality under the requirements in IRS publication 1075, "Tax Information Security Guidance for Federal, State, and Local Agencies" (http://www.irs.gov/pub/irs-pdf/p1075.pdf) as well as with the Census Bureau's own confidentiality standards. These data can be used, for instance, to tabulate covariates that can aid in the estimation of poverty rates across different geographic groups.

Federal tax records have the advantage that the laws governing them are uniform across the country and hence there is consistency in the definition of tabulations obtained from them across different geographic areas. However, these records alone cannot be used to generate reliable estimates of poverty for two reasons. First, low income households are not required to file tax returns in the U.S. Second, the pseudo poverty rate that can be derived from IRS records differs in definition from that which can be obtained by household surveys (National Research Council 2000). Nonetheless, covariates derived from IRS records serve as very valuable predictors for estimating poverty in the U.S. Census Bureau's Small Area Income and Poverty Estimates (SAIPE) Program, which estimates poverty rates for several age groups in the U.S. at the state, county, and school district levels. This is because covariates for use in small area models do not need to cover the entire target population or directly measure the quantity of interest. For a covariate to be useful in a small area estimation model it need only be strongly related to the quantity of interest and consistently defined across domains.

Inconsistencies in what data represent from place to place can lead to measurement error and to incorrect inference. When such inconsistencies are severe and cannot be adequately adjusted for, data from administrative records may not be suited for inclusion in a small area model. For instance, SAIPE considered the use of data from the National School Lunch Program, which provides free and reduced-price lunches (FRPLs) for children in schools, for estimating poverty rates at the school district level. However, due to a high level of incompleteness of the data for many states, and inconsistencies about what quantities were reported by each state, the data were not included in estimation at the school district level. Although there were particular states for which the FRPL data were more accurate and complete, they could not be used for those states because of concerns about using different methodologies for different states, both due to the difficulty in implementation and perceived fairness (National Research Council 2000). Additional reasons for not including FRPL in SAIPE modeling are cited in Cruse and Powers (2006).

A key consideration in preparing administrative records data for use in small area estimation models is the level to which the auxiliary data can be obtained. This often depends on confidentiality concerns. It may be possible to obtain the data at the unit level (i.e. person, family, household, or firm level), or it may only be possible to obtain aggregate summaries. It should be noted that most unit-level small area-level models require that the unit-level covariates be known for both the sampled and non-sampled units in the population, so that just having unit-level covariates for the sample will typically not suffice. In the area-level case, when the auxiliary variables come in the form of estimates from other surveys, care must be taken to account for their sampling error – see Section 10.3.1.

Linking or matching data from administrative records to the survey data poses challenges. Exact matching involves linking two records from the same unit, whereas statistical matching involves linking files based on similar characteristics. The error due to linking can be hard to ascertain, though a measure of error is available for many statistical and exact matching techniques (Winkler 2007). The error due to linkage is typically not incorporated in small area models.

For area-level models, which model aggregate data summaries for the domains of interest rather than unit-level data, it is not necessary to match at the unit level. However, computing such summaries sometimes involves allocating units to domains. This can be challenging for smaller areas of aggregation. For instance, geocoding involves identifying addresses with geographic locations. Addresses from the IRS need to be geocoded so that they can be allocated to the small areas of interest to produce summary statistics at the state, county, or school district levels for small area models. Geocoding of tax records from the IRS to sub county areas can be difficult, especially in some rural areas. This is primarily because some rural addresses are not in city-style format – i.e. street number and street name. At the inception of the SAIPE program, the Census Bureau had not yet developed accurate geocoding for subcounty areas, so the IRS administrative records were not used for the estimation of poverty at the school district level (National Research Council 2000). Starting with the 2005 data, SAIPE altered its methodology at the school district level to incorporate IRS data (Bell et al. 2016). There were still many tax exemptions that could not be geocoded, so Maples and Bell (2007) developed an algorithm to allocate these among school districts. Maples (2008) developed a methodology for estimating associated coefficients of variation.

Timing plays a big role in the inclusion of administrative records into small area models. One important consideration is the frequency with which the administrative records are produced, and whether that meets the needs of the small area estimation program. Usually, administrative records are released periodically – often annually or even less frequently – and sometimes data are updated later to make additions and/or corrections. Ideally, data from administrative records should cover a similar time frame as the survey of interest.

However, this is not always possible, since the reference periods of both sources are often different. For instance, for the current SAIPE methods for estimating poverty at the state, county, and school district levels, the primary data source, the American Community Survey (ACS), collects data over the course of a year and asks individuals about their income in the 12 months preceding the time of response, spanning 23 months of income overall. IRS records, on the other hand, refer to the income in a particular calendar year (Luery 2010).

Time delays in the release of administrative records can affect the timing of the release of the related small area estimates and/or the choice of which time period to use for the covariate from administrative records. In some cases, a decision is made to use lagged administrative records in order to expedite production of small area estimates. The Census Bureau's Small Area Health Insurance Estimates (SAHIEs) Program estimates numbers and proportions of health insurance coverage by counties and states. The SAHIE program uses Medicaid records from the Centers for Medicare and Medicaid Services (CMS) as well as Children's Health Insurance Program (CHIP) participation counts obtained from states and counties. Historically, SAHIE models had a one to two-year time lag in their Medicaid/CHIP covariates. In 2013 and 2014, however, many states expanded their Medicaid eligibility due to the enactment of the Patient Protection and Affordable Care Act (ACA). In response, for its 2014 estimates, SAHIE started projecting the administrative records to the year of estimation. For comparability, SAHIE rereleased estimates for 2013 using the same methodology (see Powers et al. 2016; Bauder et al. 2018).

Once some potential sources of administrative records have been identified for inclusion in small area models, relevant covariates can be derived, typically from aggregate summaries or transformations of the data. For instance, from the IRS data the SAIPE program computes a tax return pseudo-poverty rate for children for each state, computed as the number of child exemptions for returns determined to be in poverty, divided by the total number of child tax exemptions. It also computes an estimate of the proportion of people who do not file taxes and are under the age of 65 (i.e. the "non-filer rate").

When there are known inconsistencies for covariates among domains, whenever possible they should be addressed. The Supplemental Nutritional Assistance Program (SNAP, formerly known as the Food Stamp Program) provides subsidies for low-income households for food purchases. The SNAP eligibility criteria are broadly the same for all states except Hawaii and Alaska, which include some individuals with higher household incomes. For these two states the Census Bureau adjusts the data to exclude recipients who would not be eligible under the other states' criteria. It makes other adjustments for monthly outliers based on time series analysis. Some of these outliers arise from issuance of emergency SNAP benefits in response to natural disasters, particularly hurricanes. SNAP data are used to derive predictors for both the SAIPE state and county models and for SAHIE models.

In some cases, many alternative possible covariates from administrative records are available, and extensive analysis is needed to find the covariates that will be used in the final model. This is usually done with model selection tools such as the Akaike Information Criterion (AIC), checks of significance of coefficients, residual analysis, and other model diagnostics. For poverty mapping for Chilean comunas, the Ministerio de Desarrollo (Ministry of Development) had many options to choose from for auxiliary variables. In order to select the variables included in the final model, the ministry used a stepwise procedure along with model diagnostics from the statistical software Stata (Casas-Cordero Valenciaa, Encina, and Lahiri 2016). Before the statistical analysis was performed, the initial pool of potential variables was first narrowed down by subject matter experts that helped determine the relevance and reliability of the auxiliary data. Timeliness of the records was also an important consideration. This illustrates an important point – it is important to ascertain the quality of the administrative records before considering them for model inclusion. Even a statistically significant covariate may have measurement error, and this can lead to errors in inference.

After covariates from administrative records have been identified and selected for use in a small area estimation model, their quality should be periodically reevaluated. External factors such as changes in legislation or program administration may affect the predictive ability of a covariate over time. An example was mentioned above related to the effect of changes of healthcare legislation on the Medicaid/CHIP administrative records. Another example relates to the use of the SNAP covariate in SAIPE state models for poverty. In 1997, the Welfare Reform Act went into effect, which among other things, gave states more freedom to administer the SNAP program (which prior to 2008 was called the Food Stamp Program). This might have led to inconsistencies in the administration of the program that reduced the comparability of the SNAP data across states. That year, the SNAP covariate became statistically insignificant in the state poverty models. SAIPE continued to monitor the significance of the covariate, and after it was also insignificant for 1998, the covariate was removed. The covariate eventually regained its significance and was reintroduced for the 2004 estimates (Bell et al. 2016).

In addition to periodically evaluating existing models, statisticians should always be looking for new sources of covariates from administrative records. We should emphasize that administrative records for use in small area estimation will always have some error. The statistician should attempt to reduce the sources of error as much as possible, and exclude covariates from administrative records that are of poor quality and/or are not consistent in what they measure across domains, as was illustrated in this section. Errors in covariates that can be estimated or modeled could be incorporated into small area estimation models.

In most cases, it is hard to estimate the error inherent in administrative records, whether the error is considered random or deterministic. This will be discussed more in Section 10.3.

10.3 Small Area Estimation Models for Combining Information

We can envision different possible situations related to the availability of administrative data. For example, data from the administrative records can be available in one or any combination of the following forms: (i) summaries available for geographical areas that contain one or more small areas, (ii) summaries for the small areas, (iii) summaries available for geographical areas fully contained in the small areas of interest, and (iv) unit-level records. Hierarchical modeling is not necessarily simple, but it is flexible in combining information from different sources such as surveys, administrative records, and census data that are available at different levels of geography. The exact nature of hierarchical modeling depends on the nature and availability of data from different sources. In this section, we discuss the modeling and estimation, and we comment on possible issue(s) associated with the use of administrative data in small area estimation.

10.3.1 Area-level Models

Area-level models improve upon direct survey estimators by assuming a relationship between the small area parameters of interest and covariates via hierarchical models. They have the advantage that only area-level aggregate summaries of the covariates from administrative records are needed – unit-level covariates either for the sample or the population need not be available. Some early applications of area-level models in small area estimation can be found in Efron and Morris (1973, 1975) and Carter and Rolph (1974). These authors used area level covariates implicitly in forming groups of similar small areas. However, they did not explicitly use any area level covariate in modeling and did not use complex survey data.

Let θ_i be the population characteristic of interest for area i, and y_i the direct survey estimate of θ_i, with sampling variance D_i, $i = 1, ..., m$. In the context of estimating per-capita income for small places (population less than 1000), Fay and Herriot (1979) considered the following generalization of the Efron-Morris and Carter-Rolph Bayesian models:

$$\text{Level 1} \quad (\text{Sampling Distribution}) : y_i \sim ind.N\left(\theta_i, D_i\right) \tag{10.1}$$

$$\text{Level 2} \quad (\text{Prior Distribution}) : \theta_i \sim ind.N\left(x_i'\beta, A\right) \tag{10.2}$$

where x_i is a $p \times 1$ vector of known auxiliary variables derived from administrative records or other sources; D_i is the known sampling variance of y_i; β is a $p \times 1$ vector of unknown regression coefficients and A is an unknown prior variance ($i = 1, \ldots, m$). The Fay–Herriot model is a special case of a multilevel or hierarchical model where the first level captures the errors of the direct survey estimates y_i due to sampling and the second level, often called the linking model, links the true small area parameters θ_i to a set of auxiliary variables. In practice, the D_i's need to be estimated, typically using survey data. This is indeed one of the most challenging problems in the application of an area level model. We will discuss estimation of D_i subsequently.

The Fay–Herriot model given by (10.1) and (10.2) is a special case of the general linear mixed model (see, e.g. equation (5.2.1), page 98, of Rao and Molina 2015), and can be expressed as

$$y_i = \theta_i + e_i = x_i' \beta + v_i + e_i, \tag{10.3}$$

where model or linking errors $\{v_i\}$ and sampling errors $\{e_i\}$ are independent with $v_i \sim ind\ N(0, A)$ and $e_i \sim ind\ N(0, D_i)(i = 1, \ldots, m)$. Note that v_i can be viewed as a leftover random effect due to area i that is not explained by the auxiliary variables.

The Best Predictor (BP) of θ_i when the parameters are known is obtained by minimizing the mean squared prediction error (MSPE) defined as $\mathrm{MSPE}(\hat{\theta}_i) = E(\hat{\theta}_i - \theta_i)^2$, where the expectation E is with respect to the linear mixed model (10.3). The BP of θ_i is given by

$$\hat{\theta}_i^{\mathrm{BP}} = (1 - B_i)\, y_i + B_i x_i' \beta, \tag{10.4}$$

with

$$\mathrm{MSPE}\left(\hat{\theta}_i^{\mathrm{BP}}\right) = (1 - B_i)\, D_i = g_{1i}(A), \tag{10.5}$$

where $B_i = D_i/(D_i + A)$ is known as the shrinkage factor. For domains or small areas with smaller sampling variances more weight is placed on the direct estimator y_i. Note that, under the squared error loss function, the Bayes estimator of θ_i, that is, the conditional mean of θ_i given y_i (known as the posterior mean of θ_i) is identical to the BP of θ_i. Moreover, the associated measure of uncertainty for the Bayes estimator, that is, the conditional variance of θ_i given y_i (known as the posterior variance of θ_i) is identical to $\mathrm{MSPE}\left(\hat{\theta}_i^{\mathrm{B}}\right)$.

When β is unknown but A is known, the Best Linear Unbiased Predictor (BLUP) of θ_i is obtained by minimizing the $\mathrm{MSPE}(\hat{\theta}_i)$ among all linear unbiased predictors, that is, predictors of the form $\hat{\theta}_i = \sum_{j=1}^{m} l_{ij} y_j$ that satisfy the unbiasedness condition: $E(\hat{\theta}_i - \theta_i)^2$, where the expectation is with respect to the linear mixed model (10.3). Using Henderson's theory (Henderson 1953), the BLUP of θ_i can be obtained as

$$\theta_i^{\mathrm{BP}} = (1 - B_i)\, y_i + B_i x_i' \hat{\beta}^{\mathrm{WLS}}, \tag{10.6}$$

where $\hat{\beta}$ is the weighted least square estimator of β given by

$$\hat{\beta}^{\text{WLS}} \equiv \hat{\beta}^{\text{WLS}}(A) = \left(\sum_{j=1}^{m} \left(1 - B_j \right) x_j x_j' \right)^{-1} \sum_{j=1}^{m} \left(1 - B_j \right) x_j y_j.$$

Note that (10.6) is identical to (10.4) except when β replaced by $\hat{\beta}^{\text{WLS}}$. It is straight-forward to show that

$$\text{MSPE}\left(\hat{\theta}_i^{\text{BLUP}} \right) = g_{1i}(A) + g_{2i}(A), \tag{10.7}$$

where $g_{2i}(A) = B_i^2 x_i' \left(\sum_{j=1}^{m} \left(1 - B_j \right) x_j x_j' \right)^{-1} x_i$ is the additional variability in BLUP that is due to the estimation of β. Under standard regularity conditions, $g_{1i}(A) = O(1)$, but $g_{2i}(A) = O(m^{-1})$, for large m. Thus, in the standard small area higher-order asymptotic sense, $g_{1i}(A)$ contributes more to $\text{MSPE}\left(\hat{\theta}_i^{\text{BLUP}} \right)$ than $g_{2i}(A)$ does.

Note that the normality of the linking and sampling models is not needed to justify (10.6) as the BLUP of θ_i, though it is required to justify (10.4) as the BP. Under normality and assuming a non-informative flat prior on $\beta, \hat{\theta}_i^{\text{BLUP}}$, and $\text{MSPE}\left(\hat{\theta}_i^{\text{BLUP}} \right)$ are identical to the hierarchical Bayes estimator of θ_i and the corresponding posterior variance, respectively.

In practice, A is unknown. In small area applications, different estimators of A such as ANOVA, the Fay–Herriot method of moments, maximum likelihood and residual maximum likelihood methods have been considered. An Empirical Best Linear Unbiased Predictor (EBLUP), say $\hat{\theta}_i^{\text{EBLUP}}$, is obtained when A is replaced by an estimator, say $\hat{A}$. Notice that $\hat{\theta}_i^{\text{EBLUP}}$, is a weighted average of the direct estimator y_i and the regression synthetic estimator $x_i'\hat{\beta}$, where $\hat{\beta} = \hat{\beta}^{\text{WLS}}(\hat{A})$. Different well-known estimators of A are equivalent in terms of mean squared error (MSE), up to the order $O(m^{-1})$. However, in the higher-order asymptotic sense, the REML estimator of A has the least bias $(o(m^{-1}))$, compared to the bias of other standard estimators $(O(m^{-1}))$.

It is well-known that the REML, ML, and method-of-moments estimators of A can yield zero estimates, in which case the EBLUP reduces to the regression synthetic estimate for all areas; in other words the direct estimate y_i does not get any weight in the EBLUP formula, even for the largest area. In order to get around this problem, Yoshimori and Lahiri (2014a), building on earlier papers by Lahiri and Li (2009) and Li and Lahiri (2010), proposed the following general class of adjusted maximum likelihood estimators of A that includes most of the standard likelihood-based estimators of A available to date:

$$\hat{A}_h = \text{argmax}_{A \in [0, \infty)} h(A) L_{\text{RE}}(A),$$

where $L_{RE}(A)$ is the residual likelihood of A and $h(A)$ is a general adjustment term. They suggested a choice of $h(A)$ that produces a strictly positive estimate of A while maintaining the same bias and MSE properties of REML, up to the order $O(m^{-1})$.

The problem of finding an accurate estimator of the MSPE of EBLUP that captures additional variability due to the estimation of A is a challenging problem. An estimator $\widehat{MSPE}_i$ of $MSPE_i$ is called second-order unbiased or nearly unbiased if $E\left(\widehat{MSPE}_i - MSPE_i\right) = o\left(m^{-1}\right)$, under regularity conditions. For the ANOVA estimator, $\hat{A}_{ANOVA}$ of A, Prasad and Rao (1990) showed that

$$MSPE\left(\hat{\theta}_i^{EBLUP}\right) = g_{1i}(A) + g_{2i}(A) + g_{3i}(A) + o\left(m^{-1}\right), \tag{10.8}$$

where $g_{3i}(A) = \left(2D_i^2/(A+D_i)^3\right) AVar\left(\hat{A}_{ANOVA}\right)$, where $AVar\left(\hat{A}_{ANOVA}\right)$ where is the asymptotic variance of $\hat{A}_{ANOVA}$, up to the order $O(m^{-1})$. Interestingly, (10.8) holds for the general class of adjusted maximum likelihood estimators of A proposed by Yoshimori and Lahiri (2014a), under regularity conditions. Prasad and Rao (1990) noticed that a second-order unbiased estimator of $MSPE(\hat{\theta}_i^{EBLUP})$ is not obtained if we substitute $\hat{A}$ for A. By correcting the bias of $g_{1i}\left(\hat{A}_{ANOVA}\right)$, up to the order $O(m^{-1})$, they obtained the following second-order unbiased estimator of $MSPE\left(\hat{\theta}_i^{EBLUP}\right)$:

$$\widehat{MSPE}\left(\hat{\theta}_i^{EBLUP}\right) = g_{1i}\left(\hat{A}\right) + g_{2i}\left(\hat{A}\right) + 2g_{3i}\left(\hat{A}\right) + o\left(m^{-1}\right), \tag{10.9}$$

The same formula works for the REML estimator of A and the adjusted maximum likelihood estimator of Yoshimori and Lahiri (2014a), but we need additional bias corrections for the ML and adjusted maximum likelihood method of Li and Lahiri (2010). Second-order unbiased estimators based on jackknife and parametric bootstrap methods have also been proposed; see Jiang and Lahiri (2006) and Rao and Molina (2015), for a comprehensive review and comparison of these MSPE estimators.

Yoshimori and Lahiri (2014b) broadened the class of adjusted maximum likelihood estimators by introducing an area specific adjustment term $h_i(A)$, and proposed a simple second-order efficient prediction interval for θ_i of the form: $\hat{\theta}_i^{EBLUP}\left(\hat{A}_{h_i}\right) \pm z_{\alpha/2}\sqrt{g_{1i}\left(\hat{A}_{h_i}\right)}$, where $z_{\alpha/2}$ is the $100\left(1-\frac{\alpha}{2}\right)$ percentile of the standard normal deviate, and $\hat{\theta}_i^{EBLUP}\left(\hat{A}_{h_i}\right)$ is the EBLUP obtained from the BLUP when A is replaced by $\hat{A}_{h_i}$. The average length of this prediction interval is always smaller than that of the confidence interval based on the direct estimator and the coverage error is of the order $o(m^{-1})$, lower than the empirical Bayes confidence interval proposed by Cox (1976). Parametric bootstrap prediction intervals of θ_i based on EBLUP are also available; see Chatterjee et al. (2008) and Li and Lahiri (2010). Such prediction intervals have the same order of coverage error, but they are computationally intensive. Moreover, it is not known if the average length of a parametric bootstrap prediction interval is smaller than that of the direct confidence interval. Hall and Maiti (2006) also proposed a parametric bootstrap confidence interval, but it is based on a synthetic estimator.

One can assign prior distributions to β and A for a hierarchical Bayesian approach. If no prior information is available, then these parameters would typically be assigned non-informative priors (see, for instance, Berger 1985). Datta et al. (2005) and Ganesh and Lahiri (2008) considered non-informative priors for A with good frequentist properties that yield a proper posterior distribution of θ_i. Note that the posterior mean of A, an estimator of A under the hierarchical Bayesian approach, is strictly positive like the adjusted maximum likelihood estimators. One advantage of the hierarchical Bayesian approach is that it can capture different sources of uncertainties. Hierarchical Bayes implementations for the Fay–Herriot model can be implemented by numerical integration, Monte Carlo, Markov chain Monte Carlo (MCMC) or by certain approximations such as Laplace approximation or Adjustment for Density maximization (ADM); see Datta (2009), Morris and Tang (2011), Rao and Molina (2015).

Although many developments in area-level models have occurred since Fay and Herriot's seminal 1979 paper, the Fay–Herriot model is still quite useful in practice and in fact it is used for the production of official small area statistics by government agencies. For instance, SAIPE uses a hierarchical Bayes implementation for the production of state level poverty estimates. SAIPE switched from a frequentist method to a Bayesian method because estimation of the model variance, which was previously done via maximum likelihood, sometimes resulted in estimates of zero, which would imply that all the weight in EBLUP is placed on the synthetic estimate for all areas, an undesirable result. For estimation at the county level, SAIPE uses a Fay–Herriot model on log-transformed estimates of poverty counts, where the estimation of the parameters is done by maximum likelihood via an iterative method which alternates between estimating A via maximum likelihood and β via weighted least squares. Poverty mapping of Chilean comunas by the Ministerio de Desarrollo is also done via a Fay–Herriot model, featuring a variance stabilizing transformation. The issue of the potential for zero estimates for the model variance can be handled by using the adjusted maximum likelihood estimator of Li and Lahiri (2010).

Area level summaries from administrative records could be potentially useful in a situation where no survey data are available for some of the areas. For example, the Current Population Survey (CPS) sample design generally does not produce any data for a majority of U.S. counties, and so EBLUP methodology as described above cannot be used to draw inference for these counties. However, from the EBLUP methodology one may derive synthetic estimates for counties with no data, using only the regression term of (10.6). Back in the 1990's, the U.S. Census Bureau found using administrative record covariates useful in small area estimation models for poverty. This permitted estimates for counties with no survey data. This implicitly assumes that regression coefficients for the linking model do not change across counties. A possible alternative solution might be to incorporate a spatial correlation into the Fay–Herriot linking model (see Vogt 2010). An estimate derived from such a model not only uses administrative data summaries but also uses survey estimates from neighboring areas. However, with this appeal of the spatial models comes the complexity in defining spatial neighborhood and in estimation of spatial correlation. More general spatial models have been used in the small area literature (see, e.g. Rao and Molina 2015). However, the potential utility of spatial models to improve small area estimation for areas with no survey data needs further evaluation.

The sampling variance D_i is assumed to be known but must be estimated in practice. Often, direct estimates of the sampling variance are available from the survey used as the primary data source. However, these direct estimates may be unreliable, or in some cases unavailable. In such cases alternative estimates must be explored. Some approaches for this are the use of generalized variance functions (GVF's, see Wolter 2007), or simpler approaches that make assumptions about uniformity over larger levels of aggregation to obtain "smoothed" sampling variance estimates for small areas. As an example of the former, at the inception of SAIPE, the primary data source for annual poverty estimates was the CPS, later to be replaced by the ACS, a survey with much larger sample size. Sampling variances of CPS estimates at the state-level were produced but were based on small sample sizes, and at the county-level direct sampling variance estimates were not produced. For this reason, SAIPE obtained sampling variances estimates via GVF's – for the state level, a GVF with random effects was developed by Otto and Bell (1995), and for the county level, a much simpler GVF was developed (Bell 2016). For poverty mapping for Chilean comunas, the smallest territorial entity in Chile, the use of a variance-stabilizing transformation eliminated the need for sampling variance estimates, but estimates of the design effect were needed to compute the effective sample sizes for each small area. These were computed at the regional level to avoid unstable estimates at smaller levels of aggregation (Casas-Cordero Valencia, Encina, and Lahiri 2016), with the underlying assumption that these estimates were also valid under lower levels of aggregation.

We note that administrative data could be potentially used in estimating D_i for small areas. For example, one can use area level summaries from administrative records as covariates in the synthetic estimators of sampling variances of small area proportions; see Liu (2009). The method first fits the following logistic model using data only from large areas to obtain stable estimates of model parameters:

$$\text{logit} \left(p_{iw} \right) = x_i' \beta + \varepsilon_i, \quad i = 1, \dots, I, \tag{10.10}$$

where for the ith large area p_{iw} is a direct estimate of proportion P_i and x_i is a vector of known covariates available in area i and $\varepsilon_i \sim iid \, N(0, \sigma^2) \, (i = 1, \dots, I)$. Using an estimate $\hat{\beta}$ of β using data from the large areas, synthetic estimates of P_i for all areas are obtained as:

$$\tilde{p}_{isyn} = \frac{\exp \left(x_i' \hat{\beta} \right)}{1 + \exp \left(x_i' \hat{\beta} \right)}, i = 1, \dots, m. \tag{10.11}$$

Finally, smoothed estimates of D_i's are obtained as:

$$\hat{D}_{i;syn} = \frac{\tilde{p}_{isyn} \left(1 - \tilde{p}_{isyn} \right)}{n_i} \text{deff}_{iw}, \tag{10.12}$$

where deff_{iw} is a suitable approximation of the design effect for area i (e.g. design effect estimate for a large area containing the small area i).

In order to capture variation due to the estimation of sampling variances D_i and at the same time to obtain smoothed sampling variances, Liu (2009) considered a number of integrated hierarchical models for proportions, including non-normal models. In the context of estimating small area proportions, one such integrated Bayesian model – an extension of the Fay–Herriot model – is given by:

$$\text{Level 1 } \left(\text{Sampling Distribution} \right) : y_i \sim ind.N \left(\theta_i, \frac{\theta_i \left(1 - \theta_i \right)}{n_i} \text{deff}_i \right) \tag{10.13}$$

$$\text{Level 2 (Prior Distribution)}: \theta_i \sim ind.N \left(x_i' \beta, A \right). \tag{10.14}$$

Assuming a non-informative prior on the hyperparameters β and A and using MCMC, one can easily produce the hierarchical Bayesian estimate of P_i and sampling variance of y_i as the posterior means of P_i and $\frac{\theta_i(1-\theta_i)}{n_i} \text{deff}_i$, respectively. Liu (2009) (also see Liu et al. 2014) evaluated design-based properties of such hierarchical Bayesian methods using Monte Carlo simulations. Ha et al. (2014) compared a number of integrated hierarchical models in the context of estimating smoking prevalences in the U.S. states. An early example of an integrated hierarchical modeling approach in small area estimation for continuous variables can be found in Arora and Lahiri (1997). For a review of variance modeling, readers are referred to Hawala and Lahiri (2018).

The research on area-level models has been extensive. We describe here a few extensions, though many others have been proposed in the literature. For more information on area-level models, see also Molina and Rao (2015) or

Pfeffermann (2013). Some extensions to area-level modeling are those that use Generalized Linear Mixed Models, which can depart from the assumptions of normality in the original Fay–Herriot model. These can be appropriate in cases where the data are discrete, skewed, or when the model errors are thought to be heteroscedastic. Small area estimation models via GLMMs have been discussed by Ghosh et al. (1998) and Rao and Molina (2015), among others. In the context of estimating poverty rates for Chilean comunas, Ha (2013) developed an ADM approximation to the posterior distribution of small area proportions using a hierarchical Binomial-Beta model that combines information from area level summary statistics derived from different administrative records and complex survey data. Franco and Bell (2015) discuss a binomial-logit-normal model with multivariate and time series extensions.

Another type of extension is subarea multilevel models, where each area is divided into subareas, and the interest is in predicting both at the area and subarea level. Fuller and Goyeneche (1998) introduced subarea-level models with an application to SAIPE, where the subareas were the counties, nested within states (areas). Subarea-level models have also been studied by Torabi and Rao (2014), Rao and Molina (2015), and Kim et al. (2018). Erciulescu et al. (2018) developed a hierarchical Bayes subarea-level model for harvested acreage of crop commodities in U.S. counties (sub-areas) and agricultural statistics districts (areas), with choices of covariates that included administrative acreage data.

One straightforward but useful extension is the multivariate version of the Fay–Herriot model. Multivariate Fay–Herriot models have been explored, for instance, by Fay (1987), Datta, Fay, and Ghosh (1991), and Bell and Huang (2012), among others. Multivariate area-level models allow for the joint modeling of related characteristics. These could be different estimates from the same survey or from different surveys. By jointly modeling related characteristics, it is possible to improve the estimation by exploiting the correlation among them. Moreover, when attempting to estimate functions of more than one of the responses, jointly modeling properly accounts for the correlations among model errors. For instance, one may want to estimate the year to year change in a poverty rate, or some other characteristic based on estimates of the same survey for two consecutive years (e.g. Arima et al. 2018).

When jointly modeling consecutive estimates from several years of data collection, one may also choose to use time series extensions to area-level models. Time series extensions of the Fay–Herriot model are discussed in Rao and Yu (1994), Ghosh et al. (1996), Datta et al. (1999), and Rao and Molina (2015). Another growing area of research within small area estimation is spatial and spatio-temporal models, which exploit the dependence among data points across space. Note that space does not necessarily need to be defined as geographic distance. See, for instance, Esteban, Morales, and Perez (2016) and Rao et al. (2016). These models may be explored to find alternative solutions to the situation when survey data are unavailable for some areas for some time points, as mentioned earlier.

In many area-level applications, summaries from different administrative databases are used as covariates. These are likely to be subject to measurement error, where such error may be defined as the difference between the summary statistic and the value it is intended to measure. Ybarra and Lohr (2008) stimulated a flurry of research activity on extending small area methodologies to account for errors in the covariates when these errors are random, as may arise when the covariates come from other surveys. In view of this recent research productivity, one may naturally ask whether these ideas can be applied to account for measurement error in covariates arising from administrative records. To discuss this question, we assume that the Fay–Herriot model (10.3) is the true model, but that we do not observe x_i, but rather a noisy estimate X_i, of x_i. Suppose that $X_i = x_i + \eta_i$, $\eta_i \sim N(0, C_i)$. Under these assumptions, two types of measurement error have been discussed in the literature: *functional* (e.g. Ybarra and Lohr 2007) and *structural* (e.g. Bell et al. 2017). The former assumes that the true x_i is a fixed but unknown quantity. The latter assumes that x_i is random and follows a model.

A naive model would simply ignore measurement error and treat the X_i s as if they were the true x_is. The resulting predictor would be of the form

$$\hat{\theta}_i^{\text{NV}} = \left(1 - \hat{B}_{i,\text{NV}}\right) y_i + \hat{B}_{i,\text{NV}} X_i' \hat{\beta}_{\text{NV}}. \tag{10.15}$$

where B_i is the same as in (10.4). The estimators of β and B_i are given by the subscripts NV because the model parameter estimates obtained assuming the naive model is true do not converge to the true β and A in (10.3).

The implied model in expression (10.15) models the relationship between the true quantities θ_i and the noisy estimates X_i of the true quantity x_i, rather than modeling the relationship between the two true quantities θ_i and x_i, which is more reasonable. Prediction results will differ between the naive models and measurement error models, except in special cases. The naive model also assumes homoscedasticity of the residuals $y_i - X_i'\beta$, which will not hold when the C_i's differ across domains. Bell et al. (2017) compare the effects on MSPE's from using the naive, functional, or structural models, assuming either the structural or functional models are true.

Expressions for EBLUPs for functional and structural measurement error models can be found, for instance, in Ybarra and Lohr (2008) and Bell et al. (2018). We now point out a few challenges in implementing these predictors when the X_i are drawn from adminstrative records. First, applying measurement error models require estimates of the C_i. When the X_i's are derived from administrative records, it is not clear how one should define and estimate C_i in the presence of a multitude of possibly unknown sources of errors. Holt (2007) noticed a clear lack of statistical theories for assessing the uncertainty of register-based statistics. Zhang (2012) discussed some possible theories for statistics based on registers but did not put forward any concrete suggestion for defining the MSE of register-based statistics. Secondly, confidentiality aspects of most administrative data in the U.S. limit the prospect for advancement of research in understanding the theory of statistics derived from administrative data. Finally, though measurement error models assume C_i is known, their optimality properties are not guaranteed when C_i is estimated.

The literature on measurement error can shed some light on what happens when there is random error in covariates derived from administrative records provided the measurement error model, whether functional or structural, is reasonable for the inherent measurement error in the application in question. Bell et al. (2017) show that the naive model discussed above, which ignores measurement error, can lead to misstated MSEs when either the structural or functional measurement error models are true. Exceptions to this arise when the structural measurement error model is true and occur for areas where $C_i = \overline{C}$, where $\overline{C}$ is the mean of the C_is, in which case the predictions are the same for the structural and naive models. It follows that if C_i is constant for all areas then the naive and the structural measurement error predictors give the same results for all areas, but in other cases the results differ. However, the measurement error models discussed above cannot currently be used to correct measurement error in the covariates drawn from administrative records even when it is reasonable to assume that a measurement error model holds, due to the difficulties in determining the C_is.

10.3.2 Unit-level Models

As noted in the last section, uncertainty due to estimation of the sampling variances of direct estimators cannot be incorporated in an area level model without additional information on sample design and within area variation information. In order to capture this additional uncertainty, we need observations within each area. Let n_{ij} be the sample size for the ith area and y_{ij} be the value of the study variable for the jth unit (e.g. person, household, farm, etc.) of the ith small area population ($i = 1, \ldots, m; j = 1, \ldots, n_j$). One possible model is the following:

$$\text{Level 1 } (\text{Sampling Distribution}): y_{ij} \mid \theta_i \sim ind\ N\left(\theta_i, \sigma_e^2\right), \tag{10.16}$$

$$\text{Level 2 (Prior Distribution)}: \theta_i \sim ind\ N\left(x_i'\beta, \sigma_v^2\right), \tag{10.17}$$

$i = 1, \ldots, m; j = 1, \ldots, n_j$ where x_i is a $p \times 1$ vector of area specific known auxiliary variables (some of them could be summaries from administrative records); σ_e^2 and σ_v^2 are within and between area unknown variance components, respectively.

The model described by Eqs. (10.16) and (10.17) can be written in the following linear mixed model form:

$$y_{ij} = x_i' \beta + v_i + e_{ij}, i = 1, \ldots, m; j = 1, \ldots, n_i, \tag{10.18}$$

where area specific random effects $\{v_i\}$ and the random error $\{e_{ij}\}$ are independent with $v_i \sim N\left(0, \sigma_v^2\right)$ and $e_{ij} \sim N\left(0, \sigma_e^2\right)$. We are interested in predicting the mixed effect $\theta_i = x_i' \beta + v_i$ Note that if σ_e^2 is known, we can equivalently use the area level model, discussed in Section 10.3.1, on the sample mean $\bar{y}_i = n_i^{-1} \sum_{j=1}^{n_i} y_{ij}$ with $D_i = \sigma_e^2/n_i$ and $A = \sigma_v^2$.

The two-level model (10.16) and (10.17) or the linear mixed model (10.18) allows for estimation of both the variance components and accounts for uncertainty due to estimation of both the variance components. One advantage of the model is that we do not need an auxiliary variable information at the unit level, which would require complex linking procedures, especially when the auxiliary information comes from administrative records.

The exchangeability assumption in the sampling error model (10.16), however, is rather strong. It is possible to relax the modeling assumption if the auxiliary variables are available for all sampled units and some summary information about the auxiliary variables is available for the non-sampled units in each small area. Such a situation was considered by Battese et al. (1988). They proposed an EBLUP method to predict areas under corn and soybean for 12 counties of north-central Iowa using the 1978 June Enumerative Survey (JES) and satellite (LANDSAT) data.

Although satellite data may not be categorized as administrative records, it will be instructive to discuss how Battese et al. (1988) linked LANDSAT data to survey data. The unit for recording the satellite information is a pixel (about 0.45 ha), a term used for "picture element," and the unit of measurement for the survey data is a farmer. Thus, the units of measurements for these two databases are different, which could often be the case when survey data are to be linked with administrative records. Before implementing an EBLUP methodology, a data preparation step was thus needed to define a common unit for which aggregates from both survey and satellite data can be obtained.

Battese et al. (1988) considered segment (about 250 ha), the primary sampling unit (PSU) of the JES, as the common unit for the two databases. The areas under corn and soybean for each sampled segment were determined by the USDA Statistical Reporting field staff by interviewing farm operators. Using USDA procedures, recordings from LANDSAT during August and June 1978 were used to classify crop cover for all pixels in 12 counties, and this information was used to obtain areas under corn and soybean for all sampled and non-sampled segments in the 12 counties. There can be, however, errors in assigning pixels to the right segment or classifying a pixel to the right crop. There could also be potential errors in compiling data at the segment level from the information obtained from the farmers in JES. Battese et al. (1988) ignored such possible errors in developing their EBLUP method.

Let N_i be the population size and y_{ij} be the value of the study variable for the jth unit of the ith small area population ($i = 1, \ldots, m; j = 1, \ldots, N_j$). Suppose we are interested in estimating the finite population mean $\overline{Y}_i = N_i^{-1} \sum_{j=1}^{N_i} y_{ij}$, or, equivalently, the finite population total $N_i \overline{Y}_i$, when N_i is known. In Battese et al. (1988), N_i is the number of segments and $\overline{Y}_i$ is the average hectare of crop per segment for county i, the parameter of interest.

Battese et al. (1988) considered the following linear mixed model, commonly referred to as a nested error regression model:

$$y_{ij} = x'_{ij}\beta + v_i + e_{ij}, \tag{10.19}$$

$i = 1, \ldots, m; j = 1, \ldots, N_j$, where x_{ij} is a $p \times 1$ column vector of known auxiliary variables; $\{v_i\}$ and $\{e_{ij}\}$ are all independent with $v_i \sim iid\, N\left(0, \sigma_v^2\right)$ and $e_{ij} \sim iid\, N\left(0, \sigma_e^2\right)$. Thus, they assumed that the model holds for all the units of the finite population.

We can also write the model as a two-level model:

$$\text{Level 1}: y_{ij} \mid v_i \sim ind\, N\left(x'_{ij}\beta + v_i, \sigma_e^2\right) \tag{10.20}$$

$$\text{Level 2}: v_i \sim iid\, N\left(0, \sigma_v^2\right), \tag{10.21}$$

$i = 1, \ldots, m; j = 1, \ldots, N_i$.

A model-dependent estimator of $\overline{Y}_i$ can be written as

$$\hat{\overline{Y}}_i = f_i\overline{y}_i + \left(1 - f_i\right)\hat{\overline{Y}}_{ir}, \tag{10.22}$$

where $f_i = n_i/N_i$, the sampling fraction, and $\hat{\overline{Y}}_{ir}$ is a model-dependent predictor of $\overline{Y}_{ir} = \left(N_i - n_i\right)^{-1} \sum_{j \,\text{not}\in s_i} y_{ij}$, with s_i being the sample for area i ($i = 1, \ldots, m$). For the Bayes or BP of $\overline{Y}_{ir}$, we have

$$\hat{\overline{Y}}_{ir} = \overline{x}'_{ir}\beta + \tilde{v}_i, \tag{10.23}$$

where $\overline{x}_{ir} = \left(N_i - n_i\right)^{-1} \sum_{j \,\text{not}\in s_i} x_{ij}$ and $\tilde{v}_i \equiv \tilde{v}_i\left(\beta, \lambda\right) = \left[1 - B_i\left(\lambda\right)\right]\left(\overline{y}_i - \overline{x}'_i\beta\right)$, with $B_i \equiv B_i(\lambda) = \lambda/(\lambda + n_i)$, and $\lambda = \sigma_e^2/\sigma_v^2$.

In an empirical Bayes (EB) setting, one would estimate the hyperparameters using a classical method. For example, one can estimate β by the weighted least squares estimator with estimated variance components and restricted (or residual, or reduced) maximum likelihood (REML) to estimate the variance components. One can then use a resampling method (e.g. Jiang et al. 2002) or Taylor series (e.g. Datta and Lahiri 1999; Das et al. 2004) method to estimate the MSE. Confidence intervals can be obtained using the parametric bootstrap method of Chatterjee, Lahiri, and Li (2008).

In a hierarchical Bayes (HB) setting, one would put a prior on the hyperparameters. Typically, enough data will be available to estimate β and σ_e^2 so that one can use any reasonable noninformative prior distribution. For example, one can assume that *a priori* β and σ_e are independent and β and σ_e have improper uniform priors in the p-dimensional Euclidean space and positive part of the real line, respectively. The prior on σ_v is less clear cut. See Gelman (2006). One suggestion is to put an improper uniform prior on σ_v. MCMC can be applied to carry out the fully Bayesian data analysis for a variety of inferential problems. See Molina and Rao (2015) for various extensions of the nested error model proposed by Battese et al. (1988) and the different estimation methods.

Heteroscedasticity in the distribution of e_{ij} is common in real applications. One solution proposed in the literature is to assume nested error regression model (10.19) on some suitable transformations (e.g. logarithm) of the data of study and auxiliary variables. As the heterogeneity problem persists even after such transformations, Molina et al. (2014) suggested to replace σ_e^2 by $k_{ij}\sigma_e^2$ in the nested error regression model, where k_{ij} is known. In many real applications, however, it is not easy to choose k_{ij}. In a poverty mapping application, Elbers et al. (2003) discussed this problem of heteroscedasticity at length and suggested certain complex modeling on the sampling variances. The nested error regression modeling assumption on the transformed study variable and/or the estimation of complex non-linear parameters such as the poverty gap and poverty severity (see Foster et al. 1984) require availability of auxiliary variables not only for the sampled units but also for all units in the population, which limits the use of such models in many applications.

In an effort to provide a solution to the heteroscedasticity problem that does not require transformations, Bellow and Lahiri (2012) considered a nested error regression model (10.19) with σ_e^2 replaced by $x_{ij}^\delta \sigma_e^2$, where x_{ij} is a size variable available in a list frame, and δ is a parameter to be estimated from the data.

Another potential solution that avoids transformation was suggested by Gershunskaya and Lahiri (2018). The predictor of $\overline{Y}_{ir}$ is derived from a model (denoted N2) that is obtained from the nested error regression model (10.19) with the distribution of e_{ij} following a mixture of two normal distributions with zero mean but different variances:

$$e_{ij} \mid z_{ij} \sim ind \ (1 - z_{ij}) \ N\left(0, \sigma_1^2\right) + z_{ij} N\left(0, \sigma_2^2\right), \tag{10.24}$$

where the mixture part indicators z_{ij} are independently identically distributed Bernoulli random binomial variables with a common success probability π (probability of belonging to part 2). Gershunskaya and Lahiri (2018) obtained the following empirical best predictor (EBP) of $\overline{Y}_{ir}$:

$$\widehat{\overline{Y}}_{ir}^{N2} = \overline{x}_{ir}'\hat{\beta}^{N2} + \hat{v}_i^{N2}, \tag{10.25}$$

where

$$\hat{\beta}^{N2} = \left(\sum_{i=1}^{m} \sum_{j=1}^{n_i} w_{ij} x_{ij} x'_{ij} \right)^{-1} \sum_{i=1}^{m} \sum_{j=1}^{n_i} w_{ij} x_{ij} \left(y_{ij} - \hat{v}_i^{N2} \right) \tag{10.26}$$

$$\hat{v}_i^{N2} = \frac{\sigma_v^2}{D_i^{N2} + \sigma_v^2} \left(\hat{\tilde{y}}_i^{N2} - \hat{\tilde{x}}_i^{N2\prime} \hat{\beta}^{N2} \right) \tag{10.27}$$

with

$$w_{ij} = \hat{\sigma}_1^{-2} \left(1 - \hat{z}_{ij} \right) + \hat{\sigma}_2^{-2} \hat{z}_{ij}, \hat{z}_{ij} = E \left(z_{ij} \mid y_{ij}, x_{ij}, \hat{\varphi} \right)$$

$$D_i^{N2} = \left(\sum_{j=1}^{n_i} w_{ij} \right)^{-1},$$

$$\hat{\tilde{y}}_i^{N2} = \left(\sum_{j=1}^{n_i} w_{ij} \right)^{-1} \sum_{j=1}^{n_i} w_{ij} y_{ij},$$

$$\hat{\tilde{x}}_i^{N2} = \left(\sum_{j=1}^{n_i} w_{ij} \right)^{-1} \sum_{j=1}^{n_i} w_{ij} x_{ij},$$

and the hyperparameters $\varphi = (\beta, \sigma_1^2, \sigma_2^2, \sigma_v^2, \pi)$ are estimated using the expectation-maximization (EM) algorithm (Dempster, Laird, and Rubin 1977).

Note that $\hat{\tilde{y}}_i^{N2}$ accounts for outliers. Although $\hat{\tilde{y}}_i^{N2}$ resembles a "direct estimator," unlike direct estimators, it depends on units from other areas through the estimates of variances and the probabilities belonging to part 2 of the mixture. Each observation has its own conditional probability $P \left(z_{ij} = 1 \mid y_{ij}, x_{ij}, \hat{\varphi} \right) = E \left(z_{ij} = 1 \mid y_{ij}, x_{ij}, \hat{\varphi} \right)$ of belonging to part 2 of the mixture so that the observations in the sample can be ranked according to these probabilities. The estimate of $\hat{\beta}^{N2}$ (thus, the synthetic part of the estimator) is outlier robust because the outlying observations would be classified with a higher probability to the higher variance part of the mixture; hence, they would be "down-weighted" according to the formula for $\hat{\beta}^{N2}$.

Gershunskaya and Lahiri (2018) proposed the following overall bias-corrected EBP of $\overline{Y}_{ir}$:

$$\hat{Y}_{ir}^{N2+OBC} = \hat{Y}_{ir}^{N2} + n^{-1} s^R \sum_{i=1}^{m} \sum_{j=1}^{n_i} \psi_b \left(e_{ij}^{N2} / s^R \right),$$

where $e_{ij}^{N2} = y_{ij} - x'_{ij} \hat{\beta}^{N2} - \hat{v}_i^{N2}$, s^R is a robust measure of scale for the set of residuals $\left\{ e_{ij}^{N2}, i = 1, \ldots, m; j = 1, \ldots, n_i \right\}$, and $n = \sum_{i=1}^{m} n_i$, the overall sample size. They considered $s^R = \text{med} \mid e_{ij}^{N2} - \text{med} \left(e_{ij}^{N2} \right) \mid /0.6745$, and ψ_b is a bounded Huber's function with the tuning parameter of $b = 5$.

One potential problem in using unit-level covariates from administrative records is that covariate information could be missing for a number of units. In such a case, one may consider a solution proposed by Bellow and Lahiri (2012) although they encountered the problem in a covariate from a list sampling frame. Following Bellow and Lahiri (2012), one can divide the administrative records into two groups – group 1 where unit-level covariates are not missing and group 2 where unit-level covariates are missing. One can then apply an EBP or Bayesian method using the nested error model (10.19) with the unit-level covariate for group 1 and apply an alternative method for group 2. For instance, one can use EBP or Bayesian approaches without the covariates that are missing units or a simpler method such as a synthetic method used by Bellow and Lahiri (2012). Estimates from these two groups can then be combined in an appropriate way. More research is needed to address the issue of missing unit-level covariates.

10.4 An Application

We have mentioned the U.S. Census Bureau's SAIPE program throughout the chapter to illustrate concepts related to data preparation and modeling. In fact, SAIPE is a good example of a successful implementation of a small area program by a government agency. In this section, we use past data similar to those used by SAIPE to show readers how one might analyze survey data using covariates from administrative records to produce small area statistics.

From Section 10.2, recall SAIPE produces poverty statistics for various age groups at different levels of geographic aggregation in the U.S. – at the state level, county level, and at the school district level. The estimates for related school-aged (aged 5–17) children in poverty at the school district level are used for the allocation of funds by the U.S. Department of Education – over $16 billion (U.S. currency) in the fiscal year 2014. "Related" here refers to children in families. The primary data source for SAIPE's area-level models are estimates from the ACS, which samples approximately 3.5 million addresses per year. SAIPE uses data from administrative records as a source of auxiliary information for all age groups and levels of geographic aggregation. The main sources of covariates from administrative records are selected tax records obtained from an interagency agreement with the IRS and data from the SNAP. For estimates for the over 65 population SAIPE uses data from the Supplemental Security Income (SSI) program instead of SNAP. Some of the issues associated with using these administrative records as covariates were discussed in Section 10.2. SAIPE also uses estimates from the 2000 Census long-form as covariates. The census long-form, a survey that used to be part of the decennial census data collection, was discontinued after 2000 and was replaced by the ACS. Research has been conducted to explore replacing the

2000 Census long-form estimates by more current estimates based on five years of ACS data collection (Huang and Bell 2012, Franco and Bell 2015).

Much has been written about SAIPE over the years. See, for instance, the recent book chapter by Bell et al. (2016), or the many other publications available at the SAIPE website: https://www.census.gov/programs-surveys/saipe/library.html. Here, we focus on how to analyze data for school-aged children in poverty at the state level using data from the CPS and associated administrative records tabulations for the year 1993. The CPS, sponsored jointly by the U.S. Census Bureau and the Bureau of Labor statistics, is primarily designed to produce monthly estimates related to labor force participation and employment, and is used to produce national unemployment rates, among other statistics. It has a multistage probability sample design. For more information about the CPS, see http://www.census.gov/programs-surveys/cps.html.

As a historical note, SAIPE used data from the CPS from its inception until the year 2004. Starting in 2005, SAIPE began using data from the ACS due to its larger sample size – the ACS sampled approximately 3 million addresses per year at the time, whereas CPS sampled approximately 100 000 addresses in 2005. In the year 1993, CPS sampled about 60 000 addresses.

The data set we use here is from Bell and Franco (2017), available at https://www.census.gov/srd/csrmreports/byyear.html. The compressed set of files included there contains the text files cps93p.txt and CEN89RES.txt, among others. The columns corresponding to the ages 5–17 in these files contain the variables we use here, at the state level, for the year 1993. All but the last variable listed below are in the cps93p.txt file. The last variable is in the CEN89RES.txt file.

- **cps93** – The direct CPS estimated poverty rates for related children ages 5–17.
- **irspr93** – The pseudo-poverty rates tabulated from IRS tax data. These are defined as the number of child tax exemptions for poor households divided by the total number of child tax exemptions.
- **irsnf93** – The tax non-filer rates tabulated from IRS tax data, defined as the difference between the estimated population and number of tax exemptions under age 65, divided by the estimated population under age 65.
- **fs93** – The Food Stamp participation proportions. As pointed out in Section 10.2, the Food Stamp Program changed its name to the SNAP in 2008. This variable is the average monthly number of individuals receiving food stamps over a 12-month period, as a percentage of the population.
- **smpsize** – The CPS sample size (number of interviewed households).
- **fnlse** – The GVF estimates of sampling standard errors from the CPS. These are computed using the GVF developed by Otto and Bell (1995), using an iterative procedure that alternates between estimation of model parameters via maximum likelihood and estimation of the sampling standard errors.

Table 10.1 Regression prediction results for model with four regressors.

Variable	Coefficient	S.E.	t	$Pr > |t|$
Intercept	−3.477	2.224	−1.564	0.118
IRS pseudo-poverty rate	0.267	0.125	2.144	0.032
IRS non-filer rate	0.509	0.156	3.261	0.001
Food stamp participation rate	1.185	0.268	4.429	<0.001
Census residuals	1.261	0.413	3.050	0.002

- **cen89rsd** – The residuals obtained by fitting a Fay–Herriot model to the estimates of children in poverty from the 1990 census, with analogous covariates to those used here but for the year 1989. These are found in the "Age 5–17" column of the file CEN89RES.

In some years, the census residuals were replaced in the SAIPE production model by the census estimates of children in poverty. See Bell et al. (2016) for more details.

We conduct analysis in the spirit of Bell et al. (2007), which perform analysis of ACS data and related covariates to produce county-level poverty estimates. We use some of the same model selection and diagnostic tools as in this technical report as an illustration of how such an analysis might be done in practice. The data set we use in this chapter is for research purposes only and may differ slightly from that used in actual SAIPE production.

We fit a Fay–Herriot model, given by Eqs. (10.1) and (10.2). Here, y_i is given by **cps93** for each state, x_i is given by an intercept term, and **irspr93**, **irsnf93**, **fs93**, **cenrsd**, and D_i are given by **fnlse^2**, again, for each state. The analysis here is done with the "sae" package in R (Molina and Marhuenda 2015), and we invite the reader to replicate the analysis as an exercise.

We first explore the relationships between the covariates and the response. Figure 10.1 plots these against each other. All covariates appear to have a positive correlation with CPS direct estimates of poverty, though these relationships appear to be stronger for the Food Stamp participation rate and the IRS pseudo poverty rate. Both of those covariates are from administrative records. Note that the true relationship between the covariates and the true poverty rate is masked by the sampling variance of the CPS estimates (Table 10.1).

We then fit the Fay–Herriot model to the CPS poverty rates using all combinations of the four covariates described above. The 15 models considered are listed in Table 10.2.

The models are fit using restricted maximum likelihood estimation to estimate the model variance. Table 10.1 shows results for the regression coefficients. Note

Table 10.2 Model comparison.

Model	Regressors	Model variance	AIC
M1	IRS pseudo poverty rate	12.803	316.638
M2	IRS non-filer rate	25.307	341.525
M3	Food Stamp participation rate	6.049	294.007
M4	Census residuals	30.631	345.437
M12	IRS pseudo poverty rate IRS non-filer rate	7.972	308.810
M13	IRS pseudo poverty rate Food Stamp participation rate	6.051	295.332
M14	IRS pseudo poverty rate Census residuals	9.741	311.074
M23	IRS non-filer rate Food Stamp participation rate	3.449	290.033
M24	IRS non-filer rate Census residuals	24.310	339.345
M34	Food Stamp participation rate Census residuals	5.468	289.568
M123	IRS pseudo poverty rate IRS non-filer rate Food Stamp participation rate	3.061	290.188
M124	IRS pseudo poverty rate IRS non-filer rate Census residuals	4.418	300.373
M134	IRS pseudo poverty rate Food Stamp participation rate Census residuals	4.850	289.749
M234	IRS non-filer rate Food Stamp participation rate Census residuals	3.257	285.264
M1234	IRS pseudo poverty rate IRS non-filer rate Food Stamp participation rate Census residuals	1.703	282.601

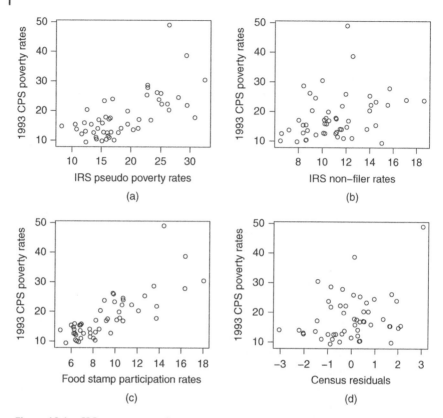

Figure 10.1 CPS poverty rates for school-aged children plotted against (a) IRS pseudo-poverty rates, (b) IRS non-filer rates, (c) Food Stamp (SNAP) participation rates, and (d) Census 1990 residuals.

that all regression coefficients are significant at the 0.05 significance level, and, as suggested in Figure 10.1, all coefficients are positive. Table 10.2 shows the estimated model variances and AIC's for all models. The lowest AIC and the lowest model variance correspond to the model including all covariates. This is not surprising since all the *t*-statistics are significant and since the model with all covariates is the one that was used for production. The largest estimated model variance is for the model with the Census residuals as the only covariate. However, the census residual covariate is intended to be used in conjunction with the other covariates. Since a model similar to that used for the year 1993 was used to produce them (using census poverty estimates and covariates corresponding to the year 1989), the residuals reflect variation in the model not explained by the administrative record covariates. This is only relevant when such covariates are in the model. The next highest model variance and AIC is for the model with only

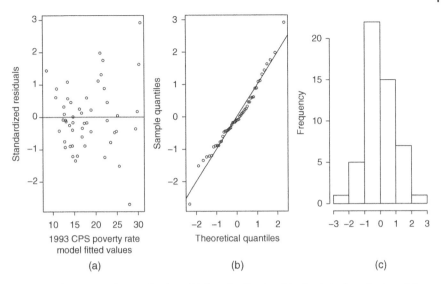

Figure 10.2 Model diagnostic plots: (a) Standardized residuals plotted against model predictions. (b) Quantile to quantile plot of standardized residuals. (c) Histogram of standardized residuals.

the IRS non-filer rate. A high model variance implies less shrinkage towards the synthetic estimator, which contains the information from the auxiliary variables, and thus a greater weight on the direct survey estimate.

In practice, one may also consider other model forms and transformations of the data, but we do not pursue them here. We now look at some model diagnostics for the model with all four covariates (M1234), which had the lowest AIC and model variance. In particular, we study the standardized residuals, defined here as:

$$r_i = \frac{\left(y_i - x_i'\hat{\beta}\right)}{\sqrt{var\left(y_i - x_i'\hat{\beta}\right)}}.$$

When the parameters are known, $var\left(y_i - x_i'\hat{\beta}\right) = D_i + A$, so we use $D_i + \hat{A}$ as a somewhat naive approximation to $var\left(y_i - x_i'\hat{\beta}\right)$. More accurate variance estimates can also be calculated under this model, but for simplicity we omit these calculations here. Figure 10.2a shows these standardized residuals plotted against the model fitted values $x_i'\hat{\beta}$. All values are between -3 and 3, indicating an absence of extreme outliers, and there does not appear to be a systematic difference between the standardized residuals and the fitted values. Under the model, these residuals should be normal and independent of the fitted values, and Figure 10.2a does not appear to contradict this. Figures 10.2b,c, a quantile to quantile plot with a normal distribution and a histogram of the residuals, also do not suggest severe deviations from the normality assumption.

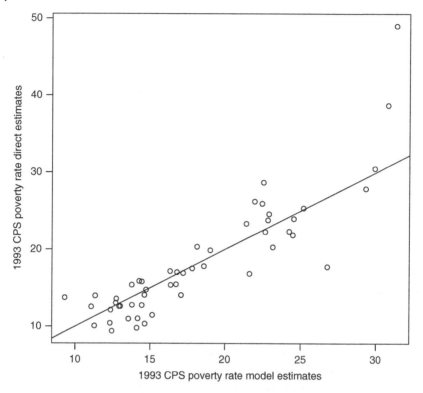

Figure 10.3 Model predictors versus direct estimates for 1993 school-aged children in poverty based on CPS data, and $y = x$ line.

We now explore the differences between the direct estimators and the model predictions. Figure 10.3 plots these against each other, along with the $y = x$ line. Note the difference between the domain and the range – the direct estimators have more extreme values. This is because shrinkage causes smoothing; for instance, the two highest values for the direct estimates correspond to smaller model predictions.

Figure 10.4a,b displays the ratios of standard errors and coefficients of variation of the modeled estimates over the direct estimates, respectively. Note large reductions in both standard errors and coefficients of variations can be achieved by modeling in this application. In fact, in all but one state, the standard errors are lower for the model estimates than for the direct estimates, with a median decrease of 57%, and range of decrease of −1% to 67%.

Of course, measurement error in the covariates, if present, would not be captured or reflected in the results under this model. Nonetheless, this application illustrates that impressive improvement in inference can be achieved by using administrative records in small area estimation models.

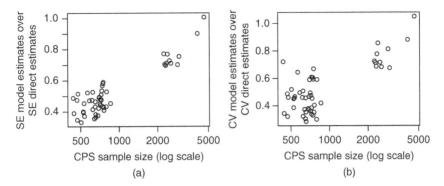

Figure 10.4 Ratios of standard errors (a) and of coefficients of variation (b) of modeled estimates over direct CPS estimates.

10.5 Concluding Remarks

We critically reviewed different statistical models and methods that can be used to improve small area estimation by utilizing information extracted from administrative records. While extensive research has been done in developing small area methodology that combines survey data with aggregate-level statistics derived from administrative records, more research is still needed to incorporate linkage errors that arise from probabilistically linking records from different databases. Some early work in this area can be found in Han and Lahiri (2018) and Han (2018). Sustained collaboration between researchers in government agencies, industry, and academia may significantly advance progress in this challenging research area.

We also need to emphasize that protecting data confidentiality associated with administrative data is vital. One way to maintain confidentiality and yet have usable auxiliary data is to develop more synthetic data that match the key properties of the real data but are not themselves confidential or sensitive.

10.6 Exercises

Data for the following exercises are from Bell and Franco (2017) at https://www.census.gov/srd/csrmreports/byyear.html. See Section 10.4 for more information on these data.

Exercise 1. Use the files cps93p.txt and CEN89RES.txt, as described in Section 10.4, to perform the following analysis:

a) Fit the full Fay–Herriot model to the data, using all of the available covariates, and using REML to estimate the model parameters. Provide summary results for the estimated model parameters. Determine whether the estimated parameters associated with the different administrative data sources are significant. These results should match those of Section 10.4.

b) Compute estimated shrinkage $B_i = D_i/(D_i + \hat{A})$, for all the states.

Exercise 2. Use the data in cps97p.txt and CEN89RES.txt for ages 5–17 to answer the following questions:

a) Fit the full Fay–Herriot model to the data, using all of the available covariates, and using REML to estimate the model parameters. The description of the variables for the 1993 and the 1997 datasets is the same and is given in Section 10.4. Provide summary results for the estimated model parameters. Determine whether the estimated parameters associated with the different administrative data sources are significant.

b) What implications does the estimate of the model variance $\hat{A}$ have on the weights placed on the direct estimates? What are the shrinkage coefficients for each of the states?

Exercise 3. Replicate the analysis in Exercises 1 and 2 using a hierarchical Bayesian approach. See Section 10.3.1 for choices of prior distributions to β and A and for choices of model fit and estimation. How do the model parameter estimates compare? In each case, provide the shrinkage coefficients.

Exercise 4. Lahiri and Pramanik (2011) computed estimates for the shrinkage coefficient, B_i based on the full Fay–Herriot models fitted to the 1993 and the 1997 data, using the following three methods:

- Exact Bayesian posterior distribution of B_i,
- The estimation method (ADM) mentioned in Section 10.3.1,
- The first-order Laplace approximation (Kass and Steffey 1989).

Table 10.3 displays the exact posterior means and variances of $\hat{B}_i$, for the four chosen states, California (CA), North Carolina (NC), Indiana (IN), and Mississippi (MS), representing both small (i.e. large D_i) and large (i.e. small D_i) states.

Use the data from cps93p.txt, cps97p.txt, and CEN89RES.txt for ages 5–17 to answer the following questions:

a) Compare your results in Exercise 1(c), Exercise 2(d), for California (CA), North Carolina (NC), Indiana (IN), and Mississippi (MS), with the posterior means of B_i in column Laplace, in Table 10.3.

b) Using the posterior means of B_i in column ADM, in Table 10.3, compute the estimated model (random effects) variance under the Li–Lahiri method.

c) Give a Taylor approximation expression to the variance of the shrinkage coefficient for the general Fay–Herriot model.

Table 10.3 Estimates of the shrinkage coefficients based on Fay–Herriot models to SAIPE state-level data.

Year	State	Posterior mean			Posterior variance		
		Exact	ADM	Laplace	Exact	ADM	Laplace
1993	CA	0.47	0.37	0.56	0.038	0.023	0.093
	NC	0.62	0.55	0.72	0.030	0.025	0.061
	IN	0.80	0.77	0.87	0.014	0.014	0.019
	MS	0.81	0.79	0.89	0.012	0.012	0.015
1997	CA	0.68	0.60	1.00	0.037	0.041	0.099
	NC	0.84	0.81	1.00	0.014	0.018	0.120
	IN	0.87	0.85	1.00	0.010	0.013	0.071
	MS	0.92	0.91	1.00	0.005	0.005	0.021

Source: Modified from Lahiri and Pramanik (2011).

d) Compute estimates of shrinkage coefficients and their estimated Taylor approximated variance using the expression in (c), evaluated at the REML point estimates for the model (random effects) variances constructed in Exercises 1 and 2, and at the ADM point estimate for the model (random effects) variances constructed in (b). How do the estimated Taylor approximated variances compare to the estimated variances in Table 10.3?

Acknowledgments

The authors thank the editors for a few constructive suggestions that led to improvement of an earlier version of the chapter. The authors also thank William Bell, Jerry Maples, and David Powers for their very useful comments during the Census Internal Review. The research of the third author was supported in part by the National Science Foundation Grant Number SES-1534413.

References

Arima, S., Bell, W.R., Datta, G.S. et al. (2018). Multivariate Fay–Herriot Bayesian estimation of small area means under functional measurement error model. *Journal of the Royal Statistical Society–Series A* 180 (4): 1191–1209.

Arora, V. and Lahiri, P. (1997). On the superiority of the Bayesian method over the BLUP in small area estimation problems. *Statistica Sinica* 7: 1053–1063.

Battese, G.E., Harter, R.M., and Fuller, W.A. (1988). An error-components model for prediction of county crop areas using survey and satellite data. *Journal of the American Statistical Association* 83: 28–36.

Bauder, M., Luery, D., and Szelepka S. (2018). Small area estimation of health insurance coverage in 2010–2016. U. S. Census Bureau. https://www2.census.gov/programs-surveys/sahie/technical-documentation/methodology/2008-2016-methods/sahie-tech-2010-to-2016.pdf (accessed 26 October 2018).

Bell, W. R., and Franco, C. (2017). Small area estimation – state poverty rate model research data files. https://www.census.gov/srd/csrmreports/byyear.html (accessed 22 October 2018).

Bell, W. R., Basel W. W., Cruse C. S., Dalzell L., Maples J. J., O'Hara B. J., Powers D. S. (2007). Use of ACS data to produce SAIPE model-based estimates of poverty for counties. https://www.census.gov/content/dam/Census/library/working-papers/2007/demo/bellreport.pdf (accessed 22 October 2018).

Bell, W.R., Basel, W.W., and Maples, J.J. (2016). An overview of the U.S. Census Bureau's small area income and poverty estimates program. In: *Analysis of Poverty Data by Small Area Estimation* (ed. M. Pratesi), 349–377. West Sussex: Wiley.

Bell, W.R., Chung, H.C., Datta, G.S., and Franco, C. (2018). Measurement error in small area estimation: functional versus structural versus naive models. *Survey Methodology* 45 (1): 61–80.

Bellow, M. and Lahiri, P. (2012). Evaluation of methods for county level estimation of crop harvested area that employ mixed models. *Proceedings of the ICES-IV*. http://www.amstat.org/meetings/ices/2012/papers/302087.pdf (accessed 22 October 2018).

Bell, W. R., Chung, H. C., Datta, G. S., & Franco, C. (2018). Measurement error in small area estimation: Functional versus structural versus naive models. *Statistics*, 06.

Bell, W. R., Basel, W. W., & Maples, J. J. (2016). An overview of the US Census Bureau's small area income and poverty estimates program. *Analysis of Poverty Data by Small Area Estimation*, 349–377.

Berger, J.O. (1985). *Statistical Decision Theory and Bayesian Analysis*. Springer-Verlag.

Brackstone, G.J. (1987). Small area data: policy issues and technical challenges. In: *Small Area Statistics* (eds. R. Platek, J.N.K. Rao, C.-E. Särndall and M.P. Singh), 3–20. New York: Wiley.

Carter, G.M. and Rolph, J.F. (1974). Empirical Bayes methods applied to estimating fire alarm probabilities. *Journal of the American Statistical Association* 69: 880–885.

Casas-Cordero Valencia, C., Encina, J., and Lahiri, P. (2016). Poverty mapping in Chilean comunas. In: *Analysis of Poverty Data by Small Area Estimation* (ed. M. Pratesi), 379–403. West Sussex: Wiley.

Chatterjee, S., Lahiri, P., and Li, H. (2008). On small area prediction interval problems. *Annals of Statistics* 36: 1221–1245.

Cox, D.R. (1976). Prediction intervals and empirical Bayes confidence intervals. In: *Perspectives in Probability and Statistics, Papers in Honor of M.S. Bartlett* (ed. J. Gani), 47–55. Academic Press.

Cruse C. S. and Powers D. S. (2006). Estimating school district poverty with free and reduced-price lunch data. https://www.census.gov/content/dam/Census/library/working-papers/2006/demo/crusepowers2006asa.pdf (accessed 22 October 2018).

Datta, G. (2009). Model-based approach to small area estimation. In: *Handbook of Statistics Sample Surveys: Inference and Analysis*, vol. 29, Part B, Pages i–xxiii (ed. C.R. Rao), 3–642. Elsevier.

Datta, G.S. and Lahiri, P. (1999). A unified measure of uncertainty of estimated best linear unbiased predictors in small area estimation problems. *Statistica Sinica* 10: 613–627.

Datta, G.S., Fay, R.E., and Ghosh, M. (1991). Hierarchical and empirical Bayes multivariate analysis in small area estimation. In: *Proceeding of the US Census Bureau 1991 Annual Research Conference*, 63–79. Washington, DC: U.S. Census Bureau.

Datta, G.S., Lahiri, P., Maiti, T., and Lu, K.L. (1999). Hierarchical Bayes estimation of unemployment rates for the U.S. states. *Journal of the American Statistical Association* 94: 1074–1082.

Dempster, A. P., Laird, N. M., & Rubin, D. B. (1977). Maximum likelihood from incomplete data via the EM algorithm. *Journal of the Royal Statistical Society: Series B (Methodological)*, 39(1), 1–22.

Das, K., Jiang, J., & Rao, J. N. K. (2004). Mean squared error of empirical predictor. *The Annals of Statistics*, 32(2), 818–840.

Datta, G. S., Rao, J. N. K., & Smith, D. D. (2005). On measuring the variability of small area estimators under a basic area level model. *Biometrika*, 92(1), 183–196.

Datta, G. S., Rao, J. N. K., & Smith, D. D. (2005). On measuring the variability of small area estimators under a basic area level model. *Biometrika*, 92(1), 183–196.

Efron, B. and Morris, C.N. (1973). Stein's estimation rule and its competitors – an empirical Bayes approach. *Journal of the American Statistical Association* 68: 117–130.

Efron, B. and Morris, C.N. (1975). Data analysis using Stein's estimator and its generalizations. *Journal of the American Statistical Association* 70: 311–319.

Elbers, C., Lanjouw, J.O., and Lanjouw, P. (2003). Micro-level estimation of poverty and inequality. *Econometrica* 71: 355–364.

Erciulescu, A.L., Cruze, N., and Nandram, B. (2018). Model-based county-level crop estimates incorporating auxiliary sources of information. *Journal of the Royal Statistical Society, Series A* 182 (1) https://doi.org/10.1111/rssa.12390. 283–303.

Esteban, M.D., Morales, D., and Perez, A. (2016). Area-level spatio-temporal small area estimation models. In: *Analysis of Poverty Data by Small Area Estimation* (ed. M. Pratesi), 349–377. West Sussex: Wiley.

Fay, R.E. (1987). Application of multivariate regression to small domain estimation. In: *Small Area Statistics* (eds. R. Platek, J.N.K. Rao, C.-E. Särndall and M.P. Singh), 91–102. New York: Wiley.

Fay, R.E. and Herriot, R. (1979). Estimation of income from small places: an application of James-Stein procedures to census data. *Journal of the American Statistical Association* 74: 269–277.

Foster, J., Greer, J., and Thorbecke, E. (1984). A class of decomposable poverty measures. *Econometrica* 52: 761–766.

Franco, C. and Bell, W.R. (2015). Borrowing information over time in binomial/logit normal models for small area estimation. *Joint Special Issue of Statistics in Transition and Survey Methodology* 16 (4): 563–584.

Fuller, W. A. and Goyeneche, J. J. (1998), *Estimation of the state variance component.* Unpublished manuscript.

Ganesh, N. and Lahiri, P. (2008). A new class of average moment matching prior. *Biometrika* 95: 514–520.

Gershunskaya, J. and Lahiri, P. (2018). Robust empirical best small area finite population mean estimation using a mixture model. *Calcutta Statistical Association Bulletin* 69 (2): 183–204. https://doi.org/10.1177/0008068317722297.

Gelman, A. (2006). Prior distributions for variance parameters in hierarchical models (comment on article by Browne and Draper). *Bayesian analysis*, 1(3), 515–534.

Ghosh, M. and Rao, J.N.K. (1994). Small area estimation: an appraisal (with discussion). *Statistical Science* 9: 55–93.

Ghosh, M., Nangia, N., and Kim, D.H. (1996). Estimation of median income of four-person families: a Bayesian time series approach. *Journal of the American Statistical Association* 91: 1423–1431.

Ghosh, M.; Rao, J. N. K. Small Area Estimation: An Appraisal. Statist. Sci. 9 (1994), no. 1, 55–76. doi:10.1214/ss/1177010647. https://projecteuclid.org/euclid.ss/1177010647.

Ghosh, M., Natarajan, K., Stroud, T. W. F., & Carlin, B. P. (1998). Generalized linear models for small-area estimation. *Journal of the American Statistical Association*, 93(441), 273–282.

Guadarrama, M., Molina, I., & Rao, J. N. K. (2016). A comparison of small area estimation methods for poverty mapping. *Statistics in Transition new series*, 1(17), 41–66.

Ha, N. (2013). Hierarchical Bayesian estimation of small area means using complex survey data. PhD dissertation, University of Maryland, College Park, MD.

Ha, N.S., Lahiri, P., and Parsons, V. (2014). Methods and results for small area estimation using smoking data from the 2008 National Health Interview Survey. *Statistics in Medicine* 33 (22). 3932–3945.

Hall, P. and Maiti, T. (2006). On parametric bootstrap methods for small-area prediction. *Journal of the Royal Statistical Society, Series B* 68: 221–238.

Han, Y. (2018), Statistical inference using data from multiple files combined through record linkage. PhD dissertation, University of Maryland, College Park.

Han, Y. and Lahiri, P. (2018). Statistical analysis with linked data. *International Statistical Review* 87 (S1) https://doi.org/10.1111/insr.12295. S139–S157.

Hawala, S. and Lahiri, P. (2018). Variance modeling for domains. *Statistics and Applications* 16 (1) (New Series of *Journal of the Indian Society of Agricultural Statistics*): 399–409.

Henderson, C.R. (1953). Estimation of variance and covariance components. *Biometrics* 9: 226–252.

Holt, T. (2007). The official statistics olympic challenge: wider, deeper, quicker, better, cheaper (with discussion). *The American Statistician* 61: 1–15.

Huang, E. T., Bell, W. R. (2012). An empirical study on using previous American Community Survey Data versus census 2000 data in SAIPE models for poverty estimates. Research Report Number RRS2012--4, Center for Statistical Research and Methodology, U.S. Census Bureau. http://www.census.gov/srd/papers/pdf/rrs2012_04.pdf. (accessed 22 October 2018).

Huang, E. T., & Bell, W. R. (2012). An empirical study on using previous American Community Survey data versus Census 2000 data in SAIPE models for poverty estimates. *Statistics*, 4.

Jiang, J., Lahiri, P., and Wan, S. (2002). A unified jackknife theory for empirical best prediction with M-estimation. *Annals of Statistics* 30: 1782–1810.

Jiming Jiang & P Lahiri (2006) Estimation of Finite Population Domain Means, Journal of the American Statistical Association, 101:473, 301-311, DOI: 10.1198/016214505000000790

Kass, R. E., & Steffey, D. (1989). Approximate Bayesian inference in conditionally independent hierarchical models (parametric empirical Bayes models). *Journal of the American Statistical Association*, 84(407), 717–726.

Kim, J.K., Wang, Z., Zhu, Z., and Cruze, N. (2018). Combining survey and non-survey data for improved sub-area prediction using a multi-level model. *Journal of Agricultural, Biological, and Environmental Statistics* 23 (2): 175–189.

Lahiri, P. and Li, H. (2009). Generalized maximum likelihood method in linear mixed models with an application in small area estimation. In *Proceedings of the Federal Committee on Statistical Methodology Research Conference*. http://www.fcsm.gov/events/papers2009.html. (accessed 22 October 2018).

Lahiri, P. and Pramanik, S. (2011). Discussion of Estimation of random effects via adjustment for density maximization. *Statistical Science* 26: 271–298. https://doi.org/10.1214/10-STS349.

Li, H. and Lahiri, P. (2010). Adjusted maximum method for solving small area estimation problems. *Journal of Multivariate Analysis* 101: 882–892. https://doi.org/10.1016/j.jmva.2009.10.009.

Liu, B. (2009). Hierarchical Bayes estimation and empirical best prediction of small area proportions. PhD dissertation, University of Maryland, College Park.

Liu, B., Lahiri, P., and Kalton, G. (2014). Hierarchical Bayes modeling of survey-weighted small area proportions. *Survey Methodology* 40: 1–13.

Luery, D. M. (2010). Small area income and poverty estimates program. Working papers, U.S. Census Bureau. https://www.census.gov/library/working-papers/ 2010/demo/luery-01.html. (accessed 22 October 2018).

Maples J.J. (2008) Calculating coefficient of variation for the minimum change school district poverty estimates and the assessment of the impact of nongeocoded tax returns. Research Report RRS2008/10, Center for Statistical Research and Methodology, U.S. Census Bureau. https://www.census.gov/srd/papers/pdf/ rrs2008-10.pdf (accessed 15–16 July).

Maples J.J. and Bell W.R. (2007). Small area estimation of school district child population and poverty: Studying the use of IRS income tax data. Research Report RRS2007/11, Center for Statistical Research and Methodology, U.S. Census Bureau. https://www.census.gov/srd/papers/pdf/rrs2007-11.pdf (accessed 15–16 July)

Molina, I. and Marhuenda, Y. (2015). sae: An *R* package for small area estimation. *The R Journal* 7 (1): 81–98.

Molina, I., & Rao, J. N. K. (2015). Small area estimation.

Molina, I., Nandram, B., & Rao, J. N. K. (2014). Small area estimation of general parameters with application to poverty indicators: a hierarchical Bayes approach. *The Annals of Applied Statistics*, 8(2), 852–885.

Morris, C. and Tang, R. (2011). Estimating random effects via adjustment for density maximization. *Statistical Science* 26: 271–287.

National Research Council (2000). *Small-Area Estimates of School-Age Children in Poverty: Evaluation of Current Methodology*, Panel on Estimates of Poverty for Small Geographic Areas, Committee on National Statistics (eds. C.F. Citro and G. Kalton). Washington, DC: National Academy Press.

Otto M.C. and Bell W.R. (1995). Sampling error modelling of poverty and income statistics for states. *Proceedings of the American Statistical Association, Social Statistics Section*, 160–165. https://www.census.gov/library/working-papers/1995/ demo/otto-01.html (accessed 25 October 2018)

Pfeffermann, D. (2013). New important developments in small area estimation. *Statistical Science* 28 (1): 40–68.

Powers, D., Bowers, L., Basel, W., and Szelepka, S. (2016). Medicaid and CHIP data methodology for SAHIE models. *Proceedings of the 2015 Federal Committee on Statistical Methodology (FCSM) Research Conference.* http://www.census.gov/ content/dam/Census/library/working-papers/2016/demo/powers-bowers-basel-szelepka-fcsm.pdf (accessed 15 July 2016).

Prasad, N.G.N. and Rao, J.N.K. (1990). The estimation of the mean squared error of small area estimators. *Journal of the American Statistical Association* 85: 163–171.

Rao, J.N.K. (2003). *Small Area Estimation*. Hoboken, NJ: Wiley.

Rao, J.N.K. and Molina, I. (2015). *Small Area Estimation*, 2e. Hoboken, NJ: Wiley.

Rao, J.N.K. and Yu, M. (1994). Small area estimation by combining time series and cross-sectional data. *Canadian Journal of Statistics* 22: 511–528.

Torabi, M. and Rao, J.N.K. (2014). On small area estimation under a subarea level model. *Journal of Multivariate Analysis* 127: 36–55.

Vogt, M. (2010). Bayesian spatial modeling: propriety and applications to small area estimation with focus on the German Census 2011. PhD dissertation, University of Trier.

Wolter, K.M. (2007). *Introduction to Variance Estimation*, 2e. New York: Springer-Verlag.

Winkler, W. E. (2007). Automatically estimating record linkage false match rates. Statistics, 5.

Ybarra, L.M.R. and Lohr, S.L. (2008). Small area estimation when auxiliary information is measured with error. *Biometrika* 95: 919–931.

Yoshimori, M. and Lahiri, P. (2014a). A new adjusted maximum likelihood method for the Fay–Herriot small area model. *Journal of Multivariate Analysis* 124: 281–294. https://doi.org/10.1016/j.jmva.2013.10.012.

Ybarra, L. M., & Lohr, S. L. (2008). Small area estimation when auxiliary information is measured with error. *Biometrika*, 95(4), 919–931.

Yoshimori, M. and Lahiri, P. (2014b). A second-order efficient empirical Bayes confidence interval. *The Annals of Statistics* 42 (4): 1233–1261. https://doi.org/10.1214/14-AOS1219.

Zanutto, E. and Zaslavsky, A. (2002). Using administrative data to improve small area estimation: an example from the U.S. Decennial Census. *Journal of Official Statistics* 18: 559–576.

Zhang, L.C. (2012). Topics of statistical theory for register-based statistics and data integration. *Statistica Neerlandica* 66: 41–63.

Zhang, L.C. and Giusti, C. (2016). Small area methods and administrative data integration. In: *Analysis of Poverty Data by Small Area Estimation* (ed. M. Pratesi), 379–403. West Sussex: Wiley.

Zhang, Li-Chun and Fosen, J. (2012) A modeling approach for uncertainty assessment of register-based small area statistics. [in special issue: Small Area Estimation] *Journal of the Indian Society of Agricultural Statistics*, 66, 91–104.

Part IV

Use of Administrative Data in Evidence-Based Policymaking

11

Enhancement of Health Surveys with Data Linkage

Cordell Golden[1] and Lisa B. Mirel[2]

[1] *U.S. National Center for Health Statistics (NCHS)*
[2] *Data Linkage Methodology and Analysis Branch, Division of Analysis and Epidemiology National Center for Health Statistics, 3311 Toledo Road, Hyattsville, MD 20782 USA*

11.1 Introduction

11.1.1 The National Center for Health Statistics (NCHS)

As the Nation's principal health statistics agency, the National Center for Health Statistics (NCHS) is responsible for collecting accurate, relevant, and timely data related to health. The mission of NCHS is to provide statistical information that can be used to guide actions and policies to improve the health of the American people (National Center for Health Statistics (NCHS) n.d.). In addition to collecting and disseminating the Nation's official vital statistics, NCHS conducts several population-based surveys, including the National Health Interview Survey (NHIS) and the National Health and Nutrition Examination Survey (NHANES), and establishment surveys of health-care facilities, including the National Hospital Care Survey (NHCS). The data collected through these surveys allow NCHS to publish widely used, reliable statistics regarding the health status of the U.S. population and selected subgroups.

The data also provide the opportunity to identify disparities in health status and use of health care services by demographic, socioeconomic status, and other population characteristics; describe experiences with the health care system; monitor trends in health status and health care delivery; and evaluate the impact of health policies and programs.

There are many questions that health surveys cannot answer on their own, in part because they often only represent a snapshot in time. In addition, most population-based health surveys rely on respondent reports and, thus, are limited by respondent recall. When these data are linked with vital or administrative

Administrative Records for Survey Methodology, First Edition.
Edited by Asaph Young Chun, Michael D. Larsen, Gabriele Durrant, and Jerome P. Reiter.

data, however, analysts can gain insight into outcomes such as mortality or health care utilization. They also can study some of the methodological issues described in previous chapters of this book, such as assessment of data quality, measuring linkage non-consent bias, and evaluating linkage algorithms. Thus, data linkages enhance the analytic capabilities and scientific value of health surveys as well as the vital and administrative data.

11.1.2 The NCHS Data Linkage Program

Over the years, NCHS has developed a data linkage program to link its health survey data with vital statistics data sources, including the National Death Index (NDI), and administrative data sources, including Centers for Medicare & Medicaid Services (CMS), Social Security Administration (SSA), the U.S. Department of Housing and Urban Development (HUD), as well as other sources in pilot projects. Although administrative data are not created for research purposes (they are created primarily for program administration; see Section VI – Analytic considerations and limitations of administrative data), the NCHS Data Linkage Program has worked extensively with partner agencies to develop data files that can be used for research.

Research with the NCHS linked data has examined disparities in mortality patterns by race and ethnicity; looked at health care utilization and expenditures by chronic health conditions (Gorina and Kramarow 2011; Honeycutt et al. 2013); followed up clinical measures through diagnosis codes obtained from claims data; and assessed health characteristics for those receiving federal housing assistance (see Section 11.5).

The NCHS Data Linkage Program, housed within the Division of Analysis and Epidemiology, began in the early 1990s. The goals of the NCHS Data Linkage Program have evolved over time. However, the primary objectives have always been to release high quality timely data that can be used to address key health topics without compromising the confidentiality of survey participants.

11.1.3 Initial Linkages with NCHS Surveys

The first linkage began with linkages of the NCHS surveys to NDI. NDI, a centralized database of death record information from state vital statistics offices, is maintained at NCHS. Many of NCHS' surveys already were collecting the identifying information needed for linkage. As a result, linkages to NDI offered a readily available opportunity to link health-related survey data with mortality for NCHS survey participants, without the complex data sharing agreements that linkages to non-NCHS data require. These early linkage projects laid the foundational framework for future linkages. These linkages also addressed requests from the research community. The NHANES I Epidemiologic Follow-up

Study (NHEFS) was the first NCHS survey to adopt linkage as a part of the data collection process. The initial and subsequent studies were designed to investigate the association between factors measured at baseline (during the interview and examination) and mortality (Loria et al. 1999).

A key policy question about the health insurance status and access to care for Social Security Disability Insurance (SSDI) beneficiaries during the 24-month waiting period prior to Medicare entitlement served as the impetus for the NCHS Data Linkage Program to expand its linkages beyond NDI and NCHS data sources (Riley 2006). The seminal interagency agreements established between NCHS, SSA, CMS, and the Office of the Assistant Secretary for Planning and Evaluation of the Department of Health and Human Services resulted in NCHS survey data being linked to Social Security and Medicare administrative records. A subset of the data from this linkage, the 1994–1996 NHIS linked with Social Security and Medicare administrative data (Golden et al. 2015; Riley 2006), was used to highlight an important policy issue, that approximately one quarter of SSDI recipients did not have health insurance in the 24 months leading up to Medicare coverage. This linkage project laid the groundwork for collaborating with other federal agencies and allowed for other key policy questions to be addressed, such as why household population surveys tended to yield lower estimates of Medicaid enrollment than the number of enrollees reported in state and national administrative data (Medicaid Undercount Project (SNACC) n.d.).

This chapter describes several NCHS health surveys that were enhanced through data linkages and the specific sources to which they were linked. It provides an overview of the NCHS Data Linkage Program, the linkage methods and processing issues, the enhancements to the survey data, as well as limitations of the linked data. It concludes with a description of plans for future developments and direction for the NCHS Data Linkage Program.

11.2 Examples of NCHS Health Surveys that Were Enhanced Through Linkage

Most NCHS population-based and establishment surveys are part of the NCHS Data Linkage Program. Below is a list of a few of these surveys with brief descriptions of their content and focus.

11.2.1 National Health Interview Survey (NHIS)

NHIS is a nationally representative, cross-sectional household interview survey that serves as an important source of information on the health of the civilian non-institutionalized population of the United States. It is a multistage sample survey with primary sampling units of counties or adjacent counties, secondary sampling units of clusters of houses, tertiary sampling units of households, and finally, persons within households. It has been conducted continuously since 1957 and the content of the survey is periodically updated (National Center for Health Statistics and National Health Interview Survey (NHIS) n.d.). NHIS has been used as the sampling frame for several NCHS surveys focusing on specialized populations including the Supplement on Aging (SOA) (National Center for Health Statistics 1984) and the Second Longitudinal Study on Aging (LSOA II) (National Center for Health Statistics n.d.-a). NHIS implemented its most recent content and structure redesign in 2019.

A subsample of households that participated in the previous year's NHIS are selected for a panel in the Medical Expenditure Panel Survey (MEPS), conducted by the Agency for Healthcare Research and Quality. MEPS is the most complete source of data on the cost and utilization of health care for the civilian noninstitutionalized population. It follows participants for two calendar years (Agency for Healthcare Research and Quality n.d.). The data linkages at NCHS further enhance the cost and utilization information collected in the MEPS household component and provide additional data for longitudinal analytic purposes.

11.2.2 National Health and Nutrition Examination Survey (NHANES)

NHANES includes an interview in the household followed by an examination in a mobile examination center (MEC). NHANES is a nationally representative, cross-sectional sample of the U.S. civilian noninstitutionalized population that is selected using a complex, multistage probability design (National Center for Health Statistics n.d.-b). Prior to becoming a continuous survey in 1999, NHANES was conducted periodically, with the last periodic survey (NHANES III) conducted between 1988 and 1994 (National Center for Health Statistics n.d.-c). NCHS also initiated a national longitudinal study, the NHEFS in collaboration with the National Institutes of Health, National Institute on Aging and other agencies of the Public Health Service. The NHEFS cohort included all persons 25–74 years of age who completed a medical examination as part of NHANES I in 1971–1975 and was comprised of a series of follow-up waves (1982–1984, 1986, 1987, and 1992) (National Center for Health Statistics n.d.-d).

11.2.3 National Health Care Surveys

The National Health Care Surveys conducted by NCHS, are a family of data collection efforts that gather information about providers of health care services and the patients they serve across the spectrum of health care settings from ambulatory and hospital care to long term care settings. Data about the providers include facility characteristics such as facility or practice size, ownership, Medicare/

Medicaid certification, services provided and specialty programs offered, and sources of revenue. For patients, data are obtained on demographic characteristics, medical conditions, and insurance status (National Center for Health Statistics n.d.-e). The encounter data include information on the types of providers seen; the services and procedures ordered, including laboratory and other diagnostic tests; and the medications prescribed. This information can be used to investigate factors that influence the use of health care resources, the quality of health care, and disparities in health care services provided to various subpopulations.

One of the National Health Care Surveys, the NHCS, integrates inpatient data formerly collected by the National Hospital Discharge Survey with emergency department, outpatient department, and ambulatory surgery center data collected by the National Hospital Ambulatory Medical Care Survey (National Center for Health Statistics n.d.-f). NHCS also incorporates substance-involved emergency department visit data previously collected by the Drug Abuse Warning Network (DAWN) conducted by the Substance Abuse and Mental Health Services Administration (SAMHSA). Previously, hospitals were asked to provide Uniform Bill (UB)-04 administrative claim data for all inpatients, emergency department visits and outpatient department visits as well as some facility-level data through a facility questionnaire. In 2014, the emphasis of NHCS' data collection shifted from UB-04 claim data to electronic health record (EHR) data.

More information about the population-based and establishment surveys described, as well as other NCHS data collection systems, can be obtained from the NCHS website: http://www.cdc.gov/nchs/index.htm. (National Center for Health Statistics (NCHS) n.d.)

11.3 NCHS Health Surveys Linked with Vital Records and Administrative Data

To date, several NCHS surveys, including NHIS and NHANES, have been linked to vital records and administrative data sources including NDI and CMS, HUD, SSA, and the United States Renal Data System (USRDS). The NCHS Data Linkage Program links health survey data to vital or administrative data, but it does not link to other surveys' data because of the inherent complications of combining survey designs and the lack of overlap across samples for the surveys. The NCHS Data Linkage Program also has conducted a number of pilot studies with state databases, including linkages of NCHS survey data to data from a state cancer registry and a single state pilot with the Supplemental Nutrition Assistance Program (SNAP) (Miller et al. 2014; Mirel et al. 2010). Some of the vital and administrative data sources that have been linked with NCHS surveys are described in this section.

11.3.1 National Death Index (NDI)

The National Vital Statistics System (NVSS) is NCHS' mechanism for collecting and disseminating the Nation's official vital statistics data. Through partnerships with the vital registration systems in the various jurisdictions, NCHS produces critical information on topics such as teenage births and birth rates, prenatal care and birth weight, risk factors for adverse pregnancy outcomes, infant mortality rates, leading causes of death, and life expectancy (National Center for Health Statistics n.d.-g).

In addition to NVSS, NCHS also houses the NDI. NDI is a centralized database of death record information from U.S. state and territory vital statistics offices dating back to 1979. NDI was established as a resource to aid health and medical researchers in determining the mortality status and other relevant information such as date of death, state of death, and death certificate number for persons in their studies (National Center for Health Statistics n.d.-h). Researchers can obtain cause of death information using the "NDI Plus" service that links the standard NDI record with information from other NCHS vital statistics resources.

11.3.2 Centers for Medicare and Medicaid Services (CMS)

CMS is responsible for administering Medicare, Medicaid, and the Children's Health Insurance Program (CHIP), and more recently, the Health Insurance Marketplace. The Medicare program provides health insurance for people aged 65 and over, people under 65 with permanent disabilities who receive SSDI, and people of all ages with end-stage renal disease (ESRD, permanent kidney failure requiring dialysis, or a kidney transplant). Medicaid is a means-tested entitlement program that provides health care coverage to some vulnerable populations in the United States, including low-income children, and the aged or disabled poor. CHIP targets low-income uninsured children and pregnant women in families with incomes too high to qualify for most state Medicaid programs (Centers for Medicare and Medicaid Services (CMS) n.d.). Both Medicaid and CHIP are jointly financed by the federal and state governments, and are administered by the states.

Data for NCHS survey participants have been linked with administrative data generated from both the Medicare and Medicaid/CHIP programs. These linkages provide a rich source of expenditure data including the amount charged and reimbursed for services billed to the Medicare and Medicaid/CHIP programs by health care providers. This allows the opportunity to study the effect that certain health conditions or risk factors may have on Medicare and Medicaid/CHIP costs (Cai et al. 2010; Hedley Dodd and Gleason 2013).

The claims records obtained from the CMS file can provide a case history of a survey participant's interaction with the health care system. In some situations the information from the administrative data may provide more detail or be more accurate than the information collected from the survey respondent, as these data come directly from an encounter between the survey participant and a health service provider. For example, the CMS claims records include procedure and diagnosis codes that allow researchers to verify health conditions reported in the NHANES interview and/or measured during the NHANES examination, identify health conditions that may not have been reported during the interview, and also examine outcomes for risk factors identified in the data. It should be noted that only claims from Fee-for-Service encounters are captured in the administrative data from CMS.

11.3.3 Social Security Administration (SSA)

The federal assistance programs/services that are administered by SSA include the Old Age, Survivors, and Disability Insurance (OASDI) and Supplemental Security Income (SSI). The OASDI program (known as "Social Security") is the largest income-maintenance program in the United States. The program provides monthly benefits designed to replace, in part, the loss of income resulting from retirement, disability, or death. OASDI has two programs: the Old Age and Survivors Insurance (OASI) program and the SSDI program. SSI is a needs-based program that provides cash assistance to the elderly, blind and disabled children and adults who have limited income and resources (Social Security Administration (SSA) n.d.). When linked with NCHS survey data, the benefit records collected from SSA's OASDI, and SSI data provide the opportunity to enhance our understanding of the specific public health challenges that face the elderly and disabled U.S. population.

11.3.4 Department of Housing and Urban Development (HUD)

HUD is the primary federal agency responsible for overseeing and managing domestic housing programs and policies. HUD manages and oversees a variety of housing and community development programs, including project-based rental subsidy programs, the Housing Choice Voucher program, community and economic development programs, Federal Housing Administration (FHA) mortgage insurance programs, and specialized programs for high-needs populations (e.g. the elderly, homeless, and disabled) (United States Department of Housing and Urban Development User n.d.). The HUD administrative data include information on the timing and receipt of the largest housing assistance programs: the Housing Choice Voucher program; public housing; and privately owned, subsidized multifamily housing. When linked with NCHS survey data, the HUD data provide the opportunity to examine relationships between housing and health.

11.3.5 United States Renal Data System and the Florida Cancer Data System

In addition to data from programs administered by federal agencies, NCHS has also linked some of its population-based health surveys to data from disease databases and registries, namely the United States Renal Data System (USRDS) and the Florida Cancer Data System (FCDS) (as a pilot project). USRDS is a national data system funded by the National Institute of Diabetes and Digestive and Kidney Diseases (NIDDK) designed to collect, analyze, and distribute information about ESRD in the United States (United States Renal Data System (USRDS) n.d.). FCDS is funded by the State of Florida and CDC's National Program of Cancer Registries. State cancer registries are responsible for collecting, managing, and analyzing data on incident cancer cases and cancer deaths. A pilot project was conducted to assess the feasibility of linking NHIS data to FCDS (Florida Cancer Data System (FCDS) n.d.).

11.4 NCHS Data Linkage Program: Linkage Methodology and Processing Issues

NCHS has developed a Data Linkage Program designed to expand the analytic utility of the Center's population-based surveys. Linked data can be used to facilitate richer analyses of the survey data by augmenting the information collected from the surveys with vital or administrative data which allows researchers to examine factors that influence disability, chronic disease, health care utilization, morbidity, and mortality. Some topics in linkage methodology and processing of files are discussed below.

11.4.1 Informed Consent in Health Surveys

Before data collected from a health survey participant can be linked with administrative records, it is necessary to obtain the participant's informed consent to be included in data linkage activities. Over time, obtaining consent from all survey participants has become an increasingly difficult task. NCHS survey participants are informed of NCHS' intent to conduct data linkage activities through a variety of informed consent procedures, which include advance letters, participant brochures, signed consent forms, and questionnaires. Each of these methods have been approved by the NCHS Research Ethics Review Board (ERB).[1] Only

---

1 The NCHS ERB, also known as an Institutional Review Board or IRB, is an administrative body of scientists and non-scientists that is established to protect the rights and welfare of human research subjects.

NCHS survey participants who have provided consent as well as the necessary personally identifiable information (PII), such as date of birth and full or partial Social Security Number (SSN), are considered *linkage-eligible.* Linkage-eligibility refers to the potential ability to link data from an NCHS survey participant to administrative data.

Prior to 2007, all NCHS survey participants who refused to provide SSN and Medicare health insurance claim number (HIC) (for linkages to CMS administrative data) were implicitly considered to have refused to participate in data linkages. The rate of linkage refusal, particularly for NHIS, steadily increased over time, reducing the number of NHIS participants eligible for record linkage (Miller et al. 2011). The reduction in the number of linkage-eligible survey participants potentially diminished the research utility of the linked data files produced by NCHS. Fewer linkage-eligible survey participants meant fewer linked participant records available in the linked data. It also increased the possibility of systematic differences between linked and unlinked records, called "consent" or "linkage" bias, (Dunn et al. 2003; Hill et al. 2002) which potentially reduced researchers' ability to make inferences to the general population.

In 2003, NCHS conducted research to assess the accuracy of NDI matches using partial SSN and other PII. The research assessed algorithms using the last four and last six digits of the SSN. The results were favorable and provided sufficient data to support changes in how NHIS collected SSN and HIC numbers for linkage (Sayer and Cox 2003). Leading up to 2007, the linkage refusal rates in NHIS had been increasing to an all-time high of about 57% in 2006 (Miller et al. 2011). Beginning in 2007, NHIS attempted to decrease linkage refusal rates by requesting only the last four (instead of the full nine) digits of SSN and HIC numbers, adding a short introduction before asking for SSN, and asking participants for their explicit permission to link to administrative records if SSN or HIC had not been provided. Following the implementation of the new consent procedures, linkage refusal rates began to decrease (Dahlhamer and Cox 2007). Although the number of linkage-eligible survey participants began to increase, linking NCHS survey data to administrative data with only partial SSNs provided a new and ongoing challenge. NCHS continues work to adapt its linkage processes and identify new strategies for linking to administrative data without full nine-digit SSN.

11.4.2 Informed Consent for Child Survey Participants

During NCHS household interview surveys, all sampled members of the household are encouraged to participate in the survey and respond to the survey questions for themselves. It is not uncommon for the interviewers to collect survey data from a knowledgeable adult (parent or guardian) for household members who are not old enough to participate in the interview (e.g. under 18 years of age).

If consent is provided by their parent or guardian and the linkage-eligibility criteria are met, NCHS survey participants under 18 years of age at the time of the survey are considered linkage-eligible. However, the consent provided by the parent or guardian does not apply once the child survey participant becomes a legal adult. As a result, in accordance with NCHS ERB guidance, NCHS only includes administrative data that were generated for program participation, claims, and other events that occurred prior to the participant's 18th birthday on the linked data files provided to researchers.

11.4.3 Adaptive Approaches to Linking Health Surveys with Administrative Data

Of the NCHS survey participants who provide consent for linkage and meet linkage eligibility criteria, the actual number of participants that are successfully linked to administrative records depends both on program enrollment and the linkage process. The linkage methods used by NCHS differ depending on the administrative or vital data source that is being used in the linkage, using a varying mix of deterministic and probabilistic linkage methods. Much of the criteria for the linkage methods are determined by the legal requirements of the agency providing the administrative records. Deterministic or rules-based approaches, which are employed for the linkages to administrative data from CMS, SSA, and HUD, assume that there is a known set of variables that can be used to link two datasets and require that the information for these variables in both datasets match exactly in order for a pair of records to be considered a match (Herzog et al. 2007). SSN along with other personal identifiers such as name, date of birth, and sex are often used as the linking variables for linkages of health survey data with administrative data from CMS, SSA, and HUD. Depending on the scope of the linkage and the quality of the data available in both datasets (survey and administrative), enhanced or modified deterministic approaches may be introduced. For instance, instead of requiring complete agreement between the information contained in the variables used for linking two datasets, the matching criteria (rules) may allow pairs of records with slightly different or similar information to be considered a match. Phonetic algorithms (i.e. Soundex (The Soundex Indexing System n.d.), NYSIIS (Taft 1970)) may be applied for the comparison of name variables to account for typographical errors and spelling variations. Similarly, to account for potential data entry errors and misreporting with date variables (i.e. date of birth) the matching criteria may be modified to allow either the month, day, or year of a date to not be equal if the other two parts of the date are in complete agreement in both datasets. These modified deterministic approaches have proven to be particularly useful as NCHS moves toward linking with four-digit SSNs.

NCHS uses a probabilistic scoring approach when linking NCHS survey data with death records from NDI. In general, probabilistic approaches take into account a larger number of potential identifiers (linking variables) and weights are computed for each identifier based on its estimated ability to correctly identify a pair of records as matches or non-matches (Gill 2001; Herzog et al. 2007). The weights are used to calculate the probability (or score) that a pair of records belongs to the same individual (survey participant). Record pairs that score above a predetermined threshold are considered to be matches, whereas pairs with scores below another predetermined threshold are considered to be non-matches; pairs that fall between both thresholds are considered to be "possible matches" and are marked for clerical review by data linkage staff. The number of record pairs marked for clerical review will vary depending on how the thresholds are set. Generally, with NCHS linkages that use a deterministic approach there is a small amount of clerical review by NCHS staff. Often this is limited to cases where a single record in one dataset (survey or administrative) matches multiple records in the other. However, with the NDI linkages there tends to be a considerably larger number of record pairs that are clerically reviewed. In an attempt to reduce the number of records marked for clerical review, data from previous linkages of NCHS surveys and NDI records have been used to establish rules that allow record pairs with certain characteristics to be systematically marked as either matches or non-matches. Although this has not completely eliminated the lengthy clerical review process, it has drastically shortened the production schedule for the linked NCHS survey-NDI data. This moves the NCHS Data Linkage Program a step closer to its goals of more timely and regularly scheduled data releases for the linked NDI data.

11.4.4 Use of Alternate Records

In an attempt to increase the likelihood of finding a correctly matching record in the linkage process, NCHS creates alternate or additional records that are included on the files used to link NCHS survey participants with administrative data. The alternate records may include different SSNs or dates of birth that have been identified through other linkages, formal names that replace shorthand or nicknames, or use variations of name parts when complex names (e.g. hyphenated) are used. Linkage submission files are created that include a record with the original information collected at the time of the survey and alternate records with every combination of original and alternate information available for a given survey participant. Additional alternate records are created for Hispanic and Asian names to account for cultural traditions regarding the use of father's surname for women and the ordering of given and family names. Although NCHS has not conducted a formal evaluation of the effectiveness of using alternate records, preliminary analyses using data from the NDI linkage have shown encouraging results (Miller et al. 2015).

11.4.5 Protecting the Privacy of Health Survey Participants and Maintaining Data Confidentiality

NCHS remains unwavering in its commitment to maintain the confidentiality and protect the PII for all of its survey participants, including those for whom records are enhanced through the NCHS Data Linkage Program. The files that are used for the NCHS data linkage process contain only the minimum number of personal identifiers required for linkage and each survey participant is assigned a randomly generated linkage identification number that does not directly link to any in-house or public-use identification. When collaborating with other agencies, all files are transferred through approved secure data transfer mechanisms. At the completion of each linkage project, NCHS requires that collaborating agencies destroy all NCHS data obtained under the interagency agreement and provide written confirmation of file destruction.

The final step in the linkage process is to delete all the PII on the file, replace the randomly generated linkage identification number with the survey identification number so that the files can be used for future research. There are two types of linked analytic data files that NCHS develops: restricted-use and if possible, public-use. Restricted-use linked analytic data files are only accessible through the NCHS Research Data Center (RDC). All researchers must submit a research proposal to gain access to these restricted data files. The proposal provides a framework which allows RDC staff to identify potential disclosure risk.

There are two types of public-use files released by the NCHS Data Linkage Program: public-use feasibility data files and public-use linked analytic data files. The data released on the two types of public-use files differ as do the purposes for which they are used. Public-use feasibility data files are developed to help interested researchers determine the maximum available sample sizes to assess the feasibility of conducting analyses utilizing the restricted-use linked files in the NCHS RDC. The public-use feasibility files do not contain any extracted administrative data such as benefits or payments; they simply include status type variables designed to provide researchers with a way to gauge whether or not their analytic samples will have a sufficient number of sample persons with linked data to warrant writing and submitting an application to access the restricted-use linked analytic files in the RDC. The public-use feasibility files are reviewed and approved by the NCHS Confidentiality Officer and the NCHS Disclosure Review Board (DRB) prior to their release.

Public-use linked analytic data files are only produced for the NDI linkages. The public-use linked mortality files contain a limited set of mortality variables (e.g.

quarter of death instead of month of death, grouped underlying cause of death codes instead of International Classification of Diseases [ICD] codes) for adult survey participants. Death information for children are not included on the public-use linked mortality files. Data perturbation techniques are used to reduce the risk of survey participant reidentification (National Center for Health Statistics 2015). As with the public-use feasibility files, the NCHS Confidentiality Officer and the NCHS DRB reviews and approves the public-use linked mortality files prior to their release.

11.4.6 Updates Over Time

Over time the Program's linkage projects have expanded to include an increasing number of health surveys and vital and administrative data sources. As noted above, many of the NCHS surveys are now linked to NDI on a routine basis. With each linkage to NDI, the number of linked NCHS surveys as well as the years of mortality follow-up information increase. For example, in 2013, 1985–2009 NHIS and 1999–2010 NHANES were linked to NDI mortality data through 31 December 2011. In 2017, the 1985–2014 NHIS and 1999–2014 NHANES were linked to NDI mortality information through 31 December 2015. Similarly, the surveys are on a routine schedule for linkage to the CMS data. In 2016, new files were released including linked Medicare data through 2013 and a new linkage was started that included Medicaid data through 2013. Previous NCHS–CMS linkages have included Medicare data through 2000 and 2007 and Medicaid data through 2009 (Golden et al. 2015). Two linkages have been conducted of several NCHS surveys to SSA benefits history data through 2000 and 2007 (Golden et al. 2015). Negotiations are underway between NCHS and SSA to update the NCHS-SSA linked data. In 2015, NCHS collaborated with HUD to link the 1999–2012 NHIS and 1999–2012 NHANES to housing assistance program data through 31 December 2014 (Lloyd et al. 2017). Although the NCHS-HUD linkage originated as a pilot project to assess the feasibility of linking health survey data to housing assistance program data and test linkage algorithms to be used in other projects, it has produced a valuable data resource that enables researchers to examine housing as an outcome or covariate associated with health. As a result, NCHS and HUD have agreed to continue their collaboration and hope to update the linked NCHS-HUD files on a routine basis. The NCHS Data Linkage Program continues to update the linked files as the NCHS survey data are released and as agreements between NCHS and collaborating agencies are updated or new agreements are implemented.

More information about the NCHS Data Linkage Program can be found on the NCHS Data Linkage website: http://www.cdc.gov/nchs/data-linkage/index.htm (National Center for Health Statistics n.d.-i)

11.5 Enhancements to Health Survey Data Through Linkage

Although the survey data collected by NCHS provide information on a wide-range of health-related topics, they often lack information on longitudinal outcomes or other key outcomes for evidence-based research. The NCHS Data Linkage Program has enabled NCHS to enhance its health survey data with information from vital and administrative data sources. These linkages provide the unique opportunity to examine factors that influence changes in health status, follow-up of clinical measures, health care utilization and expenditures, morbidity, and mortality. Beyond enhancing survey data, record linkage also reduces the cost burden of collecting additional information, given the expense of active follow-up of survey participants. Surveys may be linked to multiple years of administrative data. Depending on the survey year, administrative data may be available for survey participants at the time of the survey, as well as before or after the survey period. Several factors may influence the alignment of the survey with administrative data, including: age of the survey participant; program eligibility; discontinuous program coverage; and residential mobility of the survey participant. As an example, the temporal alignment between NHIS and Medicaid administrative data is illustrated in Figure 11.1 (Golden et al. 2015). The illustration shows the years of NHIS data (1994–2005) that have been linked with 1999–2005 Medicaid administrative data. Based on this example, Medicaid administrative data may be available for a linked 2005 NHIS participant at the time of survey (2005) as well as before (1999–2004) and after (2006–2009) the survey period.

Linking surveys to administrative data sources also can be used to inform improvements in surveys. Capturing additional information through linkages can decrease the number of items collected during the survey and reduce the burden on the survey respondent. This in turn could be used as a resource to address the decline in response rates for national surveys (Marton and Karberg 2011). Decreasing the number of survey questions also reduces a survey's implementation costs. Administrative data may also provide information that might be relevant for the imputation of missing survey data due to nonresponse. Information can be gained related to the characteristics of the survey sampling frame and used to inform survey redesign efforts. Administrative data can be used to expand the scope or content of a survey by providing information that survey respondents may have difficulty providing themselves. Detailed information on eligibility, enrollment status, and participation in state and federal programs can be obtained. Administrative data can be used to expand the analytic time window, providing a longitudinal dimension to cross-sectional surveys. Although many types of administrative data exist, the analytic utility of health survey data is most enhanced with information from the following types of administrative data:

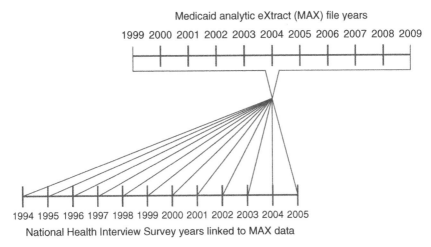

Figure 11.1 Illustration of the temporal nature of the linked data using the linked NHIS-Medicaid data as an example. Sources: CDC/NCHS, National Health interview Survey, 1994–2005; and Centers for Medicare & Medicaid Services, Medicaid Analytic eXract, 1999–2009.

- Expenditures for receipt of health services;
- Diagnosis and procedure codes generated from encounters with the health care system;
- Enrollment and participation in state and federal assistance programs;
- Information from disease databases and registries.

The linkages conducted by NCHS capture detailed information about the experiences of NCHS survey participants in federally sponsored benefit programs. Many health surveys include questions that capture information about the receipt of federal assistance and participation in federally sponsored programs. The information obtained from administrative data sources such as SSA and HUD can be used to assess the quality and examine the reliability of the survey information. The survey information can be used in health policy analyses, such as predicting the number of people who will require certain federal or state-provided services based on survey-reported health conditions; and determining whether current funding levels for entitlement programs will be adequate in the future. The linked survey and administrative data can also be used to examine health disparities between people enrolled in federal or state programs and those who are not enrolled.

The linkage of health survey data with data from disease databases and registries can produce a valuable tool for research related to disease prevention and control. For survey participants interviewed before their diagnosis, researchers are able to examine characteristics and risk factors associated with future diagnoses and survival time. For survey participants interviewed after a diagnosis, researchers are able to examine issues related to treatment effectiveness. As with administrative data generated from health services claims and program participation, data obtained from disease databases and registries can be used to assess respondent-reported diagnoses from the survey.

The linked mortality files have been used to highlight important public health issues related to disease and mortality. In 2007 Flegal et al. examined the association between underweight, overweight and obesity on cause-specific excess deaths, using the linked NHANES and NDI data (Flegal et al. 2007). In 2011 Pratt et al. assessed excess mortality due to depression and anxiety, using the linked NHIS and NDI data (Pratt et al. 2016).

Two recent publications utilized the linked NHANES data. The first, by Looker, examined the relationship between femur neck bone mineral density as measured by the dual-energy X-ray absorptiometry in the NHANES MEC and risk of incident major osteoporotic fracture (hip, spine, radius, and humerus) in older U.S. adults from NHANES III (1988–1994), using both linked mortality and Medicare data (Looker 2013). The second, Ahrens et al. utilized the linked NHANES-HUD data to compare blood lead levels among children by receipt of federal housing assistance (Ahrens et al. 2016). In addition, the linked files have offered insight into survey design and methodology. A National Health Statistics Report examined the rate of concordance between survey report of Medicaid enrollment and linked Medicaid administrative records in two national studies. This report found an almost 90% rate of concordance between the self-report of Medicaid/CHIP receipt as reported in the survey and what was found in the administrative database (Mirel et al. 2014).

Lastly, previous research has compared characteristics of beneficiaries within a program, such as comparing health-related characteristics for managed care plans (or Medicare Advantage) and fee-for-service beneficiaries in both Medicare and Medicaid (Lloyd et al. 2015; Miller et al. 2016; Mirel 2012).

11.6 Analytic Considerations and Limitations of Administrative Data

It is important to keep in mind that administrative data are collected primarily for programmatic purposes. The data are collected to support the implementation of a program or service and may not be intended for research. The data are not always standardized and there can be inconsistencies in how they are coded. Data entry is typically conducted by staff whose primary responsibilities are to provide treatment or administer services (e.g. health service providers, social workers, and facility or program administrators) and who may be less concerned with ensuring data quality. It is also important for researchers to consider the timeliness of the administrative data. The coverage period of the administrative data available for linkage may not correspond with the time period of interest to the researcher. Often it takes several months beyond the end of the coverage period to fully collect and process the administrative data in order to prepare datasets that are suitable for research. Researchers also should be aware of the policies that apply to the coverage period of the administrative data used in their analyses, as the policies related to the program's eligibility and participation may change over time. The program's eligibility policies may also result in gaps in coverage for beneficiaries. There may be discontinuities in the information of interest to researchers for beneficiaries who move in and out of a program due to eligibility requirements. Another important analytic consideration when using the linked files is the distinction between linkage-eligibility and program-eligibility. Program-eligibility is defined depending on whether an individual meets federal and state-specific eligibility criteria for a specific government-administered or -funded program. Linkage-eligibility is defined by whether the survey participant provided sufficient PII to be linked and/or provided consent for linkage. This distinction is important for several reasons. Not everyone who meets a program's eligibility requirements is included in the survey sample and if included, they may have refused to provide the necessary PII or consent to be considered linkage eligible. Conversely, a survey participant may have provided the necessary PII and consent to be considered linkage eligible, but may not meet the program's eligibility requirements to receive benefits or services.

11.6.1 Adjusting Sample Weights for Linkage-Eligibility

NCHS continues to work to find ways to improve the research utility of the NCHS linked data products. For its population-based health surveys, NCHS creates sample weights based on probabilities of selection into the surveys, with adjustments for non-response and post-stratification to annual population totals for specific population domains to provide nationally representative estimates. Although users of the NCHS surveys are encouraged to use the survey-provided sample weights in their calculations for most analyses, the properties of the survey-provided weights for linked data files with incomplete linkage due to ineligibility for linkage and non-matches are unknown and can be problematic. In addition, methods for using the survey weights for some longitudinal analyses require further research. As this is a central and complex methodological topic, ongoing works at NCHS and elsewhere are examining the use of survey weights for linked data in multiple ways.

One approach is to analyze linked data files using adjusted sample weights. This allows researchers to construct sample weights adjusted for linkage eligibility that are suitable for their specific analysis. As inferences may depend on the approach and statistical software packages used to develop weights, NCHS recommends that researchers seek assistance from a mathematical statistician for guidance on their particular project. Some statistical software packages, such as SUDAAN (Research Triangle Institute 2008), allow auxiliary information to be used to adjust the statistical weights for non-response. The sample weights available on NCHS population-based health survey data files can be adjusted for incomplete linkage and non-matches using standard weighting domains based on the sample design of the survey to reproduce population counts within certain domains (e.g. sex, age, and race and ethnicity subgroups) (Curtin et al. 2013; Parsons et al. 2014). These counts are called "control totals" and are estimated from the full survey sample. Results from an evaluation of this approach for adjusting survey weights for the linked files are encouraging (Judson et al. 2013; Larsen et al. 2012). NCHS continues to investigate alternate approaches for addressing issues related to missing data, including the use of multiple imputation techniques (Zhang et al. 2016).

11.6.2   Residential Mobility and Linkages to State Programs and Registries

Residential mobility can be a major challenge for linkages involving national health survey data and administrative data from state-based programs and disease registries (e.g. FCDS). While the current weighting strategy (creating weights to the year of the survey) is valid, it may result in a nontrivial reduction in sample size by giving a zero weight to linked survey participants who were interviewed in one state but participated in a non-federal program or diagnosed with a disease in another state. NCHS is investigating alternative weighting strategies to reduce the loss of sample sizes.

Similarly, Medicaid is a state-based program and analysts need to be aware that different states may offer different types of coverage. States are permitted to cover many services that federal law designates as optional, including dental services, prescription drugs, case management, and hospice services. State variation in Medicaid coverage, with regard to both program eligibility and covered services, results in state differences in enrollment rates and expenditures. Other factors, including the age distribution, the poverty rate, and the Medicaid provider reimbursement rates, also contribute to variation among states in enrollment, service use, and costs. As a result, Medicaid operates more than 50 distinct programs – one in each state, the District of Columbia, and each of the territories.

11.7   Future of the NCHS Data Linkage Program

NCHS' long-standing data collection and data linkage programs fulfill its mission as a federal statistical agency, to develop and disseminate data and information to improve the nation's health. The linked data program supports the Foundations for Evidence-Based Policymaking Act of 2018 and several directives on the use of administrative records issued by the Office of Management and Budget (OMB) of the Executive Office of The President, including:

- *M-11-02*: Sharing Data While Protecting Privacy (Office of Management and Budget, Executive Office of the President 2010)
- *M-13-13*: Open Data Policy – Managing Information as an Asset (Office of Management and Budget, Executive Office of the President 2013)
- *M-13-17*: Next Steps in the Evidence and Innovation Agenda (Office of Management and Budget, Executive Office of the President 2013)
- *M-14-06*: Guidance for Providing and Using Administrative Data for Statistical Purposes (Office of Management and Budget, Executive Office of the President 2014)

The linked data result in resources that can be used by researchers and policy makers for analytical and statistical purposes to enhance the information from health surveys alone.

As the NCHS Data Linkage Program continues to move forward it will face many challenges, but also many opportunities. Ensuring confidentiality and protection for all participants' personally identifiable information will remain at the forefront of the NCHS Data Linkage Program. Some of the challenges will include: addressing methodological issues such as declining survey response rates (Marton and Karberg 2011); respondents becoming less willing to provide the PII necessary for eligibility and accurate matches (Miller et al. 2011); assessing alternative methods to address incomplete linkage due to ineligibility for linkage and non-matches; exploring differences in survey participants who agree to linkage compared with participants who do not; and developing matching algorithms that will rely on the collection of more limited PII (e.g. four-digit SSN). Similarly, as surveys modify or update their designs (e.g. the upcoming redesign of NHIS and the collection of EHRs for NHCS), the NCHS Data Linkage Program will be impacted in terms of who is linkage-eligible and how to proceed with the linkage algorithms. In addition, legal and policy barriers remain between federal agencies, which create challenges for developing and outlining the terms of interagency agreements for linkage. The OMB M-14-06 directive clearly identifies these challenges and lays out a structure for moving forward particularly when the linked data products will be used for statistical purposes. OMB defines using the data for statistical purposes as "the description, estimation, or analysis of the characteristics of groups, without identifying the individuals or organizations that comprise such groups" (Office of Management and Budget, Executive Office of the President 2014).

As for opportunities, the NCHS Data Linkage Program will continue to focus on releasing more timely data through improvements in methodology and linkage algorithms. Improvements to the linkage algorithms will include assessing different strategies such as using alternate records, incorporating some probabilistic components into the methods, and maximizing the use of the PII that is currently being collected. In addition, future changes to administrative data may improve the linkage capabilities to the administrative records. One example of such an improvement is with the Medicaid and CHIP data. CMS is in the process of redesigning its system for the collection and management of Medicaid and CHIP data. Once fully implemented, the Transformed Medicaid Statistical Information System (T-MSIS) will provide the research community with a complete, accurate, and timely national database of detailed Medicaid and CHIP information. This new format may aid in future data linkage projects (Centers for Medicare and Medicaid Services n.d.-a).

As the linked data files are released, they are and will continue to be accompanied with detailed data documentation and summary reports in the form of Data Briefs, National Health Statistics Reports, or other NCHS publications. To aid researchers in the use of the linked data, the NCHS Data Linkage Program teamed with the NCHS Division of Health and Nutrition Examination Surveys to create a web tutorial for researchers willing to use the linked CMS data (National Center for Health Statistics n.d.-j). The NCHS Data Linkage Program plans to continue this type of outreach to data users in the form of tutorials as well as workshops at national conferences.

The linked data files are crucial to help inform public health policy. As the NCHS Data Linkage Program continues to move forward, a goal will be to make the data more available for researchers while protecting confidentiality. For example, the NCHS linkage program will consider developing visual dashboards of linkage rates, enrollment in plans (e.g. Medicare advantage and traditional Medicare), and causes of death by key survey variables (e.g. reported chronic conditions and access to care). These dashboards could be made available through the NCHS data visualization gallery, which was established to make NCHS data accessible interactively by policy analysts, researchers, and the general public (National Center for Health Statistics n.d.-k). In addition to data visualization efforts, the NCHS Data Linkage Program will explore the release of public use synthetic data files for the surveys that are linked to other administrative data beyond the NDI. While ensuring confidentiality and protection of PII remain the program's utmost priority, the program recognizes the need for public use files that allow access to at least some of the broad features of the data outside of the RDC. For example, public use files could contain summarized variables, such as high level aggregates for total Medicare dollars spent, summary variables of utilization (e.g. hospital stays, physician visits, and home health visits), and prevalence of chronic conditions as summarized by the Medicare claims. The release by CMS of their first-ever Medicare Current Beneficiary Survey public-use file has laid the groundwork for a public-use file that contains both survey and aggregated administrative data (Centers for Medicare and Medicaid Services n.d.-b).

Another avenue for the NCHS linked data program will be to inform survey design. Linked data can inform questionnaire development, inform imputation methodologies and, with more timely data, could potentially help with responsive and adaptive design techniques. The program also will continue research into adjusting sampling weights for linkage eligibility to ensure that the linked data can be used for nationally representative estimates.

The NCHS Data Linkage Program is exploring new avenues for linked data products. The first is the use of EHRs with data linkage. As health service providers are increasingly being encouraged to transition to the use of EHRs, linkage projects involving health survey data and EHRs will likely be a part of the future agenda for the NCHS Data Linkage Program. In October 2015, it was announced that eligible hospitals and health care professionals could fulfill the CMS EHR Incentive Program public health reporting objectives also known as Meaningful Use (MU), through the submission of data to NCHS' National Health Care Surveys (Centers for Medicare and Medicaid Services n.d.-c). As a result, NCHS is anticipating a high volume of EHR data from MU registrants that can be used for linkage across hospital settings within a sampled hospital system (linking multiple claims to single visits or discharges) as well as other data sources (i.e. NDI). A second avenue is investigating the linkage of health surveys to a potential database containing a federal registry of cancer surveillance. Lastly, the program is looking into methods utilizing the linked data to inform follow-up studies of survey participants.

The prospects and benefits for enhancing the data collected by the NCHS surveys with vital and administrative records are extensive, but these need to be balanced by the evolving challenges of protecting confidentiality and ensuring data integrity, as well as data accuracy, relevance, and timeliness.

11.8 Exercises

1 What are some of the limitations of using administrative data for research?

2 What are the benefits to using alternative records for data linkage?

3 List some enhancements to national health surveys by using linked vital or administrative records data.

4 Describe how privacy concerns play an important role in the linking of survey data with vital and administrative records.

Acknowledgments

The authors would like to thank Christine S. Cox, Kimberly A. Lochner, and Jennifer D. Parker for their helpful feedback and comments on the topics discussed in this chapter.

Disclaimer

The views expressed in this paper are those of the authors and no official endorsement by the Department of Health and Human Services or the Centers for Disease Control and Prevention (CDC) is intended or should be inferred.

References

Agency for Healthcare Research and Quality. (n.d.). Medical expenditure panel survey (MEPS). https://meps.ahrq.gov/mepsweb/.

Ahrens, K.A., Druss, B.G., Manderscheid, R.W., and Walker, E.R. (2016). *Housing assistance and blood lead levels in children in the United States, 2005–2012. Am J Public Health* 39: e1–e8.

Cai, L., Lubitz, J., Flegal, K.M., and Pamuk, E.R. (2010). The predicted effects of chronic obesity in middle age on medicare costs and mortality. *Medical Care* 48 (6): 510–517.

Centers for Medicare and Medicaid Services. (n.d.-a). Transformed Medicaid Statistical Information System (T-MSIS). https://www.medicaid.gov/medicaid/data-and-systems/macbis/tmsis/index.html.

Centers for Medicare and Medicaid Services. (n.d.-b). Medicare Current Beneficiary Survey (MCBS). https://www.cms.gov/Research-Statistics-Data-and-Systems/Research/MCBS/index.html.

Centers for Medicare and Medicaid Services. (n.d.-c). Medicare and Medicaid Electronic Health Records (EHR) Incentive Programs. https://www.cms.gov/Regulations-and-Guidance/Legislation/EHRIncentivePrograms/.

Centers for Medicare and Medicaid Services (CMS). (n.d.). http://www.cms.gov/.

Curtin, L.R., Mohadjer, L.K., Dohrmann, S.M. et al. (2013). National Health and nutrition examination survey: sample design, 2007–2010. National Center for Health Statistics. *Vital Health Stat* 2 (160), https://www.cdc.gov/nchs/data/series/ sr_02/sr02_160.pdf.

Dahlhamer, J.M. and C.S. Cox, (2007). Respondent consent to link survey data with administrative records: results from a split-ballot field test with the 2007 National Health Interview Survey. *Paper presented at the 2007 Federal Committee on Statistical Methodology Research Conference*, Arlington, VA.

Dunn, K., Jordan, K., Lacey, R.J. et al. (2003). Patterns of consent in epidemiologic research: evidence from Over 25,000 responders. *American Journal of Epidemiology* 159 (11): 1087–1094.

Evidence-Based Policymaking Commission Act of 2016, H.R. 1831, 114th Cong. (2015-2016).

Flegal, K.M., Graubard, B.I., Williamson, D.F., and Gail, M.H. (2007). Cause-specific excess deaths associated with underweight, overweight, and obesity. *JAMA* 298 (17): 2028–2037.

Florida Cancer Data System (FCDS). (n.d.). http://fcds.med.miami.edu/inc/welcome .shtml.

Gill, L. (2001). *Methods for Automatic Record Matching and Linkage and their Use in National Statistics*. National Statistics: London, UK.

Golden, C., Driscoll, A.K., Simon, A.E. et al. (2015). Linkage of NCHS Population Health Surveys to administrative records from social security administration and centers for medicare & medicaid services. National Center for Health Statistics. *Vital Health Stat* 1 (58), https://www.cdc.gov/nchs/data/series/sr_01/sr01_058.pdf.

Gorina, Y. and Kramarow, E.A. (2011). Identifying chronic conditions in medicare claims data: evaluating the chronic condition data warehouse algorithm. *Health Services Research* 46: 1–18.

Hedley Dodd, A. and P.M. Gleason (2013). Using the MAX–NHANES merged data to evaluate the association of obesity and Medicaid costs. MAX Medicaid Policy Brief #16. Washington, DC: Mathematica Policy Research.

Herzog, T.N., Scheuren, F.J., and Winkler, W.E. (2007). *Data Quality and Record Linkage Techniques*. New York, NY: Springer.

Hill, S., J. Atkinson, and T. Blakely. (2002). Anonymous record linkage of census and mortality records: 1981, 1986, 1991, 1996 Census Cohorts. NZCMS Technical Report No. 3.

Honeycutt, A.A., Segel, J.E., Zhuo, X. et al. (2013). Medical costs of CKD in the Medicare population. *J Am Soc Nephrol* 24 (9): 1478–1483.

Judson, D.H., Parker, J.D., and Larsen, M.D. (2013). *Adjusting Sample Weights for Linkage-Eligibility Using SUDAAN*. Hyattsville, MD: National Center for Health Statistics.

Larsen, M.D., M. Roozeboom, and K. Schneider (2012). Nonresponse adjustment methodology for NHIS-medicare linked data. *Proceedings of the American Statistical Association Joint Statistical Meetings.*

Lloyd, P.C., A.E. Simon, and J.D. Parker, (2015). Characteristics of Children in Medicaid Managed Care and Medicaid Fee-forservice, 2003–2005. Natl Health Stat Report. In JSM Proceedings, Survey Research Methods Section. Alexandria, VA: American Statistical Association, (80): 1–15.

Lloyd, P.C., Helms, V.E., Simon, A.E. et al. (2017). Linkage of 1999–2012 National Health Interview Survey and National Health and Nutrition Examination Survey Data to U.S. Department of Housing and Urban Development Administrative Records. *Vital Health Stat* 1 (60).

Looker, A. (2013). Femur neck bone mineral density and fracture risk by age, sex, and race or Hispanic origin in older US adults from NHANES III. *Archives of Osteoporosis* 8 (141). https://pubmed.ncbi.nlm.nih.gov/23715737/.

Loria, C.M., Sempos, C.T., and Vuong, C. (1999). Plan and operation of the NHANES II Mortality Study, 1992. National Center for Health Statistics. *Vital Health Stat* 1 (38), https://www.cdc.gov/nchs/data/series/sr_01/sr01_038.pdf.

Marton, K. and J.C. Karberg, (2011). The future of federal household surveys: summary of a workshop. Committee on National Statistics; Division of Behavioral and Social Sciences and Education; Research Council.

Medicaid Undercount Project (SNACC). (n.d.). https://www.census.gov/did/www/snacc/.

Miller, E.A., et al. (2014). Linkage of 1986–2009 National Health Interview Survey with 1981–2010 Florida Cancer Data System. National Center for Health Statistics. *Vital Health Stat* 2 (167), https://www.cdc.gov/nchs/data/series/sr_02/sr02_167.pdf.

Miller, D.M., R. Gindi, and J.D. Parker, (2011). Trends in record linkage refusal rates: Characteristics of National Health Interview Survey participants who refuse record linkage. Presented at Joint Statistical Meetings 2011. Miami, FL (30 July–4 August).

Miller, E.A., et al. (2015). Use of alternate information to improve linkage with the National Death Index (NDI). *Presented at: Annual Conference of North American Association of Central Cancer Registries 2015*, Charlotte, NC (16–18 June).

Miller, E.A., Decker, S.L., and Parker, J.D. (2016). Characteristics of medicare advantage and fee-for-service beneficiaries upon enrollment in medicare at age 65. *J Ambul Care Manage* 39 (3): 231–241.

Mirel, L.B. (2012). *Health Characteristics of Medicare Traditional Fee-for-Service and Medicare Advantage Enrollees: 1999–2004 National Health and Nutrition Examination Survey Linked to 2007 Medicare Data*. National health statistics reports; no 53. Hyattsville, MD: National Center for Health Statistics.

Mirel, L.B. et al. (2010). Linking the National Health and Nutrition Examination Survey to Supplemental Nutrition Assistance Program Administrative Data: A one state pilot study. In: *Proceedings of Statistics Canada Symposium 2010, Social Statistics: The Interplay among Censuses, Surveys and Administrative Data*. Ottawa: Statistics Canada.

Mirel, L.B., Simon, A. E., Golden, C., et al. (2014). Concordance between survey report of Medicaid enrollment and linked Medicaid administrative records in two national studies. National Health Statistics Reports; no 72. Hyattsville, MD: National Center for Health Statistics.

National Center for Health Statistics. (1984). Supplement on aging (SOA). http://www.cdc.gov/nchs/lsoa/soa1.htm.

National Center for Health Statistics. (2015). Comparative analysis of the NHIS public-use and restricted-use linked mortality files: 2015 public-use data release. February. Hyattsville, MD.

National Center for Health Statistics. (n.d.-a). Second longitudinal study on aging (LSOA II). http://www.cdc.gov/nchs/lsoa/lsoa2.htm.

National Center for Health Statistics. (n.d.-b) National Health and Nutrition Examination Survey (NHANES). http://www.cdc.gov/nchs/nhanes/index.htm.

National Center for Health Statistics. (n.d.-c) Third National Health and Nutrition Examination Survey (NHANES III). http://www.cdc.gov/nchs/nhanes/nhanes3.htm.

National Center for Health Statistics. (n.d.-d) NHANES I Epidemiologic Followup Study (NHEFS). http://www.cdc.gov/nchs/nhanes/nhefs/nhefs.htm.

National Center for Health Statistics. (n.d.-e) National Health Care Surveys. http://www.cdc.gov/nchs/nhcs/index.htm.

National Center for Health Statistics. (n.d.-f) National Hospital Care Survey. http://www.cdc.gov/nchs/dhcs/index.htm.

National Center for Health Statistics. (n.d.-g) National Vital Statistics System (NVSS). http://www.cdc.gov/nchs/nvss/index.htm.

National Center for Health Statistics. (n.d.-h) National Death Index (NDI). http://www.cdc.gov/nchs/ndi/index.htm.

National Center for Health Statistics. (n.d.-i). NCHS data linkage activities. http://www.cdc.gov/nchs/data-linkage/index.htm (accessed 17 August 2020)

National Center for Health Statistics. (n.d.-j) *NHANES-CMS Linked Data Tutorial*. http://www.cdc.gov/nchs/tutorials/NHANES-CMS/index.htm.

National Center for Health Statistics. (n.d.-k) *NCHS Data Visualization Gallery*. https://blogs.cdc.gov/nchs-data-visualization/.

National Center for Health Statistics (NCHS). (n.d.). http://www.cdc.gov/nchs/index.htm (accessed 17 August 2020).

National Center for Health Statistics and National Health Interview Survey (NHIS). (n.d.). http://www.cdc.gov/nchs/nhis/index.htm.

Office of Management and Budget, Executive Office of the President. 2010: M-11-02: Sharing Data While Protecting Privacy. Washington, D.C.

Office of Management and Budget, Executive Office of the President. 2013: M-13-13: Open Data Policy – Managing Information as an Asset. Washington, D.C.

Office of Management and Budget, Executive Office of the President. 2013: M-13-17: Next Steps in the Evidence and Innovation Agenda. Washington, D.C.

Office of Management and Budget, Executive Office of the President. (2014). M-14-06: Guidance for Providing and Using Administrative Data for Statistical Purposes. Washington, D.C.

Parsons, V.L. et al. (2014). Design and estimation for the National Health Interview Survey. *Vital Health Stat* 2 (165).

Pratt, L.A., Druss, B.G., Manderscheid, R.W., and Walker, E.R. (2016). Excess mortality due to depression and anxiety in the United States: results from a nationally representative survey. *Gen Hosp Psychiatry* 39: 39–45.

Research Triangle Institute. (2008). *SUDAAN Language Manual Release 10.0.* Research Triangle Institute: Research, Triangle Park, NC.

Riley, G. (2006). Health insurance and access to care among social security disability insurance beneficiaries during the medicare waiting period. *Inquiry* 43 (3): 222–230.

Sayer, B. and C.S. Cox. (2003). How many digits in a handshake? National Death Index matching with less than nine digits of the social security number. *Proceedings of the American Statistical Association Joint Statistical Meetings.*

Social Security Administration (SSA). (n.d.). http://www.ssa.gov/.

Taft, R.L., (1970). Name search techniques, New York State identification and intelligence system. Special Report No. 1, Albany, NY.

The Soundex Indexing System. (n.d.). National Archives and Records Administration. 2007-05-30. http://www.archives.gov/research/census/soundex.html.

United States Department of Housing and Urban Development User. (n.d.). https://www.huduser.gov/portal/home.html.

United States Renal Data System (USRDS). (n.d.). http://www.usrds.org/.

Zhang, G., Parker, J.D., and Schenker, N. (2016). Multiple imputation for missingness due to nonlinkage and program characteristics: a case study of the National Health Interview Survey linked to Medicare claims. *J Surv Stat Methodol* 4 (3): 319–338.

12

Combining Administrative and Survey Data to Improve Income Measurement

Bruce D. Meyer[1,2] and Nikolas Mittag[3]

[1] *NBER, AEI and U.S. Census Bureau, 4600 Silver Hill Road, Washington, DC 20233, USA*
[2] *Harris School of Public Policy Studies, University of Chicago, 1307 E. 60th Street, Chicago, IL 60637, USA*
[3] *CERGE-EI, Politických vězňů 7, Prague 1 110 00, Czech Republic*

12.1   Introduction

Large, nationally representative surveys are among the key sources of information for both academics and policy makers. Household surveys are the basis of official rates of poverty, unemployment and inflation, and many other official statistics in the U.S. Survey estimates not only provide information that guides policy choices, but are also directly used to allocate federal funds. Policy makers often rely on detailed household characteristics from surveys for purposes such as improving the targeting of policies. Such information is also essential to academic researchers studying household decisions and the effects of policies. This situation makes the collection of large, detailed, and accurate data indispensable. However, recent research has documented an alarming and often growing extent of survey error in key measures, such as income, education, and health, in important household surveys. See Bound, Brown, and Mathiowetz (2001) for a review of earlier studies and Meyer, Mok, and Sullivan (2015) for an overview of the recent literature on income and transfer programs.

This chapter reviews how combining administrative and survey data can improve survey accuracy. Building on an overview of the growing applied literature that links administrative records to survey data, we argue that linking administrative records to household surveys can be used to measure the importance of each error source and suggest ways to reduce them. Recent projects have demonstrated these potential benefits of data linkage by linking administrative microdata to U.S. household surveys, such as the Current Population Survey (CPS), the American Community Survey (ACS), and the Survey of Income and Program

Administrative Records for Survey Methodology, First Edition.
Edited by Asaph Young Chun, Michael D. Larsen, Gabriele Durrant, and Jerome P. Reiter.
© 2021 John Wiley & Sons, Inc. Published 2021 by John Wiley & Sons, Inc.

Participation (SIPP). Databases that contain large parts or the entire survey population of interest are increasingly becoming available from administrative records, private companies, or public sources such as the Internet. Advances in computation and methods to link data sources have not only greatly reduced the cost of data linkage, but also increased the accuracy of the linked data. These advances make it possible to implement linkage broadly and frequently enough to continuously evaluate and improve surveys. Thereby, linking administrative records to survey data offers many promising ways to alleviate the problem of survey error and provide us with the accurate and detailed data necessary to design good policies. We specifically emphasize recent research on government income transfer programs and discuss the advantages of administrative data and record linkage for data users in academia and policy. Thereby, we complement early chapters of this book, which put more emphasis on the perspective of data producers and statisticians.

Linking administrative records to surveys at the household or individual level allows us to add variables to the survey. This process can add information on questions that were not asked or eliminate the need for some existing questions, reducing respondent burden. Here, we are concerned with measuring and improving the accuracy of survey responses, so we focus on cases where data linkage is used to add a second measure of a variable that is included in the survey data, but measured with error. If the administrative data contain accurate measures that are comparable to the survey definitions and the data sources can be linked with little error, then the variable from the linked administrative records can be considered a measure of truth. We focus on such cases, which heavily rely on high quality administrative records and linkages as discussed in Section 12.3 of this book, as well as on researchers to ensure and assess the accuracy of the linked data as described in Section 12.2 of this volume. See Celhay, Meyer and Mittag (2019a), Courtemanche, Denteh, and Tchernis (2018), and Meyer and Mittag (2019a) for evidence on data accuracy for specific linked data sets. Many of the methods we survey can still be used as long as errors in the linked administrative variable are sufficiently rare that it provides a reasonable approximation to the truth. In Section 12.7, we discuss this issue further and refer to methods for cases in which both survey and administrative measures are subject to error.

We focus on the case of U.S. government transfers, where the linked administrative measure can often provide a measure of true program receipt, including the amount received. We do not mean to argue that administrative records or transfer program records are more accurate in general. Niehaus and Sukhtankar (2013) provide an impressive example to the contrary. However, by virtue of being paid from government resources, transfer payments usually leave an accurate paper trail to ensure accountability. The administrative records are usually scrutinized to avoid fraud both on behalf of the government agency administering the program and the individuals who receive the transfers. The records often contain official identifiers, which facilitate linkage. Therefore, linking program records can provide us with an accurate measure of a survey concept. Past research on government transfer programs, such as Medicaid (Davern et al. 2008), the Supplemental Nutrition Assistance Program (SNAP, formerly The Food Stamp Program, see Bollinger and David, 1997; Celhay, Meyer, and Mittag 2018a; Cerf Harris 2014; Kirlin and Wiseman 2014, Marquis and Moore 1990; Meyer, Mittag, and Goerge forthcoming; Taeuber et al. 2004), cash welfare (Celhay, Meyer, and Mittag 2018a; Lynch et al. 2007, and social security and pensions (Nicholas and Wiseman 2009, 2010; Gathright and Crabb 2014; Bee and Mitchell 2017a,b), have used data linkage to demonstrate substantial survey error that has important effects on survey estimates. Both policy makers and academics heavily rely on the data in which these studies document sizeable error. Transfer programs are key tools of the government to combat poverty and inequality, so understanding their consequences is important to academics examining the distribution of income, work incentives, and the well-being of recipients.

We use the case of transfer receipt to show how data linkage can improve the accuracy of survey data. We first show how linked data can be used to estimate total survey error and decompose it into key components, such as generalized coverage error (the combination of coverage error, unit nonresponse error, and errors from weighting), item nonresponse error, and measurement error. Knowing the magnitude of different errors should be the first step in a plan to reduce them. We then discuss each of these three error sources in Sections 12.3–12.5. For each source, we summarize how linked data can be used to examine the nature and consequences of errors, review the evidence on the extent of error and discuss how data linkage can contribute to improving the problem. We provide an illustration of how data linkage can improve measures of poverty and the income distribution in Section 12.6. In Section 12.7, we discuss likely effects of errors in the linked variable and review ways to assess and improve data accuracy.

12.2 Measuring and Decomposing Total Survey Error

Recent research has documented survey error from non-sampling sources such as nonresponse and measurement error that is often more problematic than sampling error. Nonetheless, survey producers routinely provide users with information on sampling error, but data users usually do not have enough information about non-sampling error to assess its impact on their estimates or account for it. To understand and address the problem, as well as to allow researchers to assess the reliability of their estimates, we need a framework to catalog and measure non-sampling error. Decomposing survey error into to its sources can provide a joint assessment of multiple sources of error in the same survey, allowing easier and more meaningful comparisons.

Understanding the importance of each error source is crucial to reduce overall survey error in a cost-effective manner. For example, survey producers often spend considerable resources to increase response rates, even though the impact on final estimates and nonresponse bias seems to be small (Groves 2006; Groves and Peytcheva 2008). Reducing nonresponse may also amplify measurement error (Sakshaug, Yan, and Tourangeau 2010; Tourangeau, Groves, and Redline 2010), which we find to be a much larger source of bias in mean transfer dollars received (Meyer and Mittag forthcoming). This observation raises the question whether the same overall error reduction could be achieved at a lower cost by shifting some resources toward the reduction of measurement error. To answer such questions, we need to understand the magnitude of each error source, as well as how the measures taken and funds spent affect the error from each source. This requires methods to measure survey error and its components on a routine basis.

To illustrate our points, we rely on a framework and examples from some of our recent work. In Meyer and Mittag (forthcoming), we argue that administrative records that cover the entire population of interest and data linkage enable us to estimate total survey error (TSE) and its components (Groves and Lyberg 2010). In particular, we define measures of total non-sampling survey error in the sum or mean of a variable as well as components of this error due to survey coverage, item nonresponse and measurement error. Accurate records on the entire population linked to the survey data can provide the measures of truth required to estimate these measures of error. Thereby, data linkage provides an inexpensive and easily implementable way to analyze and summarize survey error.

The unlinked population data enable us to calculate the target amount that the survey should yield. Administrative records usually include individuals that the survey does not intend to cover and records that cannot be linked to the survey data due to linkage problems, so estimating the target amount likely requires some adjustment. For example, in Meyer and Mittag (forthcoming), the statistic of interest is an estimate of average dollars received from a welfare program. The surveys that we use do not include individuals in group quarters or experiencing homelessness. We adjust for this intentional difference by subtracting payments to the two groups from the total amount paid according to the unlinked administrative data. We adjust for linkage problems such as linkage failure in a similar way. Both adjustments are likely typical in that they contain components that can be calculated from the administrative data (such as payments to individuals experiencing homelessness) and components that need to be estimated from other surveys (such as payments to individuals in group quarters). We estimate the TSE in average dollars of transfer payments, $\varepsilon_{\mathrm{TSE}}$, as the difference between the estimate of average dollars received according to the survey reports in the linked data and average dollars paid according to the unlinked administrative records adjusted for intentional coverage differences and linkage problems.

We then decompose TSE into generalized coverage error, item nonresponse error and measurement error:

$$\varepsilon_{TSE} = \varepsilon_{GCE} + \varepsilon_{INR} + \varepsilon_{ME}.$$

Generalized coverage error, ε_{GCE}, is the difference between average dollars actually received by those in the survey population and the target amount of the survey in our case. More generally it is the difference between the statistic of interest, such as the mean of a variable, in the population the survey sample actually captures and the same statistic in the population that the survey intends to represent. We can estimate generalized coverage error as the difference between the weighted average of the administrative variable in the linked data and the target amount for the survey.

Item nonresponse error, ε_{INR}, is the contribution to TSE from the difference between the estimate of interest according to imputed and accurate values of nonrespondents. We estimate item nonresponse error from the linked data as the weighted sum of the difference between the imputed survey responses and the accurate values from the administrative data for item nonrespondents divided by the population size.

Finally, measurement error, ε_{ME}, is the contribution to TSE from the difference between the statistic of interest according to survey reports and true values among respondents. Just as item nonresponse error, we can estimate it from the linked data as the weighted sum of the difference between survey reports and the accurate administrative values among respondents divided by the population size. We further decompose item nonresponse and measurement error into two parts due to misclassification, false positives and false negatives, and a part due to errors in amounts among those correctly classified. This division is useful for the many important variables that combine a continuous part with a mass point at zero, such as income or expenditure and their components. The decomposition can be extended to other TSE components such as processing error and other measures of TSE, such as mean squared error.

In Meyer and Mittag (forthcoming), we estimate and decompose survey error in average transfer dollars received from SNAP and Public Assistance (cash welfare) in the New York State sample of the ACS, CPS, and the SIPP. TSE is substantial both in absolute and relative terms in all cases except for SNAP in the SIPP. All three surveys understate transfer dollars received, with larger overall error for Public Assistance than for SNAP. TSE ranges from missing 4.3% of dollars received (SNAP in the SIPP) to missing almost two out of every three dollars of Public Assistance in the CPS. Figure 12.1 presents the results of our TSE decomposition. The size of the vertical bars indicates survey error as a fraction of the target amount. The numbers in each error band provide the error due to each component as a percentage of TSE for the respective survey and program. TSE is negative in all cases, but some error components are positive and thus reduce overall error. The sign and the magnitude of the error components vary across surveys and programs, which makes the composition of TSE differ substantially across the columns of Figure 12.1. The only component that leads surveys to understate dollars received in all cases is measurement error. For transfer dollars received, measurement error is by far the largest source of error. Further decomposing it into binary and continuous parts shows that it is mainly due to underreporting of binary receipt. Our estimates of generalized coverage error and item nonresponse error are much smaller and often bias estimates in opposing directions. The TSE due to these two components combined is less than 10% of the target amount in all cases.

12.3 Generalized Coverage Error

Generalized coverage error is a combination of several types of error. The sampling frame may not include all units in the population of interest (frame error) and some sampled units may not respond to the survey at all (unit nonresponse error). In addition, survey processing, such as adjusting the sample weights to correct for unit nonresponse, may introduce additional error. We focus on studies that link microdata from surveys to administrative records to assess and improve the problem of survey coverage. If the administrative records include the entire population of interest and data linkage is accurate or one can adjust for missed links, then the linked data are representative of the population of interest. Then, data linkage enables us to observe the same variable for the population and the survey sample. Therefore, we can directly compare survey estimates to their population values without any concerns for comparability of what the variables in the two data sources measure. Thereby, data linkage provides us with a simple, but powerful, tool to analyze survey representativeness and coverage in the aggregate. Note that this does not require the variable in the administrative data to be an accurate measure. If the variable from the administrative data is measured with error, we can still estimate the population size of linked units and compare it to the actual population size in the administrative data. The same idea can be applied to subpopulations of interest, such as single parent households or the elderly as long as they can be identified in the administrative records.

A few recent studies have gone even further by linking the administrative records to the sampling frame or all sampled units rather than survey participants only. Linking directly to the sampling frame is more complicated, because usually little information beyond addresses is available for units that were not sampled or did not respond. However, this type of data linkage allows us to assess the problem of survey representativeness in much more detail. First, it makes it possible to isolate the different sources of error that lead to generalized coverage error. For example, frame error is the difference between the statistic of interest according to the administrative variable linked to the survey frame and according to the unlinked population data. Unit nonresponse error can be estimated from the records linked to unit nonrespondents. The effect of the weight adjustment is the difference in the statistic of interest estimated using the adjusted and the original sample weights. Second, it allows researchers to study these sources of error at the household or individual level. For example, Bee, Gathright, and Meyer (2017) link the 2011 CPS Annual Social and Economic Supplement (CPS ASEC) frame to internal revenue service (IRS) Form 1040 data for tax year 2010 using address information from both sources. They use the resulting linked data to compare characteristics of respondents and nonrespondents obtained from the tax records. These characteristics include income, self-employment status, marital status, presence and number of children, and the receipt of pensions and certain government benefits.

The preliminary results from a few recent studies that have used data linkage to examine survey coverage are encouraging both in terms of the usefulness of this approach and in terms of survey accuracy. To our knowledge, Meyer and Mittag (forthcoming) is the only study of this kind that examines transfer programs. As the results in Figure 12.1 show, they find that the coverage of program recipients in all three surveys is high, at least compared to the magnitude of other sources of error. Bee, Gathright, and Meyer (2017) focus on unit nonresponse and assess the consequences for measures of income in the CPS ASEC. They are unable to reject that the distribution of household income among nonrespondents is the same as that of respondents, though the groups differ on other characteristics such as marital status and number of dependents. Quite remarkably, an analysis of the SIPP by Mattingly et al. (2016) using the same approach of linking tax records to respondent and nonrespondent addresses produces nearly the same result of no significant differences between respondents and nonrespondents at a wide range of percentiles of income, but does not examine other characteristics of the two groups. Brummet et al. (2017) use a similar approach to examine income in the Consumer Expenditure Survey.

Linking administrative records both to the sample frame and to respondents only likely directly points to methods to improve survey coverage. The problem of survey coverage is notoriously difficult to study, so that even descriptions of who is missing from the survey can help to address the problem. The methods described above yield a detailed description of the geographic distribution of units that are missing from the survey. If the administrative data include individual or household characteristics, then they also provide estimates of the demographic characteristics of those missing from the survey. This information can be used to improve the

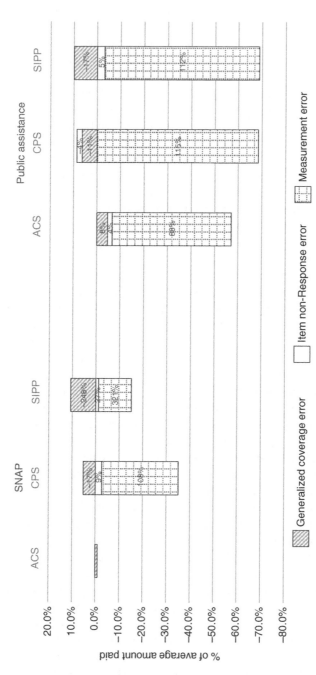

Figure 12.1 TSE components in percent of average dollars paid and as shares of TSE. *Notes*: The size of the bars indicates the error amount as a percentage of average dollars received per household by the relevant population (which varies across surveys). The numbers in the bars provide the error due to each component as a percentage of total survey error for the respective survey and program. All estimates are based on the linked New York sample using survey weights adjusted for Protected Identification Key (PIK) probability. The ACS does not ask for SNAP amounts, so we can only estimate coverage error. The SIPP is not claimed to be representative of New York state, so the estimates of generalized coverage error for the SIPP should be interpreted with caution. Source: Meyer and Mittag (forthcoming).

survey at the design stage, by improving the sampling frame or increased efforts to include groups that are known not to respond. It can also be used ex post by adjusting the sample weights to make information from the two data sources match.

12.4 Item Nonresponse and Imputation Error

Item nonresponse is the problem that some survey respondents may refuse to answer some questions or sections of the survey, so that some information is missing for these observations. Survey agencies often fill in these values using imputation methods such as the hot deck, see e.g. Little and Rubin (2002) for an overview. Applied researchers usually deal with the problem of item nonresponse either by using only respondents with complete interviews or by including the imputed values of item nonrespondents. For consistency, both strategies assume that item nonresponse is conditionally random (missing at random [MAR]). Data linkage provides an accurate measure of the response for both respondents and nonrespondents in a survey. Thereby, it allows us to test MAR, examine how respondents differ from item nonrespondents as well as how imputed values differ from accurate responses, and to directly estimate the bias from each strategy.

Item nonresponse is often found to be related to other observable characteristics. However, the crucial question for the consistency of estimates is whether respondents differ in their true response conditional on the observed characteristics in the model of interest. See Heitjan and Rubin (1991) and Heitjan (1994) for discussions. The availability of an accurate measure for the entire sample in linked data makes it possible to examine how item nonrespondents and respondents differ in terms of the true response. For instance, Bollinger et al. (2019) and Chenevert, Klee, and Wilkin (2016) compare linked tax records of respondents and nonrespondents to the question on earned income. Both reject the assumption that item nonresponse is conditionally random.

In Celhay, Meyer, and Mittag (2018a), we use administrative records from New York State on SNAP and Public Assistance linked to the ACS, CPS, and SIPP to examine nonresponse to the question on program receipt. We conduct two tests of MAR: first, we test whether the probability of program receipt among item nonrespondents differs from the overall population after conditioning on covariates typically included in models of program receipt. Second, we test whether the conditional distribution of program receipt differs between nonrespondents and the entire population. Both tests provide evidence that nonrespondents differ systematically from the entire population, even conditional on many covariates. Thus, excluding item nonrespondents will bias not only unconditional population statistics, but also estimates of models that condition on these covariates. More specifically, the results in Celhay, Meyer, and Mittag (2018a) show that item nonrespondents are more likely to be program recipients than the overall population both unconditionally and conditional on key covariates. As such, estimated receipt rates are likely to be biased when excluding item nonrespondents. We also find that item nonrespondents and respondents differ in how variables, such as income, relate to program receipt. We show that coefficient estimates associated with these variables are biased in models of program receipt that do not include item nonrespondents.

Imputation allows researchers to include item nonrespondents in such analyses. This solution avoids the sample selection bias from excluding item nonrespondents but may introduce new bias due to imputation error. Unfortunately, little is known about the accuracy of imputations and the merits of different imputation procedures, see e.g. Andridge and Little (2010) and Little and Rubin (2002). Recent studies provide evidence that imputation can induce substantial error in survey data (Meyer, Mittag, and Goerge forthcoming; Celhay, Meyer, and Mittag 2018a) and in estimates derived from them (Lillard, Smith, and Welch 1986; Hirsch and Schumacher 2004; Bollinger and Hirsch 2006). Celhay, Meyer, and Mittag (2018a) use linked data to examine how imputations of transfer receipt differ from the accurate administrative variable. They reject the hypothesis that imputed values correctly reproduce the joint distribution of program receipt and covariates commonly used in studies of transfer programs. In line with the results in Hirsch and Schumacher (2004) and Bollinger and Hirsch (2006), the differences are particularly severe for variables that were not used in the imputation procedure.

These results raise the question whether data users should use or exclude the imputed observations. There is considerable disagreement on this issue, see Angrist and Krueger (2001) for a discussion. Our results from Celhay, Meyer, and Mittag (2018a) above imply that neither including imputations nor excluding item nonrespondents yields consistent estimates. We use the linked data to directly estimate and compare the bias from each strategy for models of transfer receipt. We find the bias to be smaller when excluding item nonrespondents than when using their imputed values. This finding suggests that it may be better to use only respondents than to include imputed values. However, the magnitude of the bias terms also depends on the model of interest and the imputation procedure, so it is not clear whether this finding generalizes. Nonetheless, the results in Celhay, Meyer, and Mittag (2018a) can provide some guidance, because one would expect the bias from nonresponse to be determined by the differences between respondents and nonrespondents and the bias from imputation error to depend on how the imputed values differ from the true values.

Linked data can help to improve the problem of item nonresponse both directly and indirectly. Linked data can provide us with the information on the extent of error among the imputed observations that is needed to make informed decisions, such as whether to use imputed values or not. The accurate measure of the response that linked data can provide also makes it possible to evaluate the accuracy of different imputation procedures. Studying the merits and disadvantages of different imputation methods can help to refine existing procedures and choose the optimal method in specific cases. The results on match bias (Hirsch and Schumacher 2004; Bollinger and Hirsch 2006) and the mounting evidence that item nonresponse is not conditionally random also indicates that it is important to incorporate information beyond key survey variables in the imputation procedure. Data linkage allows survey producers to incorporate information from the accurate administrative records in the imputation procedure. See Davern, Meyer, and Mittag (2019), Mittag (2019), or Blackwell, Honaker, and King (2017) for further discussion. Linked data may even make it possible to replace the missing values by those from the administrative data. See Section 12.6 for an illustration.

12.5 Measurement Error

Measurement error in surveys is defined as the difference between the value of a variable of interest obtained from survey responses and the true value of the same variable. Such survey response errors have been documented for many different variables, such as income (Abowd and Stinson 2013; Bound and Krueger 1991; Bollinger 1998; Dahl, DeLeire, and Schwabish 2011), education (Black, Sanders, and Taylor 2003) or drug use by teenagers (Johnson and Fendrich 2005). The main challenge of studying measurement error in surveys is the availability of a variable's true value. Unlike studies that compare aggregate numbers from survey data and official records (Meyer, Mok, and Sullivan 2015), linking survey data to administrative records allows studying measurement error at the individual level, significantly expanding the scope of analysis. Contrary to aggregate comparisons, studies that use linked data at the individual level can separate the measurement error component from other sources of survey error such as frame, coverage, or survey nonresponse error (Meyer, Mok, and Sullivan 2015; Meyer and Mittag forthcoming). An accurate measure of the response at the individual level also makes it possible to explore the nature and consequences of these errors in greater detail, for example by examining their distribution and relation to other variables. Finally, data linkage enables us to study the extent of both over- and underreporting in addition to net reporting. For the example of binary variables such as transfer receipt, these data make it possible to observe program recipients who do not report participation (false negatives) and non-recipients who report participation (false positive). Using linked data, researchers can estimate how much of the differences between surveys and official records are due to under-reporting of government transfers and how much underreporting is offset by overreporting. This simple distinction is relevant since the two types of error may cancel out in aggregate comparisons of survey receipt, clouding the magnitude of measurement errors in surveys. In addition, the sum of the two error rates rather

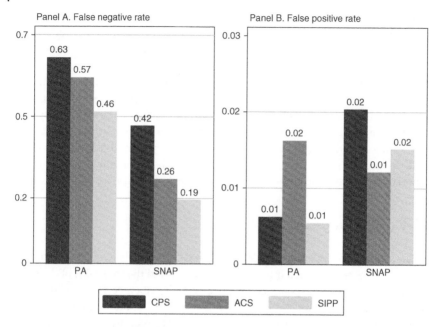

Figure 12.2 Error rates in reported receipt of SNAP and public assistance in the CPS, ACS, and the SIPP. *Notes:* Rates of misreporting program receipt among recipients (Panel A) and non-recipients (Panel B) of Food Stamps (FSP or SNAP) and Public Assistance (PA) according to the linked administrative program receipt variable. Source: New York State data for 2007–2012 from Celhay, Meyer, and Mittag (2018a).

than the net error rate determines the bias in common models – see Hausman, Abrevaya, and Scott-Morton (1998) for an example.

As pointed out in the introduction, studies of various welfare programs have found shockingly high error rates in the report of program receipt in different surveys and programs. For example, 60% of welfare recipients in the CPS and the same share of pension recipients in the ACS fail to report receipt (Meyer and Mittag 2019b; O'Hara et al. (2016). Celhay, Meyer, and Mittag (2018a) link multiple surveys to administrative records from New York State covering 2007–2012. They report that the probability of a false negative response of SNAP receipt varies from 19% in the SIPP to 42% in the CPS, while the false positive rate varies from just above 0.5% for cash welfare in the SIPP to just above 2% for SNAP in the ACS. Figure 12.2 summarizes their results.

These results are problematic for estimates of aggregate statistics of receipt, such as take-up rates or total dollars received. These biases can be large compared to other well-studied sources of survey error such as imputation or nonresponse (Celhay, Meyer, and Mittag 2018a). Even worse, the studies mentioned above show that response errors are not independent of other respondent characteristics or the true value of the variable, so that they likely bias both causal and descriptive estimates obtained from survey data (e.g. Bollinger and David 1997; Bound, Brown, and Mathiowetz 2001; Meyer, Mittag, and Goerge forthcoming; Meyer and Mittag 2017, forthcoming). For example, Meyer, Mittag, and Goerge (forthcoming) find that survey error leads to biased estimates of the determinants of program receipt. Using linked data, their results indicate that the survey data understate participation by single parents, non-whites, and the elderly and how participation declines as incomes rise.

Using linked data also enables researchers to study the nature of measurement error by exploring the determinants of errors in surveys. To understand and address the problem of measurement error it is important to examine how respondent characteristics, behavior, and survey design affect response accuracy. Inaccurate responses may arise from cognitive problems, such as recall, or respondent cooperation. Survey features related to measurement error include the recall period, choice of survey mode, the role of interviewers, and data post processing. Reviews of the main theories of misreporting can be found in Sudman and Bradburn (1973), Sirken (1999), and Bound, Brown, and Mathiowetz (2001). In Celhay, Meyer, and Mittag (2018b), we use administrative records linked to the ACS, CPS, and SIPP to examine several hypotheses related to respondents' cognition, their cooperation and survey design.

The literature on respondents' characteristics usually discusses how cognitive aspects of respondents affect survey response (e.g. Tourangeau 1984). An important source of error arises from memory or recall error, when respondents are asked about past events (Groves et al. 2009, p. 231). In Celhay, Meyer, and Mittag (2018b) and Meyer and Mittag (2019a), we show that the length of time since last program receipt substantially increases measurement error in reported program receipt. Such recall error is among the most important sources of measurement error we find, but this relationship may depend on the question at stake and survey design. Celhay, Meyer, and Mittag (2018b) also provide evidence of survey error arising from confusion regarding the dates of program receipt. This confusion causes some respondents to report program receipt that happened either before or after the reference period (telescoping). We find that households that are more dependent on government transfers are better reporters on average, which supports the hypothesis that salience improves reporting.

The quality of survey data is also affected by respondents' motivation or attitudes toward surveys. Linked data enable us to test whether respondents who are more cooperative are less likely to misreport in surveys. For example, Bollinger and David (2001) use SIPP data linked to administrative records to show that reporting errors predict later attrition. Extending their analysis in Celhay, Meyer, and Mittag (2018b), we find that respondents who have higher imputation rates in other sections of a survey are more likely to misreport participation in questions of program receipt. Another frequently discussed reason for misreporting is the unwillingness to provide socially undesirable answers. In the case of government transfers, Celhay, Mittag, and Meyer (2018c) provide evidence that stigma associated with participation in welfare programs can explain why people fail to report participation to interviewers. Finally, many survey design features are hypothesized or known to be related to reporting errors, although the empirical evidence is often mixed (Groves 2004). Linked data can be used to examine how survey design features affect response accuracy. However, this requires exogenous (conditionally random) variation in survey design features.

Linked data can help to improve the problem of survey error in various ways. The insights gained from linked data can improve the way we use surveys and how we interpret survey estimates. Linked data allow us to describe the extent and predictors of errors and provide estimates of the bias for specific cases. This can help survey users assess the accuracy of their estimates and the likely biases, particularly in the absence of theoretical results. If formulas for (asymptotic) bias from measurement error are available, estimating the bias is helpful to gauge the applicability of the formulas to finite samples and whether the magnitudes and directions of the bias conform to what the patterns of survey error suggest (Meyer and Mittag 2017). For example, given the high error rates in reports of transfer receipt found in linked data, survey users should be skeptical of survey estimates of receipt rates. Yet, results from linked data on the bias in models of program receipt suggest that strongly significant estimated associations in a given direction are unlikely to be reversed even when the dependent variable suffers from high and systematic misclassification (Meyer and Mittag 2017; Celhay, Meyer, and Mittag 2018a).

Understanding the nature of survey errors not only helps to assess the bias, but also puts us in a better position to correct it. Several bias corrections and consistent estimators have been proposed in the literature. However, they require additional assumptions on the errors, such as conditional independence. Linked data can inform us whether these assumptions are plausible and hence which correction methods are likely to improve estimates. For example, Meyer and Mittag (2017) find that corrections for misclassification of the dependent variable in binary choice models that require misclassification to be unrelated to covariates can easily make estimates worse if this assumption does not hold. On the other hand, corrections that incorporate information on these correlations from linked data can substantially improve estimates. The insights on the nature of errors and the bias discussed above are particularly useful if results from linked data are available for similar variables in similar surveys. By providing insights on the reasons for misreporting, linked data can also help to gauge the reliability of survey data more generally. Studies such as Celhay, Meyer, and Mittag (2018b) point to conditions under which survey errors are particularly problematic that likely hold more generally.

Linked data can also provide us with a better understanding of how survey design affects measurement error. Linking surveys to administrative records on a more routine basis would shed more light on the effects of (changes in) survey design on response accuracy. A better understanding of the causes and predictors of misreporting that can be gained from linkage studies would also allow survey producers to provide users with more information on the likely accuracy of a given response. For example, we find the share of an individual's responses that required imputation to predict survey errors in the individual's other responses (Celhay, Meyer, and Mittag 2018b). Investigating which paradata predicts response errors and publishing the results could provide data users with better means to assess the likely prevalence of measurement error in the samples they use. An emerging literature also examines how such information can be used to improve the accuracy of estimates derived from error-ridden data (e.g. Da Silva, Skinner, and Kim 2016). Finally, linking more accurate measures of existing variables in surveys can also be used to improve survey accuracy directly. Survey producers could replace the misreported variable by a more accurate administrative variable or combine survey responses and administrative data in synthetic variables if direct substitution is infeasible or undesirable. See Davern, Meyer, and Mittag (2019) and Mittag (2019) for further discussion. We review an example of direct substitution in Section 12.6.

12.6 Illustration: Using Data Linkage to Better Measure Income and Poverty

The sections above describe ways in which data linkage can help us to understand and measure the errors in survey data. Data linkage can also improve survey quality more directly. If an accurate measure of a survey variable can be linked to the data, one can replace the error-prone survey variable with the more accurate linked variable. We use our work that corrects analyses of poverty and the income distribution to illustrate that such direct substitution is feasible and can alter important conclusions obtained from the contaminated survey data. See Bee and Mitchell (2017a) for similar analyses that use direct substitution.

In Meyer and Mittag (2019b), we combine the New York State sample of the 2008–2013 CPS ASEC – the government's source of official poverty and inequality statistics – with administrative records for SNAP, Public Assistance, and housing assistance to better measure payments made, and the impact of the programs on the income distribution and poverty. The linked data confirm that misreporting, particularly underreporting of program receipt, is severe. More importantly, we show that replacing the error-ridden survey variable by the accurate linked variable in several prototypical analyses of poverty and transfer programs sharply alters the picture of well-being at the bottom of the income distribution and the effects of transfer programs. Using the administrative variables, poverty and inequality are lower than officially reported, program effects are larger, and fewer individuals have been missed by the combination of programs. Incomes below the poverty line are substantially understated in the CPS ASEC. This is particularly severe for incomes below half the poverty line where more than US$ 1400 per person from the programs we examine alone are missing in the survey. These missing transfer dollars exceed cash income reported by this group. Although underreporting as a share of income becomes smaller as income rises, substantial dollars are missed even toward the middle of the income distribution.

Underreporting of transfer receipt also makes government anti-poverty policies appear much less effective: the four programs moved a much larger fraction of people out of poverty than the CPS suggests. Including transfers from these programs in the income definition reduces the poverty rate by 2.8% according to the CPS reports. This reduction rises to 5.3% when using the more accurate administrative measures. The difference is particularly pronounced for housing assistance, where correcting for underreporting triples the poverty reduction. Both the understatement of household income and the poverty reducing effect in the survey are even more pronounced for some subpopulations that are at particular risk of deprivation. The understatement is particularly large for single mothers: correcting for survey errors increases their overall poverty reduction due to the four programs by 11 percentage points, amplifying the poverty reducing effect of public assistance more than sixfold and that of housing assistance more than 10-fold. In addition, we find that the fraction of non-working single mothers missed by government transfers is much lower than previously reported. This underlines that the coverage of the safety net is better than that suggested by the survey.

These results show that survey errors, mainly the misreporting of government transfer receipt and amounts, but also nonresponse and inaccurate imputation, lead to a greatly distorted view of the situation of those with the fewest resources and the effects of transfer programs. However, the results also underline that enhancing survey data with administrative records can provide improved answers to questions of relevance for both policy makers and academics. These analyses were conducted on the most recent vintage of the data at the time of initial writing, so such enhanced data products could be made available on a timely enough basis to inform policy.

12.7   Accuracy of Links and the Administrative Data

The analyses above consider the linked administrative variables to be accurate versions of the survey concepts. In many cases one will not be able to exactly match the survey concepts or the data linkage will not be error-free. In such less than idealized circumstances, many of the benefits of record linkage we discuss above still apply. See Courtemanche, Denteh, and Tchernis (2018) and Meyer and Mittag (2019a) for discussions of the merits and perils of examining the accuracy of a specific survey using linked data in less-than-ideal circumstances.

For the analyses of survey coverage, one mainly needs the administrative records to be complete and reliably linked to the survey, as we discuss in Section 12.3. Similarly, analyses of unit nonresponse remain meaningful as long as data linkage provides a measure of the survey response that does not differ systematically between respondents and nonrespondents. If response status does not predict errors in the linked variable, then statistics such as differences in means or tests of whether the distribution of the linked variable differs by response status remain valid. This condition may not hold, for example if the links are (partly) made based on information provided only by respondents. By the same logic, many analyses of item nonresponse remain valid as long as item nonresponse status does not predict errors in linkage. This condition may often seem more plausible for the case of item nonresponse, because contrary to the case of unit nonresponse, the response to the question being examined is usually not used in the linkage process.

Studying measurement error at the household or individual level requires a measure of "truth" at the same level. Therefore, linkage errors cause bias in such analyses. See Meyer, Mittag, and Goerge (forthcoming) for a discussion of the likely consequences for estimated misreporting rates. If the rates of linkage error are negligible compared to the error rates in the survey responses, comparing the administrative and the survey variable may still yield valuable insights about survey error. If the error rates in the linked variable are too high to only have a negligible effect on estimates, one can still analyze many of these problems by using methods that rely on two error-ridden variables. These methods are beyond the scope of this chapter; see Abowd and Stinson (2013), Kapteyn and Ypma (2007), Oberski et al. (2017), or Meijer, Rohwedder, and Wansbeek (2012) for applications. What can researchers do to assess the extent to which their data deviates from this gold standard and how can they address problems they identify? A growing methodological literature develops methods to examine and address inaccuracies in the administrative data and linkage errors. Chapters 4–7 of this book discuss methods to assess the accuracy of administrative records and summarize the conditions under which administrative records are likely accurate. We only provide key references and focus on strategies commonly used in the literature on income and transfer programs.

A common way to examine the overall accuracy of the unlinked administrative records is to compare them to control totals that the administrative records should match in aggregate. For government programs, it is usually possible to obtain aggregate statistics, such as the total number of recipients or the total amount paid out, from independent sources. For example, the total amount paid out by a government program is often published by the agency that administers the program, such as the Department of Health and Human Services or the Department of Agriculture in the U.S. Meyer, Mok, and Sullivan (2015) provide information on the sources of aggregate payments for many U.S. transfer programs, and discuss how these numbers can be made comparable to survey estimates.

Validating the survey response also requires matching the concept of the survey question. For the case of transfer reporting, a key issue is matching the reference period of the survey. This matching is possible if the reference period in the survey is clearly defined and the administrative data contain detailed information on the timing of receipt. See Meyer and Mittag (2019a) for a discussion. This issue is less straightforward when validating more complex concepts such as income. One problem is that the two data sources may define income in different ways. Tax records, for example, often contain net income after some tax deductions such as pension contributions, while surveys often do not exclude tax deductions and may ask for gross income. A second issue is that the tax records may include only taxable sources of income (but not nontaxable or in-kind income) and miss informal sources that may be captured in the survey. See Abowd and Stinson (2013) and Bollinger et al. (2019) for discussions of income sources that tax records may not capture.

Finally, obtaining an accurate measure from the administrative data requires data linkage to be error free. If data linkage is deterministic based on official identifiers, as is often the case with administrative records in Europe, such errors can only arise from mis-assigned official identifiers. However, most record linkages in the U.S. use probabilistic record linkage. Chapter 8 provides an overview; see NORC (2011) and Wagner and Layne (2014) for descriptions of the Person Identification Validation System of the U.S. Census Bureau, which was used to link the data in many of the studies discussed so far. An advantage of the Person Identification Validation System is that it links both the administrative and the survey records to a third data source from which common identifiers are obtained. Linking both data sources to such a population register lets the researcher observe whether a record was not linked to the other source because it was not included in the data (e.g. because the household did not receive the program), or because the record was unlinkable. In the former case, the unit has an identifier from the population register, but this identifier was not present in the other data source. This information makes it likely that the record would have been linked if it were in the other data source, particularly if linkage failure in the other data source is rare. For example, administrative records on transfer receipt often contain linkable identifiers for almost all records. Thus, not finding a survey unit with a linkable identifier in the administrative data makes it very likely that the unit did not receive the transfer. However, if a survey or administrative record is not linkable, no identifier from the population register is attached to it. This situation makes it possible to examine the extent of linkage failure in both data sources and can often be used to improve or solve the problem of linkage error.

Survey units that cannot be linked present a missing data problem: if they are not missing completely at random, the linked sample is not representative of the survey population. Adjusting for this problem is only necessary if one needs a representative sample of the population and linkage failure predicts the outcome of interest. Otherwise, one may be able to use the linked sample without further problems. If the remaining records are correctly linked, one can still examine survey error in the linked data. However, if one is interested in population estimates or concerned that survey errors are related to linkage failure, one can often use the detailed information in the survey to examine or even address this problem. One can estimate binary choice models where the dependent variable is an indicator for whether or not the survey unit is linkable to examine whether linkage failures are random and, if not, how they vary with characteristics of the unit. This estimation is often informative about likely biases. If linkage failure is random conditional on the variables in the survey, one can use methods for missing data to address it. A common approach is inverse probability weighting (Horvitz and Thompson 1952; Wooldridge 2007). However, the assumption that linkage failure is conditionally random is not testable, because the administrative variable is not observed for the survey units that cannot be linked. One way to obtain evidence on whether the procedure yields consistent estimates is to examine whether the linked data reproduce the distribution of other survey variables, most importantly, of survey reports of the variable of interest.

There are usually far fewer unlinkable administrative records than survey units, and, in some of the applications discussed above, this problem is negligible. For administrative records that could not be linked, we still observe the value of the administrative variable of interest. Therefore, we can examine the extent and importance of this problem, for example by calculating the share of individuals or payments that are missing from the linked data due to linkage failure in the administrative data. It is also often possible to adjust aggregate numbers, for example as in Meyer and Mittag (forthcoming).

12.8 Conclusions

Recent research that links administrative and survey microdata has demonstrated the potential of data linkage to remedy the problem of survey inaccuracy. As the work we review here underlines, data linkage can provide us with evidence on the extent of survey error and its key sources. Understanding the extent of error is important for survey users to gauge the reliability of their estimates. Measuring the error and its components is crucial for survey producers to cost-effectively increase accuracy by spending resources on the error sources that are important and on measures to address them that have been shown to reduce the error. Linked data can also be used to study specific sources of error, such as the key components of generalized coverage error, item nonresponse error and measurement error that we survey here. By adding an accurate measure of the variable of interest to the survey data, data linkage puts researchers in the unique position to study errors at the level of the unit of observation. As the literature on survey errors in transfer receipt we review above shows, this addition provides unprecedented detail on the nature of errors, which often points to potential remedies to the identified problems.

However, data linkage has yet to reach its full potential. The methods and benefits outlined in this chapter extend beyond studies of government transfers and other income components, because they are applicable whenever an accurate measure can be obtained from another data source. Many extensions are possible. The analyses could be extended to study further error components, such as processing error. The existing studies mainly document survey error in means or totals, but the same or similar methods could be used to study other statistics such as variances or possibly regression coefficients. Most studies in this review are case studies that used a short time period from a few surveys, mainly because most studies linked a convenience sample based on data availability. The full benefits of such analyses will accrue with more routine application. With data linkage becoming more common, methods to assess the accuracy of linked data and to understand and address the problems that arise will likely improve further. Linking data on the entire U.S. population or at least multiple states will help to examine geographic heterogeneity in survey error and provide larger sample sizes. As the results in Celhay, Meyer, and Mittag (2018a) underline, many studies would likely yield more informative results if they were able to link larger samples. Finally, routinely linking surveys to administrative data would allow survey producers to monitor how survey errors change over time and whether the measures taken to reduce them yield the desired results. As the study of poverty and the income distribution in Section 12.6 shows, linking the data on a timely basis would also allow government agencies and policymakers to rely on more accurate information.

12.9 Exercises

1 How does data combination allow us to measure and decompose total survey error? Define key error components and discuss how data combination would allow us to estimate them.

2 For one or two of the error components you defined, what are survey and administrative data sources that provide comparable information?

3 Discuss what kind of data you need to examine survey coverage. What are the minimal requirements? What are the benefits of additional information?

4 Describe direct and indirect ways in which linked data can improve the problems of item nonresponse and measurement error.

5 Suppose your linked administrative variable is subject to error. For a specific case (e.g. one error source or a specific application), discuss (i) what you can still learn from the data and how, and (ii) what cannot be learned from the data due to the errors in the administrative variable.

Acknowledgments

Any opinions and conclusions expressed here are those of the authors and do not necessarily represent the views of the U.S. Census Bureau. All results have been approved for disclosure by the Census Bureau's Disclosure Review Board (approvals dated 1/12/2017, 10/6/2017, and authorization numbers CBDRB-FY18-106 and CBDRB-FY18-14). We would like to thank the Alfred P. Sloan, Russell Sage, Charles Koch, and Menard Family Foundations for their support.

References

Abowd, J.M. and Stinson, M.H. (2013). Estimating measurement error in annual job earnings: a comparison of survey and administrative data. *Review of Economics and Statistics* 95 (5): 1451–1467.

Andridge, R.R. and Little, R.J.A. (2010). A review of hot deck imputation for survey non-response. *International Statistical Review* 78 (1): 40–64.

Angrist, J.D. and Krueger, A.B. (2001). Empirical strategies in labor economics. In: *Handbook of Labor Economics*, vol. 3A (eds. O. Ashenfelter and D. Card), 1277–1366. Amsterdam: Elsevier.

Bee, C.A. and Mitchell, J. (2017a). The hidden resources of women working longer: evidence from linked survey-administrative data. In: *Women Working Longer: Increased Employment at Older Ages* (eds. C. Goldin and L.F. Katz). Chicago, IL: University of Chicago Press.

Bee, C.A. and Mitchell, J. (2017b). Do older Americans have more income than we think? U.S. Census Bureau SESHD Working Paper #2017-39.

Bee, C.A., Gathright, G.M.R., and Meyer, B.D. (2017). *Bias from Unit Non-response in the Measurement of Income in Household Surveys. Unpublished paper.* University of Chicago.

Black, D.A., Sanders, S., and Taylor, L. (2003). Measurement of higher education in the census and current population survey. *Journal of the American Statistical Association* 98 (463): 545–554.

Blackwell, M., Honaker, J., and King, G. (2017). A unified approach to measurement error and missing data: overview and applications. *Sociological Methods & Research* 46 (3): 303–341.

Bollinger, C.R. (1998). Measurement error in the Current Population Survey: a nonparametric look. *Journal of Labor Economics* 16 (3): 576–594.

Bollinger, C.R. and David, M.H. (1997). Modeling discrete choice with response error: food stamp participation. *Journal of the American Statistical Association* 92 (439): 827–835.

Bollinger, C.R. and David, M.H. (2001). Estimation with response error and nonresponse: food-stamp participation in the SIPP. *Journal of Business and Economic Statistics* 19 (2): 129–141.

Bollinger, C.R. and Hirsch, B.T. (2006). Match bias due to earnings imputation: the case of imperfect matching. *Journal of Labor Economics* 24 (3): 483–519.

Bollinger, Christopher R., Barry T. Hirsch, Charles M. Hokayem, and James P. Ziliak. 2019. "Trouble in the Tails? What We Knowabout Earnings Nonresponse Thirty Years after Lillard, Smith, and Welch." *Journal of Political Economy*, 127(5): 2143–2185. https://doi.org/10.1086/701807

Bound, J. and Krueger, A.B. (1991). The extent of measurement error in longitudinal earnings data: do two wrongs make a right? *Journal of Labor Economics* 9 (1): 1–24.

Bound, J., Brown, C., and Mathiowetz, N. (2001). Measurement error in survey data." (Chapter 59. In: *Handbook of Econometrics*, vol. 5 (eds. J.J. Heckman and E. Leamer), 3705–3843. Amsterdam: Elsevier.

Brummet, Q., Flanagan-Doyle, D., Mitchell, J. and Voorheis, J. (2017). Investigating the use of administrative records in the consumer expenditure survey. Center for Administrative Records Research and Applications, U.S. Census Bureau.

Celhay, P., Meyer, B.D. and Mittag, N. (2018a). Errors in reporting and imputation of government benefits and their implications, unpublished manuscript.

Celhay, P., Meyer, B.D. and Mittag, N. (2018b). What leads to measurement error? Evidence from reports of program participation in three surveys, unpublished manuscript.

Celhay, P., Meyer, B.D. and Mittag, N. (2018c). Stigma in welfare programs, unpublished manuscript.

Chenevert, R.L., Klee, M.A. and Wilkin, K.R. (2016). Do imputed earnings earn their keep? Evaluating SIPP earnings and nonresponse with administrative records. U.S. Census Bureau SEHSD-Working Paper 2016-18.

Courtemanche, C., Denteh, A., and Tchernis, R. (2018). Estimating the associations between SNAP and food insecurity, obesity, and food purchases with imperfect administrative measures of participation, unpublished manuscript.

Da Silva, D.N., Skinner, C., and Kim, J.K. (2016). Using binary paradata to correct for measurement error in survey data analysis. *Journal of the American Statistical Association* 111 (514): 526–537.

Dahl, M., DeLeire, T., and Schwabish, J.A. (2011). Estimates of year-to-year volatility in earnings and in household incomes from administrative, survey, and matched data. *Journal of Human Resources* 46 (4): 750–774.

Davern, M., Call, K.T., Ziegenfuss, J. et al. (2008). Validating health insurance coverage survey estimates: a comparison of self-reported coverage and administrative data records. *Public Opinion Quarterly* 72 (2): 241–259.

Davern, M., Meyer, B.D., and Mittag, N. (2019). Creating improved survey data products using linked administrative-survey data. Journal of Survey Statistics and Methodology. 7(3): 440-463.

Gathright, G.M.R., and Crabb, T.A. (2014). Reporting of SSA program participation in SIPP. Working Paper, U.S. Census Bureau.

Groves, R.M. (2004). *Survey Errors and Survey Costs*. New York: Wiley.

Groves, R.M. (2006). Nonresponse rates and nonresponse bias in household surveys. *Public Opinion Quarterly* 70 (4): 646–675.

Groves, R.M. and Lyberg, L. (2010). Total survey error: past, present, and future. *Public Opinion Quarterly* 74 (5): 849–879.

Groves, R.M. and Peytcheva, E. (2008). The impact of nonresponse rates on nonresponse bias: a meta-analysis. *Public Opinion Quarterly.* 72: 167–189.

Groves, R.M., Fowler, F.J., Couper, M.P. et al. (2009). *Survey Methodology (Wiley Series in Survey Methods)*. New York: Wiley.

Hausman, J.A., Abrevaya, J., and Scott-Morton, F.M. (1998). Misclassification of the dependent variable in a discrete-response setting. *Journal of Econometrics* 87 (2): 239–269.

Heitjan, D.F. (1994). Ignorability in general incomplete – data models. *Biometrika* 81: 701–708.

Heitjan, D.F. and Rubin, D.B. (1991). Ignorability and coarse data. *Annals of Statistics* 19: 2244–2253.

Hirsch, B.T. and Schumacher, E. (2004). Match bias in wage gap estimates due to earnings imputation. *Journal of Labor Economics* 22 (3): 689–722.

Horvitz, D.G. and Thompson, D.J. (1952). A generalization of sampling without replacement from a finite universe. *Journal of the American Statistical Association.* 47: 663–685.

Johnson, T. and Fendrich, M. (2005). Modeling sources of self-report bias in a survey of drug use epidemiology. *Annals of Epidemiology* 15 (5): 381–389.

Kapteyn, A. and Ypma, J.Y. (2007). Measurement error and misclassification: a comparison of survey and administrative data. *Journal of Labor Economics* 25 (3): 513–551.

Kirlin, J.A., and Wiseman, M. (2014.) Getting it right, or at least better: improving identification of food stamp participants in the national health and nutrition examination survey, Working Paper.

Lillard, L., Smith, J.P., and Welch, F. (1986). What do we really know about wages? The importance of nonreporting and census imputation. *The Journal of Political Economy* 94 (3): 489–506.

Little, R.J.A. and Rubin, D.B. (2002). *Statistical Analysis with Missing Data*, 2e. New York: Wiley.

Lynch, V., Resnick, D.M., Stavely, J. and Taeuber, C.M. (2007). Differences in estimates of public assistance recipiency between surveys and administrative records. U.S. Census Bureau working paper.

Marquis, K.H. and Moore, J.C. (1990). Measurement errors in SIPP program reports. In: *Proceedings of the 1990 Annual Research Conference*, 721–745. Washington, DC: U.S. Bureau of the Census.

Mattingly, T., Choi, J., Finney, T., Hoop, R., Hornick, D., Nieman, D., Rothhaas, C., Westra, A. and White, M. (2016). Results of a nonresponse bias analysis using survey of income and program participation (SIPP) addresses matched to internal revenue service (IRS) data. Memorandum, U.S. Census Bureau.

Meijer, E., Rohwedder, S., and Wansbeek, T. (2012). Measurement error in earnings data: using a mixture model approach to combine survey and register data. *Journal of Business & Economic Statistics* 30 (2): 191–201.

Meyer, B.D. and Mittag, N. (2017). Misclassification in binary choice models. *Journal of Econometrics* 200 (2): 295–311.

Meyer, B.D. and Mittag, N. forthcoming. An empirical total survey error decomposition using data combination. Journal of Econometrics.

Meyer, B.D. and Mittag, N. (2019a). Misreporting of government transfers: how important are survey design and geography? Southern Economic Journal, 86 (1): 230–253.

Meyer, B.D. and Mittag, N. (2019b). Using linked survey and administrative data to better measure income: implications for poverty, program effectiveness and holes in the safety net. *American Economic Journal: Applied Economics* 11 (2): 176–204.

Meyer, B.D., Mok, W.K.C., and Sullivan, J.X. (2015). Household surveys in crisis. *Journal of Economic Perspectives* 29 (4): 199–226.

Meyer, B.D., Mittag, N. and Goerge, R. forthcoming. Errors in survey reporting and imputation and their effects on estimates of food stamp program participation. Journal of Human Resources.

Mittag, N. (2019). Correcting for misreporting of government benefits. American Economic Journal: Economic Policy, 11 (2): 142-164.

Nicholas, J. and Wiseman, M. (2009). Elderly poverty and supplemental security income. *Social Security Bulletin* 69 (1), https://www.ssa.gov/policy/docs/ssb/v70n2/v70n2p1.html.

Nicholas, J. and Wiseman, M. (2010). Elderly poverty and supplemental security income, 2002–2005. *Social Security Bulletin* 70 (2).

Niehaus, P. and Sukhtankar, S. (2013). Corruption dynamics: the golden goose effect. *American Economic Journal: Economic Policy* 5 (4): 230–269.

NORC. (2011). Assessment of the US Census Bureau's person identification validation system. NORC at the University of Chicago Final Report presented to the US Census Bureau.

Oberski, D.L., Kirchner, A., Eckman, S., and Kreuter, F. (2017). Evaluating the quality of survey and administrative data with generalized multitrait-multimethod models. *Journal of the American Statistical Association* 112 (520): 1477–1489.

O'Hara, A., Bee, C.A. and Mitchell, J. (2016). Preliminary research for replacing or supplementing the income question on the American Community Survey with administrative records, 2015. American Community Survey Research and Evaluation Report Memorandum Series #ACS16-RER-6, U.S. Census Bureau.

Sakshaug, J.W., Yan, T., and Tourangeau, R. (2010). Nonresponse error, measurement error, and mode of data collection: tradeoffs in a multi-mode survey of sensitive and non-sensitive items. *Public Opinion Quarterly* 74 (5): 907–933.

Sirken, M. (1999). *Cognition and Survey Research*, vol. 322. Wiley-Interscience.

Sudman, S. and Bradburn, N.M. (1973). Effects of time and memory factors on response in surveys. *Journal of the American Statistical Association* 68 (344): 805–815.

Taeuber, C., Resnick, D.M., Love, S.P., Stavely, J. Wilde, P. and Larson, R. (2004). Differences in estimates of food stamp program participation between surveys and administrative records, U.S. Census Bureau working paper.

Tourangeau, R. (1984). Cognitive science and survey methods. In: *Cognitive Aspects of Survey Methodology: Building a Bridge Between Disciplines*, 73–100. Washington, DC: The National Academies Press. https://doi.org/10.17226/930, https://www.nap.edu/catalog/930/cognitive-aspects-of-survey-methodology-building-a-bridge-between-disciplines.

Tourangeau, R., Groves, R.M., and Redline, C.D. (2010). Sensitive topics and reluctant respondents: demonstrating a link between nonresponse bias and measurement error. *Public Opinion Quarterly* 74 (3): 413–432.

Wagner, D., and Layne, M. (2014). The person identification validation system (PVS): applying the center for administrative records research and applications' (CARRA) record linkage software. U.S. Census Bureau.

Wooldridge, J.M. (2007). Inverse probability weighted estimation for general missing data problems. *Journal of Econometrics* 141 (2): 1281–1301.

Bollinger, Christopher R., Barry T. Hirsch , CharlesM. Hokayem, and James P. Ziliak. 2019. "Trouble in the Tails? What We Knowabout Earnings Nonresponse Thirty Years after Lillard, Smith, and Welch." Journal of Political Economy, 127(5): 2143-2185. https://doi.org/10.1086/701807

13

Combining Data from Multiple Sources to Define a Respondent: The Case of Education Data

Peter Siegel, Darryl Creel, and James Chromy

RTI International

13.1 Introduction

As the use of administrative data in education surveys increases, we need to think through and deal with the associated issues. Some of these issues include using these data to derive variables, impute data, and compute weights. However, we can also use administrative data to think of a survey respondent in a different way from the typical interview respondent. A usable case rule can be developed to identify key information items or combinations of these items needed to qualify as a unit respondent. If data are available from various sources, then it is possible to define a respondent (usable case) based on having enough data from any source for items that are important for analysis. The usable case rule can sometimes be satisfied with data from only a subset of the sources, and a respondent may not have completed an interview. A response rate is computed based on how a respondent is defined. Computing a response rate based on usable data may inflate the unit response rate while providing more cases for analysis.

The focus of this chapter will be on how best to define a study respondent (unit respondent) in the presence of a significant amount of administrative data accompanying the survey data. There are two implications of this definition. First, how will the response rate be computed, and, second, what cases will be included on the analysis file?

The discussion in this chapter assumes that the administrative data are accessible, reliable, and timely. While these assumptions are currently not true for many surveys, research is being conducted to improve the use of administrative data in surveys, as discussed in earlier chapters. Also, the definition and reference time of a variable may vary between the survey question and other data sources, and this could potentially introduce systematic bias. This needs to be considered

Administrative Records for Survey Methodology, First Edition.
Edited by Asaph Young Chun, Michael D. Larsen, Gabriele Durrant, and Jerome P. Reiter.

when determining whether a source should be used for defining a respondent. As administrative data continue to be more available for supplementing survey data, this chapter helps those conducting surveys to rethink how a survey respondent is defined.

The rest of the introduction provides more information on defining unit respondents. Then, the chapter will cover a literature review and methodological details. Research done for the National Center for Education Statistics (NCES) using data from the National Postsecondary Student Aid Study (NPSAS) will be provided as an example of defining a unit respondent when survey data are accompanied by administrative data from several sources. The chapter concludes with a discussion of the advantages and disadvantages of two approaches to define a unit respondent, followed by practical implications.

13.1.1 Options for Defining a Unit Respondent When Data Exist from Sources Instead of or in Addition to an Interview

There are at least two approaches to defining a unit respondent when data from administrative sources exist in addition to survey responses. One approach is to define a unit respondent as a sample member completing the survey interview under some rule for determining what constitutes "complete." This approach is most common, and unit response rates computed this way follow the American Association for Public Opinion Research (AAPOR) definition (The American Association for Public Opinion Research 2016) and the NCES Statistical Standards (U.S. Department of Education 2012). Nonresponse weight adjustments are used to compensate for the nonrespondents and to reduce the potential for unit nonresponse bias. The data collected from other sources may then be used to fill in or impute any missing data item values for these respondents. Another approach is to define a unit respondent as a sample member with sufficient data from any source to be judged complete. Filling in data using other sources, when available, and using logical and statistical imputations are all used to compensate for missing data and to reduce the potential for item nonresponse bias. The NCES Statistical Standards require a nonresponse bias analysis when unit or item response rates are less than 85%.

In this chapter, we will compare the two approaches. Of particular interest are the approaches' different use of weight adjustment and imputation to compensate for nonrespondents and reduce potential nonresponse bias. While weighting and imputation have been compared in the past, this chapter will examine this comparison in the context of education data when administrative data are available, allowing more imputation and less weight adjustment.

Nonresponse weighting adjustments are designed to reduce potential unit-level nonresponse bias. For surveys with a response rate less than 85%, the potential for nonresponse bias can usually not be ignored. Adjusting the weights of respondents to compensate for nonrespondents ideally produces respondent distributions that are similar to full population distributions, thus reducing the potential nonresponse bias. Predictor variables used in the adjustment are chosen that are thought to predict response status, reduce nonresponse bias, and be correlated with key outcomes.

Imputation is designed to reduce item-level nonresponse bias. Imputation reduces nonresponse bias by replacing missing items with statistically plausible values. For items with a response rate less than 85%, the potential for nonresponse bias when not replacing missing item values usually cannot be ignored. Ideally, replacing missing item data with reasonable values produces imputed sample distributions that resemble full population distributions, thus reducing the potential for nonresponse bias. The imputation method and variables used are designed to ensure that imputed data are plausible, and that the nonresponse bias is small.

13.1.2 Concerns with Defining a Unit Respondent Without Having an Interview

There are potential concerns with the adoption and use of a unit respondent definition without requiring an interview. Such an approach fails to address any bias among the interview nonrespondents because imputations are done on an item-specific basis and include true item nonresponse mixed with the unit nonresponse. When a large share of the interview nonrespondents are not locatable, it raises a concern that they may have some shared characteristics that are not necessarily captured in the item imputations. There could be other group differences due to the unit-level nonresponse bias analysis not being performed on the interview. Focusing the nonresponse bias analysis on a narrowly defined unit respondent rate also ignores the potential bias that may be due to missing data from other data sources for the item that is being analyzed. To the extent these items come from one source versus another, any systematic bias associated with that individual source may not be addressed in the individual item imputations.

The over-riding concern is that the purpose of conducting nonresponse bias analysis is to better understand, and ultimately adjust for, measurable sources of nonresponse bias in a data collection. Combining the results from multiple sources to get a high response rate does not properly address the potentially significant bias issues associated with the low response/data availability rates for the items coming from each contributing information component. Similarly, correcting for these missing data using item imputation does not necessarily capture the systematic bias that may be associated with participation or the availability of information from any one of the information sources.

A third concern is based on ethical and privacy issues because a non-interview nonrespondent may not know that his/her data are being obtained from administrative sources and has not explicitly given consent to do this. However, the U.S. Family Educational and Privacy Act (FERPA) of 1974 allows the disclosure of educational information of students without prior consent but under a set of conditions.

A fourth concern is the impact that nonresponse weight adjustments versus item-level imputation may have on the variance in the resulting estimates.

13.2 Literature Review

There are not many surveys that define a unit respondent based on something other than having an interview. A comparison of weighting and imputation may help decide which unit respondent definition is preferable for a study. There are various weighting and imputation methods to handle missing data due to survey nonresponse. The appropriate methods to use may be study-specific and depend on many things, including amount of missing data, type of nonresponse (complete, partial, or item), pattern of missingness (e.g. missing at random or not), and data available for weighting or imputation. Michaud (1986) concluded that imputing a lot of data and weighting by the inverse of the probability of selection often leads to similar estimates and that the choice between them is dictated by resource constraints, the number of records to be processed, and the type of information required. Hughes and Peitzmeier (1989) found that a multiple imputation (MI) technique produced estimates with less bias than weighting approaches, but due to complexities with MI recommended proceeding with a weighting approach. Rässler and Schnell (2003) encourage the use of MI for unit-nonresponse in survey practice, assuming the data are missing at random, because it produced similar estimates to weighting and reduced the nonresponse bias.

International studies provide good examples of using administrative and survey data together to improve data quality, compute weights, impute data, and derive variables. Countries that have registers often combine the register data with survey data. The Nordic countries produce some statistics by using register data only, but for other statistics take advantage of survey data to supplement register data and vice versa (United Nations Economic Commission for Europe 2007). Wallgren and Wallgren (2014) discuss the use of weighting and imputation to supplement register data and quality issues with register surveys. Quality of administrative data must be determined as discussed by Daas et al. (2009) before they can be combined with survey data. As we see in international studies, there are various uses of administrative data combined with survey data, and this chapter takes another step to show how the combination of data can be used to define a study respondent.

13.3 Methodology

When an interview respondent is used, there is the potential for a large amount of unit nonresponse bias due to interview nonresponse, if the response rate is low. When a usable case respondent is used, there is the potential for a large amount of item nonresponse bias due to missing items, especially when a subset of items is only available from one source, such as the interview. Therefore, weight adjustment for the former and imputation for the latter are important tools for reducing potential nonresponse bias.

13.3.1 Computing Weights for Interview Respondents and for Unit Respondents Who May Not Have Interview Data (Usable Case Respondents)

The weighting approach is similar regardless of how respondents are defined. For the interview respondent approach, the following steps can be used:

- Define a respondent as a sample member who completes the interview.
- Compute a nonresponse weight adjustment to account for the potentially large amount of nonresponse. Use weighting classes or a response propensity model with variables known for respondents and nonrespondents that are correlated with interview response.
- Compute a poststratification weight adjustment to correct for population undercoverage. Use control totals from known population totals for key outcomes, when available.

For the usable case respondent approach, the following steps can be used:

- Define a respondent as a sample member who has sufficient data to be considered a unit respondent.
- Compute a nonresponse weight adjustment to account for nonresponse, which may be low. Use weighting classes or a response propensity model with variables known for respondents and nonrespondents that are correlated with study response. These variables may be the same as those used in the interview response or a subset or superset of those variables.
- Compute a poststratification weight adjustment to correct for population undercoverage. Use control totals from known population totals for key outcomes, when available. The control totals will likely be the same for either approach.

13.3.1.1   How Many Weights Are Necessary?

The number of weights to compute for a study depends on the analysis plan and how analysts will use the data. Typically, the fewer the weights the better because it causes less confusion among the data users. Cross-sectional studies frequently have just one weight, and longitudinal (panel) studies typically have multiple weights to examine different sets of respondents across the panel. Although a respondent is usually defined based on interview response, either cross-sectionally or longitudinally, a respondent can be defined based on interview response, study response, or both. This chapter will show that one weight may be sufficient to cover both interview and usable case respondents. Nevertheless, there may be studies for which both weights could be computed, allowing analysts to decide which they prefer, based on the potential for bias with each weight and subsequent imputation.

13.3.2   Imputing Data When All or Some Interview Data Are Missing

Missing data can be ignored when analyzing survey data. However, this can potentially introduce bias into the analysis if the data are not missing at random. Also, the number of cases used for analysis may be small if there are a lot of missing data. Alternatively, imputation can be done for surveys when there are missing items. When a sample member completes the interview but does not answer all the questions that he/she is eligible to answer, these missing items can be replaced using statistical or logical imputation. Similarly, when an entire data source is missing, the missing items can also be replaced with imputed data. Although there are many imputation methods, including regression, hot deck, cold deck, mean value, and variants involving multiple imputation, the empirical analyses in Section 13.4 will use a weighted sequential hot deck imputation procedure (Cox 1980) among all usable case respondents. Hot deck is chosen because the study used as an example in Section 13.4 has a large sample size and many cases to choose from as donors for the hot deck imputation. The imputation sequence of the variables needs to consider the level of missingness, pattern of missingness, and logical associations among variables. For the hot deck procedure, imputation classes are needed. Generally, variables with lower levels of missingness are imputed first and then can be used as classes for the imputation of variables with higher levels of missingness. Classes can be formed based on knowledge of the data but are commonly created using a tree-based prediction model, and then the weighted sequential hot deck is performed within these imputation classes. Sample members who were legitimately skipped for certain items are typically excluded from the imputations for those items, and they are not counted as item nonrespondents.

13.3.3 Conducting Nonresponse Bias Analyses to Appropriately Consider Interview and Study Nonresponse

The bias in an estimated mean based on respondents, $\bar{y}_R$, is the difference between this mean and the target parameter, π (i.e. the mean that would be estimated if a complete census of the target population were conducted and everyone responded). This bias can be expressed as follows:

$$B\left(\bar{y}_R\right) = \bar{y}_R - \pi.$$

The estimated mean based on nonrespondents, $\bar{y}_{NR}$, can be computed if data for the particular variable are available for virtually all the nonrespondents. The true target parameter, π, can be estimated for these variables as follows:

$$\hat{\pi} = (1 - \eta)\bar{y}_R + \eta\bar{y}_{NR},$$

where η is the weighted item (or unit) nonresponse rate. For the variables that are from the frame, rather than from the sample, π can be estimated without sampling error. The bias can then be estimated as follows:

$$\hat{B}\left(\bar{y}_R\right) = \bar{y}_R - \hat{\pi}$$

or, equivalently,

$$\hat{B}\left(\bar{y}_R\right) = \eta\left(\bar{y}_R - \bar{y}_{NR}\right).$$

This formula shows that the estimate of the nonresponse bias is the difference between the mean for respondents and that for nonrespondents, multiplied by the weighted nonresponse rate. Note that this is equivalent to the difference between the mean for respondents and that for the full sample:

$$\hat{B}\left(\bar{y}_R\right) = \bar{y}_R - \bar{y}_{(R+NR)}.$$

The estimated relative bias is calculated as:

$$R\hat{B}\left(\bar{y}_R\right) = \hat{B}\left(\bar{y}_R\right)/\hat{\pi}$$

We cannot measure item-level bias after imputation, so we indirectly evaluate how well the imputation works in reducing bias. The item estimates before and after imputations are compared to determine whether the imputation changed the biased estimate to suggest a possible reduction in bias. An alternative method to evaluate nonresponse bias after imputations is to randomly divide the respondents into three groups and compare the three populations for a set of key variables. See the NCES Statistical Standards Appendix B (U.S. Department of Education 2012) for more information on measuring bias.

13.4 Example of Defining a Unit Respondent for the National Postsecondary Student Aid Study (NPSAS)

13.4.1 Overview of NPSAS

Data from the 2007–08 National Postsecondary Student Aid Study (NPSAS:08) are used to illustrate the methodology and show examples of results. NPSAS is conducted for the U.S. Department of Education's National Center for Education Statistics (NCES) and collects comprehensive data on how students and their families pay for postsecondary education. NPSAS:08 employed a two-stage sample selection process. First, 1960 institutions were selected with probability proportional to size (PPS), and 1730 provided student enrollment lists, which resulted in an overall institutional participation rate of 90% (weighted). Second, a sample of about 138 000 students was selected from the lists provided by participating institutions. The students of analytical interest are those enrolled in postsecondary education in the United States and Puerto Rico at any time between 1 July 2007 and 30 April 2008. Students completed an interview either on the web or on the phone, with approximately 132 800 eligible sample members, and 95 360 (71%, weighted) completing the student interview. In parallel, additional data about sample students were obtained from participating institutions' records, and about 1670 institutions provided institutional record information for about 130 410 students. Approximately 126 620 of these student records received were considered complete student records. The institution- and student-level weighted response rates for record abstraction were both 96%. Data were also collected from the Department of Education's Central Processing System (CPS) and National Student Loan Data System (NSLDS), as well as from the National Student Clearinghouse (NSC), ACT, and SAT.

Key variables were identified across the various data sources to determine the minimum requirements to support the analytic needs of the study. Sample members who met these minimum requirements were classified as unit respondents.[1] To be a unit respondent, a sample member had to have data for three critical variables – student type (undergraduate, graduate, or first-professional student), gender, and date of birth. In addition, unit respondents had to have data – from any source – for at least 8 variables out of a set of 15 additional prespecified variables. These variables include dependency status, income, expected family contribution (EFC), class level, and race. See Appendix 14.A for the full unit respondent definition. About 127 700 of the 132 800 eligible sample members (96%) were classified as

---

[1] In NPSAS:08, unit respondents were called study respondents. This terminology later changed to study members because the sample members may not actually respond themselves. This example will continue the use of the terms unit respondents and useable case respondents to be consistent with the rest of the chapter.

NPSAS:08 unit respondents. For most unit respondents (approximately 88%), data were available from at least two of the major data sources (e.g. student interview, institutional student records, or the CPS). Details about the NPSAS:08 study can be found in the 2007–08 National Postsecondary Student Aid Study (NPSAS:08) Full-scale Methodology Report (Cominole et al. 2010).

To investigate the question of which type of respondent is preferred and, in particular, which method best reduces bias, we looked at NPSAS:08 data to compare the methods. First, we looked at using a usable case respondent definition, and for this approach we defined a respondent as a sample member who has sufficient data from any source and did the following:

- Computed a weight, adjusted for nonresponse.
- Imputed 20 interview items (with low item response rates).
- Computed item-level bias estimates for the 20 imputed items.
- Computed estimates (e.g. proportions and means) and standard errors using the weight and imputed data for the 20 items.

Second, we looked at using an interview respondent approach, and for this approach we defined a respondent as a sample member who completes the interview and did the same activities as described above for the usable case respondent approach. Additionally, we computed student interview-level bias estimates.

The 20 NPSAS variables used in the analyses below were chosen from the set of variables for which the data were mainly available from the interview rather than from other data sources. The variables were chosen to get a variety of types of data and amount of data to be imputed. Some related variables were chosen, so that a vector approach to imputation could be used. Many of the variables are only applicable for a subset of the students. The 20 variables are:

- *ATTENDMR*: Main reason for attending school;
- *CSTCMPTR*: Cost of special equipment (student reported);
- *DEPCARE*: Dependents: children in daycare;
- *DISMOBIL*: Disability: mobility impairment;
- *DISOTHER*: Disability: other long-lasting condition;
- *DISSENSR*: Disability: sensory impairment;
- *DISTWK*: Job: distance from NPSAS school to work;
- *DISTYPES*: Disability: main type of condition or impairment;
- *EVER2PUB*: Ever attended community college;
- *FINAIDA*: Financial aid decisions: compared lender options;
- *FINAIDB*: Financial aid decisions: discussed with family/friends;
- *FINAIDC*: Financial aid decisions: researched on Internet;
- *FINAIDD*: Financial aid decisions: talked with staff;

- *HSTYPE*: Type of high school attended;
- *JOBNUM*: Job: number (excluding work-study/assistantship);
- *LNREPAY*: Expect help with repaying student loans;
- *RAINDTRB*: Race: American Indian or Alaska Native recognized tribe;
- *SJASST*: School job: assistantship;
- *SJWKST*: School job: work-study job; and
- *SPINCOL*: Spouse attending college;

Table 13.1 shows the item nonresponse rates for usable case respondents for the 20 variables overall and by interview respondents and nonrespondents.

Table 13.1 Item nonresponse rates among usable case respondents.

	Nonresponse rate		
Variable	Overall	Interview respondents	Interview nonrespondents
EVER2PUB	16.53	0.82	58.20
RAINDTRB	25.81	9.33	100.00
ATTENDMR	32.81	3.11	98.31
SPINCOL	33.41	13.80	100.00
JOBNUM	33.77	10.08	100.00
HSTYPE	34.21	10.78	96.21
FINAIDA	34.29	10.79	100.00
FINAIDB	34.29	10.79	100.00
FINAIDC	34.29	10.79	100.00
FINAIDD	34.29	10.79	100.00
DISSENSR	34.32	10.83	100.00
DISMOBIL	34.35	10.87	100.00
DISOTHER	34.36	10.88	100.00
SJWKST	36.99	14.30	100.00
SJASST	36.99	14.30	100.00
LNREPAY	38.68	18.58	100.00
DEPCARE	41.35	17.99	100.00
CSTCMPTR	43.37	23.12	100.00
DISTWK	55.79	34.84	100.00
DISTYPES	84.17	56.16	100.00

Note: The table is sorted by the overall nonresponse rate.

13.4.2 Usable Case Respondent Approach

The usable case respondent approach involves using the NPSAS:08 analysis weight for usable case respondents and then imputing missing data among unit respondents. The 20 variables listed above were imputed in NPSAS:08 for unit respondents. The standard practice for NPSAS imputations is to impute undergraduate and graduate students separately, so we continued this partition and further divided these two groups into interview nonrespondents and interview respondents. Consequently, we had a total of four groups that were imputed separately. Cominole et al. 2010 provides details about the general process.

To measure the bias due to item-level nonresponse, we computed item-level bias estimates before and after imputation for the 20 re-imputed items using the procedures and formulas described in Section 13.3.3. We considered performing the item-level nonresponse bias analysis separately for item nonrespondents among interview respondents and for interview nonrespondents. However, computing the bias is not meaningful for interview nonrespondents when there are few or no item respondents. Therefore, we performed the item-level nonresponse bias analysis overall, which is sufficient for comparison with the interview weight approach described in Section 13.4.3.

The bias estimates were computed for each of the 20 variables before imputation based on the following covariates from the Integrated Postsecondary Education Data System (IPEDS), NPSAS enrollment lists, and administrative data that are available for all sample members:

- institution type;
- region;
- institution enrollment from IPEDS file (categorical);
- student type;
- Pell Grant receipt (yes/no);
- Pell Grant amount (categorical);
- Stafford Loan receipt (yes/no);
- Stafford Loan amount (categorical);
- federal aid receipt (yes/no);
- CPS record indicator (yes/no);
- age group (three levels); and
- gender.

The Taylor series variance estimation procedure (Woodruff 1971) is used to compute variances of point estimates used below for comparing the usable case respondent and interview respondent approaches.

13.4.2.1 Results

Table 13.2 shows the results of the item-level nonresponse bias analysis before imputation as well as a comparison of estimates after imputation. The mean and

Table 13.2 Summary of item nonresponse bias analysis for usable case respondents.

	Mean absolute value of relative bias	Median absolute value of relative bias	Percent of covariate categories with significant bias	Percent difference in pre-imputation and post-imputation means
ATTENDMR	9.05	2.60	54.76	0.04[a]
CSTCMPTR	7.95	4.39	77.78	0.00
DEPCARE	11.12	6.25	64.44	0.02
DISMOBIL	7.59	3.07	77.78	0.03[a]
DISOTHER	7.58	3.17	75.56	0.02[a]
DISSENSR	7.60	3.11	75.56	0.03[a]
DISTWK	9.01	5.95	64.44	0.00
DISTYPES	11.80	5.82	46.67	0.04[a]
EVER2PUB	5.21	2.92	92.86	0.11[a]
FINAIDA	7.50	3.08	75.56	0.01
FINAIDB	7.50	3.08	75.56	0.01
FINAIDC	7.50	3.08	75.56	0.01[a]
FINAIDD	7.50	3.08	75.56	0.02[a]
HSTYPE	6.61	3.49	71.43	0.20[a]
JOBNUM	7.54	3.37	75.56	0.02[a]
LNREPAY	8.37	6.29	82.22	0.00
RAINDTRB	5.54	3.68	22.22	0.13[a]
SJASST	5.39	3.61	72.5	0.00
SJWKST	5.39	3.61	72.5	0.03[a]
SPINCOL	9.03	5.40	55.56	0.00

a) The difference between the pre- and post-imputation means is significant at the 0.05 level.
Note: Small values that are significant are due to large sample sizes.

median relative biases shown in the table were computed from the absolute value of the relative biases across the covariates listed above. Bias exists for these items, and the differences in the before and after imputation means were significantly different for over half of the variables, which suggests a possible bias reduction for these variables. See Appendix 14.B for an example of detailed nonresponse bias analysis results and Cominole et al. 2010 for more details of the item-level bias analysis.

Table 13.3 suggests that 8 of the 20 variables have remaining bias after imputations using the alternative method to evaluate nonresponse bias after imputations.

Table 13.3 Summary of alternative item nonresponse bias
analysis for usable case respondents after imputation.

Variable	Significant difference in estimates between random groups
ATTENDMR	Yes
CSTCMPTR	No
DEPCARE	Yes
DISMOBIL	No
DISOTHER	No
DISSENSR	No
DISTWK	No
DISTYPES	Yes
EVER2PUB	Yes
FINAIDA	Yes
FINAIDB	No
FINAIDC	No
FINAIDD	Yes
HSTYPE	Yes
JOBNUM	No
LNREPAY	Yes
RAINDTRB	No
SJASST	No
SJWKST	No
SPINCOL	No

The after imputation results in Table 13.3 are not always consistent with the after imputation results in Table 13.2.

We also computed estimates (e.g. proportions and means) and standard errors created with the study weight and imputed data for the 20 items. These estimates are shown graphically in Figures 13.1 and 13.2, respectively, in Section 13.4.4 in comparison with estimates from the interview weight approach.

13.4.3 Interview Respondent Approach

The interview respondent approach involves creating a NPSAS:08 interview weight for interview respondents using methodology following the NPSAS:08

Table 13.4 Summary of student interview nonresponse bias analysis.

	Mean absolute value of relative bias	Median absolute value of relative bias	Percent of covariate categories with significant bias
Before weight adjustments	5.11	4.34	79.17
After nonresponse adjustment	0.24	0.00	4.17

procedures and then imputing missing data among interview respondents. Cominole et al. 2010 provides details about the general process.

To measure the bias due to interview nonresponse, we computed student interview-level nonresponse bias estimates before weight adjustments and after the nonresponse weight adjustment. We used the same procedures as described in Section 13.3.3 for the item-level nonresponse bias analysis, and the bias estimates were computed based on the same covariates as listed in Section 13.4.2 for the item-level nonresponse bias analysis.

We imputed the 20 variables listed above for all interview respondents using the weighted sequential hot deck procedure, and we computed item-level nonresponse bias estimates among interview respondents.

13.4.3.1 Results

Table 13.4 shows the results of the student interview-level nonresponse bias analysis. The nonresponse adjustments worked to reduce nonresponse bias. See Appendix 14.B for an example of detailed nonresponse bias analysis results.

As shown in Cominole et al. 2010, bias also exists due to institution nonresponse. However, the institution response rate was above 85% overall and for seven of the nine institution sectors (strata), so institution-level nonresponse bias analysis was only conducted for the two sectors with a response rate less than 85%. Weight adjustments reduced the bias, so that the amount of remaining significant bias was small. Given the small amount of institution-level bias, the institution response rates above 85%, and the small amount of significant bias after student-level nonresponse adjustments, the confounding between institution-level bias and student-level bias appears to be minimal. However, possible confounding does need to be kept in mind when interpreting the results here.

See Table 13.5 for the results of the item-level bias analysis. Bias exists for these items, despite the relatively low nonresponse rates. The differences in the before and after imputation means were significantly different for 8 of the 20 variables, which suggests a possible bias reduction for these variables. See Appendix 14.B for an example of detailed nonresponse bias analysis results.

Table 13.5   Summary of item nonresponse bias analysis for interview respondents.

Variable	Item respondents versus item nonrespondents			
	Mean absolute value of relative bias	Median absolute value of relative bias	Percent of covariate categories with significant bias	Percent difference in pre-imputation and post-imputation means
ATTENDMR	5.91	0.91	52.38	0.02[a]
CSTCMPTR	4.75	2.01	53.33	0.03[a]
DEPCARE	6.45	3.97	48.89	0.01
DISMOBIL	4.17	1.53	60.00	0.01
DISOTHER	4.16	1.57	60.00	0.00
DISSENSR	4.18	1.54	60.00	0.01
DISTWK	6.37	3.54	57.78	0.01
DISTYPES	10.76	6.76	31.11	0.03[a]
EVER2PUB	0.31	0.30	52.38	0.01[a]
FINAIDA	3.95	1.57	57.78	0.01[a]
FINAIDB	3.95	1.57	57.78	0.00
FINAIDC	3.95	1.57	57.78	0.00
FINAIDD	3.95	1.57	57.78	0.01[a]
HSTYPE	3.72	1.54	54.76	0.05[a]
JOBNUM	4.11	1.46	62.22	0.00
LNREPAY	4.35	2.36	64.44	0.01
RAINDTRB	2.48	1.08	22.22	0.04[a]
SJASST	2.48	1.18	62.50	0.01
SJWKST	2.48	1.18	62.50	0.01
SPINCOL	6.14	4.54	53.33	0.02

a)   The difference between the pre- and post-imputation means is significant at the 0.05 level.

Table 13.6 suggests that about 4 of the 20 variables have remaining bias after imputations using the alternative method to evaluate nonresponse bias after imputations. The after imputation results in Table 13.6 are not always consistent with the after imputation results in Table 13.5.

We also computed estimates (e.g. proportions and means) and standard errors created with the interview weight and imputed data for the 20 items. These estimates are shown graphically in Figures 13.1 and 13.2 in Section 13.4.4 in comparison with estimates from the usable case respondent approach.

Table 13.6 Summary of alternative item nonresponse bias analysis for interview respondents after imputation.

Variable	Significant difference in estimates between random groups
ATTENDMR	Yes
CSTCMPTR	No
DEPCARE	No
DISMOBIL	No
DISOTHER	No
DISSENSR	No
DISTWK	Yes
DISTYPES	No
EVER2PUB	No
FINAIDA	No
FINAIDB	No
FINAIDC	No
FINAIDD	No
HSTYPE	Yes
JOBNUM	No
LNREPAY	Yes
RAINDTRB	No
SJASST	No
SJWKST	No
SPINCOL	No

13.4.4 Comparison of Estimates, Variances, and Nonresponse Bias Using Two Approaches to Define a Unit Respondent

Computing item-level bias estimates for the usable case respondent approach and student-level bias estimates for the interview respondent approach makes sense since the bias due to interview nonresponse is adjusted with imputations in one approach and with nonresponse weight adjustments in the other approach. However, comparing the bias estimates across the two approaches is tricky.

As shown in Section 13.4.3, the nonresponse adjustment for the interview weight does a good job in reducing bias due to student interview nonresponse. When using the study weight in Section 13.4.2, the imputation reduces item nonresponse bias, but it appears that bias may still remain for about half of the

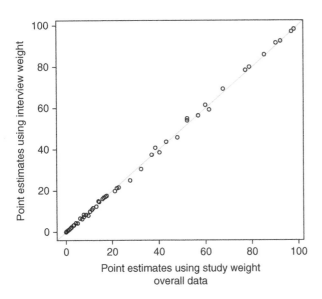

Figure 13.1 Comparison of point estimates using the interview weight and the study weight.

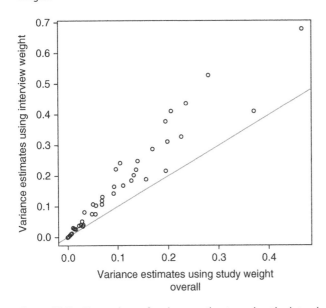

Figure 13.2 Comparison of variance estimates using the interview weight and the study weight.

variables analyzed. However, with one method to analyze the bias, bias also appears to remain after imputation when using the interview weight, but the alternative method of analyzing the bias shows that there may be less item-level bias when using the interview weight.

A comparison of the point estimates computed with the study weight and with the interview weight in Figure 13.1 shows that the estimates are similar for the 20 variables analyzed. While many of the variance estimates shown in Figure 13.2 are higher with the interview weight, the variance due to imputation is not accounted for, which would likely increase the variance estimates with the study weight more so than with the interview weight. Also, there are more cases associated with the study weight than with the interview weight, so that also causes estimates using the interview weight to have larger variances. A couple of outliers were removed from the figures, but the conclusions were identical.

These results indicate that using a usable case respondent definition should yield similar estimates and item-level bias as having an interview weight.

13.5 Discussion: Advantages and Disadvantages of Two Approaches to Defining a Unit Respondent

Weighing the advantages and disadvantages of each approach to defining unit respondents can help decide which approach is desirable for a given study. There are numerous advantages and disadvantages to an approach defining unit respondents as interview respondents, using weight adjustments to reduce nonresponse bias, and an approach defining unit respondents as usable case respondents, using imputation to reduce nonresponse bias. The advantages and disadvantages vary under different scenarios, such as the amount of data available and sample size.

13.5.1 Interview Respondents

Advantages

- An interview weight would be adjusted for interview nonresponse to properly account and adjust for interview nonresponse, as well as to reduce the bias due to interview nonresponse.
- The NCES Statistical Standards and AAPOR definitions are written in terms of interview respondents, so the current Standards and definitions would apply without any revision.
- A nonresponse weight adjustment could be more effective than imputation if the reasons why the student did not complete the interview are not related to item nonresponse and are not included in the imputation. Nonresponse due to not locating a student may be an example.

Disadvantages

- With an interview respondent approach, there could be a study weight in addition to an interview weight. However, two weights may be confusing to the data users, i.e. they may not understand which weight to use for which analyses.
- Users may also intentionally choose the wrong weight, so that they have a larger number of cases in their analyses. That is, using the interview weight will yield fewer cases for analysis than using the study weight.
- Weighting is a "one size fits all" approach to adjusting for nonresponse. That is, weight adjustments are done globally across all items, and computing weight adjustments for nonresponse uses data at the unit level and does not take into account item-specific data and relationships. This is especially true when there is an abundance of variables to use in the adjustment for missing data.
- An interview weight could add unequal weighting effects, which would cause an increase in variance.
- Multiple weights can cause problems for analysts using multilevel modeling (MLM), e.g. hierarchical linear modeling (HLM), because of the missing data and fewer cases to analyze.
- When analyzing subgroups that are not considered in computing the weights, there could be large weight variation in such subgroups which would inflate the variance.

13.5.2 Usable Case Respondents

Advantages

- Imputation is designed to reduce bias due to item nonresponse.
- Imputation with a usable case respondent definition allows one to take advantage of multiple data sources rather than ignoring these data when there is interview nonresponse.
- Imputation maintains the important relationships between variables. This can be done by identifying these relationships prior to imputation and checking for consistencies after imputation. The order in which variables are imputed also takes the relationships into account. Furthermore, once a variable is imputed using the weighted sequential hot deck procedure, it can then be used to form an imputation class for a later variable imputation.
- An imputed dataset is rectangular, as opposed to a file with missing values, and easy to use. Imputation will provide more cases for analysis and maybe better precision than a file with missing data.
- Imputation can be thought of as a weight for an individual. Imputation generally works best when there are numerous variables that can be brought to bear in the adjustment for missing data.

Disadvantages

- Analysts may not be getting an accurate measure of how much bias due to interview nonresponse exists in the variables they are using.
- The single imputation approach does not account for the variance due to imputation, and therefore the variance is underestimated. When a large amount of data is imputed, this variance should not be ignored.
- Imputation adds variance to estimates, and this variance increases as the amount of missing data and number of variables increase.

13.6 Practical Implications for Implementation with Surveys and Censuses

As administrative data are used more frequently to supplement survey data and issues with using administrative data are overcome, we need to rethink how we define a unit respondent. Some education surveys use data from multiple data sources, and this trend will only continue as survey response rates generally tend to decline. It will be important to collect data from other sources to minimize the burden on sample members. Additionally, there has been a vast increase in the amount of administrative data, in general, and in U.S. education specifically. Schools typically have data from student records and transcripts that can be used, and extant data sources also contain relevant data for students. The State Longitudinal Data Systems (SLDS) may be a future source of administrative data. The NCES Statistical Standards for unit response rates and AAPOR unit response rate guidelines are written in terms of responding to a data collection, e.g. interview. Therefore, it is important to think through issues that will be raised as data from multiple sources are used to define respondents and compile analysis files.

A usable case rule can be developed to identify key information items or combinations of these items needed to qualify as a unit respondent. Studies with reliable, accessible, and timely administrative data available should determine if the typical interview respondent definition is appropriate, or if a usable case definition would make better use of the data. The best approach when using data from multiple sources will vary based on the study.

Given the results above using NPSAS data and the advantages and disadvantages of the unit respondent definitions, it makes sense for NPSAS to use a usable case respondent definition with one weight, imputing for missing data, and conducting nonresponse bias analysis at the institution, usable case respondent, survey, and item levels. Additional variables could be identified to use for imputation classes that may not be based on interview data, such as an indicator of whether a person was located. Also, additional covariates perhaps could be added to the item-level nonresponse bias analyses.

For your study, think through the advantages and disadvantages of each approach. Think about what would be best for analysts using the data. In this age of a lot data being available, the typical interview respondent approach may no longer be the best choice. If you have the luxury of time and money, try both methods, and see which is a better fit.

13.A Appendix

13.A.1 NPSAS:08 Study Respondent Definition

Student-level data for NPSAS:08 were collected from various sources: student records by means of computer-assisted data entry (CADE); student interviews; and administrative federal and private databases such as the Central Processing System (CPS), the National Student Loan Data System (NSLDS), National Student Clearinghouse (NSC), ACT files, and SAT Reasoning Test (SAT) files. For NPSAS:08, key variables were identified across these data sources to determine the minimum requirements to support the analytic objectives of the study. Sample members who met these minimum requirements were classified as *study respondents*. Specifically, a study respondent was any sample member who was determined to be study eligible (see Cominole et al. 2010), and, at a minimum, had valid data from any source for the following:

- student type (undergraduate or graduate/first professional);
- date of birth or age;
- gender; and
- at least 8 of the following 15 variables:
    - dependency status;
    - marital status;
    - any dependents;
    - income;
    - expected family contribution (EFC);
    - degree program;
    - class level;
    - baccalaureate status;
    - months enrolled;
    - tuition;
    - receipt of federal aid;
    - receipt of nonfederal aid;
    - student budget;
    - race; or
    - parent education.

13.B Appendix

Table 13.B.1 Student interview nonresponse bias before and after nonresponse weight adjustments for selected variables.

Variable	Before nonresponse adjustment						After nonresponse adjustment			
	Unweighted respondents	Unweighted nonrespondents	Respondent mean weighted	Nonrespondent mean weighted	Estimated bias	Relative bias	Overall mean, before adjustments	Overall mean, after adjustments	Estimated bias	Relative bias
CPS record available										
Yes	61 280	20 280	58.02	47.71	3.01[a)]	0.05	55.01	55.01	0.00	0.00
No	34 090	17 130	41.98	52.17	−2.98[a)]	−0.07	44.96	44.99	−0.03	0.00
Federal Aid Status										
Received	58 550	19 030	54.18	42.96	3.28[a)]	0.06	50.91	50.91	0.00	0.00
Did not receive	36 820	18 410	45.82	57.04	−3.28[a)]	−0.07	49.09	49.09	0.00	0.00
Pell Grant status										
Received	33 630	12 170	27.20	24.96	0.65[a)]	0.02	26.54	26.54	0.00	0.00
Did not receive	61 730	25 260	72.80	75.04	−0.65[a)]	−0.01	73.46	73.46	0.00	0.00
Total Pell amount received[b)]										
US$ 1616 or less	8140	3420	6.42	7.15	−0.22[a)]	−0.03	6.63	6.63	0.00	0.00
1617 to 2840	8140	3230	6.77	6.92	−0.04	−0.01	6.81	6.81	0.00	0.00
2841 to 4309	7880	2550	6.27	5.09	0.34[a)]	0.06	5.93	5.93	0.00	0.00
4310 or more	9460	2970	7.73	5.79	0.57[a)]	0.08	7.16	7.16	0.00	0.00

Table 13.B.1 (Continued)

| Variable | Before nonresponse adjustment | | | | | | After nonresponse adjustment | | | |
	Unweighted respondents	Unweighted nonrespondents	Respondent mean weighted	Nonrespondent mean weighted	Estimated bias	Relative bias	Overall mean, before adjustments	Overall mean, after adjustments	Estimated bias	Relative bias
Stafford Loan status										
Received	45 260	14 580	41.37	32.35	2.63[a]	0.07	38.73	38.73	0.00	0.00
Did not receive	50 100	22 850	58.63	67.65	−2.63[a]	−0.04	61.27	61.27	0.00	0.00
Total Stafford amount received[b]										
US$ 3500 or less	12 030	4190	11.66	9.99	0.49[a]	0.04	11.17	11.17	0.00	0.00
3501 to 5500	15 390	4440	12.43	8.92	1.02[a]	0.09	11.40	11.39	0.01	0.00
5501 to 7510	6250	2590	5.40	4.92	0.14[a]	0.03	5.26	5.27	−0.01	0.00
7511 or more	11 600	3360	11.87	8.53	0.98[a]	0.09	10.90	10.90	0.00	0.00

a) Bias is significant at the 0.05 level.
b) Pell Grant amount and Stafford Loan amount categories were defined by quartiles.
Note: Numbers may not sum to totals due to rounding. CPS = Central Processing System.

Table 13.B.2 Item nonresponse bias before imputation for ATTENDMR for usable case respondents.

Variable	Unweighted respondents	Unweighted nonrespondents	Respondent mean weighted	Nonrespondent mean weighted	Estimated bias	Relative bias
Age group						
15–23	15 720	8130	51.34	52.15	−0.27	−0.53
24–29	6070	3300	19.83	21.17	−0.46	−2.24
30 or above	8830	4160	28.84	26.68	0.73	2.59
CPS record available						
Yes	19 210	9570	62.73	61.42	0.44[a]	0.71
No	11 420	6010	37.27	38.58	−0.44[a]	−1.17
Federal aid status						
Received	17 450	8530	56.96	54.74	0.75[a]	1.33
Did not receive	13 180	7050	43.04	45.26	−0.75[a]	−1.70
Pell Grant status						
Received	13 110	6680	42.81	42.84	−0.01[a]	−0.03
Did not receive	17 510	8910	57.19	57.16	0.01[a]	0.02
Total Pell amount received[b]						
US$ 1616 or less	3570	2030	11.65	13.03	−0.47[a]	−3.86
1617 to 2840	3390	1860	11.07	11.95	−0.30[a]	−2.60
2841 or more	6150	2780	20.07	17.86	0.75[a]	3.86
Stafford Loan status						
Received	10 490	5440	34.28	34.88	−0.20[a]	−0.59
Did not receive	20 120	10 150	65.72	65.12	0.20[a]	0.31
Total Stafford amount received[b]						
US$ 3500 or less	3970	1960	12.97	12.60	0.13[a]	0.98
3501 to 5500	2330	1210	7.60	7.78	−0.06	−0.76
5501 to 7512	3000	1620	9.80	10.39	−0.20	−1.99
7513 or more	1190	640	3.88	4.11	−0.08	−2.02

a) Bias is significant at the 0.05 level.
b) Pell Grant amount and Stafford Loan amount categories were defined by quartiles. Two quartiles were collapsed for Pell amount.
Note: Numbers may not sum to totals due to rounding. CPS = Central Processing System.

Table 13.B.3 Item nonresponse bias before imputation for ATTENDMR for item nonresponse among interview respondents.

Variable	Unweighted respondents	Unweighted nonrespondents	Respondent mean weighted	Nonrespondent mean weighted	Estimated bias	Relative bias
Age group						
15–23	15 590	300	51.35	34.64	0.47[a]	0.91
24–29	6020	230	19.82	26.12	−0.18[a]	−0.88
30 or above	8750	340	28.83	39.24	−0.29[a]	−0.99
CPS record available						
Yes	19 030	370	62.67	42.92	0.55[a]	0.88
No	11 330	500	37.33	57.08	−0.55[a]	−1.45
Federal aid status						
Received	17 280	340	56.94	38.55	0.51[a]	0.91
Did not receive	13 070	530	43.06	61.45	−0.51[a]	−1.17
Pell Grant status						
Received	12 980	190	42.76	22.09	0.58[a]	1.36
Did not receive	17 370	680	57.24	77.91	−0.58[a]	−0.99
Total Pell amount received[b]						
US$ 1620 or less	3720	50	12.25	5.29	0.19[a]	1.61
1621 to 2840	3420	60	11.25	6.79	0.12	1.12
2841 or more	5840	90	19.24	10.01	0.26[a]	1.35
Stafford Loan status						
Received	10 390	250	34.26	29.11	0.14	0.42
Did not receive	19 940	620	65.74	70.89	−0.14	−0.22
Total Stafford amount received[b]						
US$ 3500 or less	3940	60	12.99	7.02	0.17[a]	1.30
3501 to 5500	2300	60	7.56	7.25	0.01	0.12
5501 to 7874	3020	70	9.95	7.83	0.06	0.60
7875 or more	1130	60	3.73	7.02	−0.09	−2.39

a) Bias is significant at the 0.05 level.
b) Pell Grant amount and Stafford Loan amount categories were defined by quartiles. Two quartiles were collapsed for Pell amount.
Note: Numbers may not sum to totals due to rounding. CPS = Central Processing System.

References

Cominole, M., Riccobono, J., Siegel, P., and Caves, L. (2010). *2007–08 National Postsecondary Student Aid Study (NPSAS:08) Full-Scale Methodology Report (NCES 2011-188)*. U.S. Department of education. Washington, DC: National Center for Education Statistics.

Cox, B. (1980). The weighted sequential hot deck imputation procedure. In: *Proceedings of the Section on Survey Research Methods, American Statistical Association*, 721–726. Alexandria, VA: American Statistical Association.

Daas, P., Ossen, S., Vis-Visschers, R., and Arends-Tóth, J. (2009). *Checklist for the Quality Evaluation of Administrative Data Sources*. The Hague/Heerlen: Statistics Netherlands.

Hughes, A.L. and Peitzmeier, F.K. (1989). Weighting and imputation for nonresponse in CPS gross flows estimation. In: *Proceedings of the Section on Survey Research Methods, American Statistical Association*, 279–285. Alexandria, VA: American Statistical Association.

Michaud, S. (1986). Weighting vs imputation: a simulation study. In: *Proceedings of the Section on Survey Research Methods, American Statistical Association*, 316–320. Alexandria, VA: American Statistical Association.

Rässler, S. and Schnell, R. (2003). *Multiple imputation for unit-nonresponse versus weighting including a comparison with a nonresponse follow-up study*. IWQW Discussion Paper Series. Germany: University of Erlangen-Nuremberg.

The American Association for Public Opinion Research (2016). *Standard Definitions: Final Dispositions of Case Codes and Outcome Rates for Surveys*, 9e. AAPOR.

U.S. Department of Education (2012). *NCES Statistical Standards (NCES 2014-097)*. Washington, DC: National Center for Education Statistics, Institute of Education Sciences, U.S. Department of Education.

United Nations Economic Commission for Europe (2007). *Register-Based Statistics in the Nordic Countries*. New York and Geneva: United Nations Publication.

Wallgren, A. and Wallgren, B. (2014). Theory and quality of register-based statistics. In: *Register-Based Statistics: Statistical Methods for Administrative Data*, 2e. Chichester, UK: Wiley, https://www.wiley.com/en-us/Register+based+Statistics%3A+Statistical+Methods+for+Administrative+Data%2C+2nd+Edition-p-9781118856000.

Woodruff, R.S. (1971). A simple method for approximating the variance of a complicated estimate. *Journal of the American Statistical Association* 66 (334): 411–414.

Index

Administrative Records for Survey Methodology, First Edition.
Edited by Asaph Young Chun, Michael D. Larsen, Gabriele Durrant, and Jerome P. Reiter.
© 2021 John Wiley & Sons, Inc. Published 2021 by John Wiley & Sons, Inc.

-

Printed and bound by CPI Group (UK) Ltd, Croydon, CR0 4YY